TOXICOLOGY OF TRACE ELEMENTS

Advances in Modern Toxicology

Editor
Myron A. Mehlman

Vol. 1, Part 1—Myron A. Mehlman, Raymond E. Shapiro, and Herbert Blumenthal — New Concepts in Safety Evaluation

Vol. 2—Robert A. Goyer and Myron A. Mehlman — Toxicology of Trace Elements

Vol. 3—H. F. Kraybill and Myron A. Mehlman — Environmental Cancer

Vol. 4—Francis N. Marzulli and Howard I. Maibach — Dermatotoxicology and Pharmacology

IN PREPARATION

Vol. 1, Parts 2 and 3—Myron A. Mehlman, Raymond E. Shapiro, and Herbert Blumenthal — New Concepts in Safety Evaluation

Gary Flamm and Myron A. Mehlman — Mutagenesis

Advances in Modern Toxicology

VOLUME 2

TOXICOLOGY
OF TRACE ELEMENTS

EDITED BY

ROBERT A. GOYER
UNIVERSITY OF WESTERN ONTARIO

MYRON A. MEHLMAN
NATIONAL INSTITUTES OF HEALTH

HEMISPHERE
PUBLISHING CORPORATION

Washington   London

A HALSTED PRESS BOOK

JOHN WILEY & SONS

New York   London   Sydney   Toronto

Hemisphere Publishing Corporation
1025 Vermont Ave., N.W., Washington, D.C. 20005

Distributed solely by Halsted Press, a Division of John Wiley & Sons, Inc., New York.

1 2 3 4 5 6 7 8 9 0   D O D O   7 8 3 2 1 0 9 8 7

Library of Congress Cataloging in Publication Data
Main entry under title:

Toxicology of trace elements.

   (Advances in modern toxicology ; v. 2)
   Includes bibliographical references.
   1. Trace elements–Toxicology. 2. Metals–
Toxicology. I. Goyer, Robert A. II. Mehlman,
Myron A. III. Series.
RA1231.T7T69   615.9'2   76-55351
ISBN 0-470-99049-X

Printed in the United States of America

CONTENTS

PREFACE

Metals are unique environmental and industrial pollutants in that they are neither created nor destroyed by humans but are only transported and transformed into various products. Often these activities result in exposure to trace metals by persons not ordinarily in contact with them, and sometimes chemical forms are created that are not usually present in nature.

The chapters in this volume are concerned with the toxicological properties of the various forms of each metal, and emphasis is given to potential adverse health effects on humans. Introductory background material regarding environmental sources and important nonhuman effects are also provided where appropriate. It is not intended that these chapters contain all that is known about the metals under consideration, but rather emphasis has been given to current interests and problems.

The chapter on compounds of mercury by Suzuki compares the metabolism of different forms of mercury at the cellular and biochemical level. The problem of removal of organic mercury compounds from target tissue continues to be a major aspect of the mercury problem. The relative effectiveness of a variety of potential therapeutic agents is reviewed. The chapter is concluded with comparisons of the clinical and cellular consequences of exposure to organic mercurials with detailed discussion of the chemical pathology of Minamata disease.

Lead continues to be of interest because of potential central nervous system effects in children with greater than normal blood lead levels and chronic excessive exposure to lead by industrial workers. There is a growing consensus that a safe blood level for young children is no higher than 40 μg/ 100 ml, whereas a permissible blood lead level among workers with occupational exposure to lead has been reduced by NIOSH to 60 μg/100 ml. These levels were determined by improved methods for detecting subclinical functional effects of lead, particularly central nervous system effects, in both children and workers. The chapter by Goyer and Mushak combines a review of the toxicology of lead and biochemical parameters of lead effect with a discussion and critique of commonly employed laboratory tests of lead effect.

The chapter by Fowler introduces the toxicology of arsenic with a brief review of sources in the environment and accumulation in living things, including humans. The effects of the multiple forms of arsenic are compared, and cellular mechanisms of arsenical toxicity are reviewed. Trivalent arsenicals are usually more toxic than pentavalent forms. It is interesting that arsenicals have been used to enhance growth, but the biological basis for these seemingly desirable effects is not known. Arsenic does inhibit cellular respiration and accumulates in mitochondria. Correlation of this cellular effect with observed toxicity is incomplete but may in part explain toxic effects of arsenic on hepatic parenchymal cells and renal tubular lining cells. Arsine toxicity is reviewed in considerable detail. Arsine poisoning produces a severe hemolysis and hemoglobinuric nephropathy, and the basis for the hemolysis may be arsine-induced reduction in the glutathione content of red blood cells.

In the chapter on copper toxicology Hill points out that the toxic potential of copper is relatively low except in two special circumstances: humans with Wilson's disease and certain domestic animals. Persons with Wilson's disease are unusually sensitive to copper because of reduced amounts of ceruloplasmin. For a number of reasons domestic animals are much more susceptible to copper toxicity than are humans. One reason is that agricultural practices include copper-containing drenches on animals and, sometimes, the feeding of high levels of copper, particularly to swine. Sheep are susceptible to copper toxicosis under conditions in which other animals would not be affected. Zinc, iron, and molybdenum reduce susceptibility to excess copper.

The chapter by Nielsen summarizes current knowledge of metabolism and toxicology of nickel, excluding carcinogenesis. The toxicity of nickel or nickel salts is relatively low, probably because gastrointestinal absorption is slight, and there is little lifetime accumulation in tissues. Specific effects of various nickel salts in different species of animals are compared. The effect of parenterally administered nickel salts on carbohydrate and insulin metabolism is particularly interesting. Sensitization of skin for nickel is a potential occupational health problem, and mechanisms involving a nickel–protein complex are presented. Nickel toxicity by inhalation, particularly inhalation of nickel carbonyl, is a serious occupational health problem. Mechanisms of toxicity and possible therapeutic measures are reviewed.

Vanadium is widely distributed in nature but rarely occurs in high concentration. Awareness that the vanadium content of certain crude oils and ores may be magnified by refining and smelting has stimulated interest in the potential toxicity of this metal. The chapter by Waters reviews the environmental distribution of vanadium, and the metabolism and potential toxic effects on humans. Although vanadium is now believed to be essential for the chick and rat, human requirements are uncertain. Inhalation of vanadium dust causes severe irritation of the respiratory tract and may produce bronchitis and bronchospasm. Vanadium measured in blood and urine serves as evidence of exposure, but systemic effects are not likely except with unusually high levels of exposure. Cysteine content of fingernails decreases with vanadium exposure. This test may be a useful index of vanadium exposure in industrial workers.

The toxicology of selenium and tellurium is less well known than many other metals. Reasons are that industrial uses are more recent and also that the multiple chemical forms of these metals make study more difficult. The chapter by Fishbein contains a detailed review of literature dealing with both human and experimental exposures. Selenium is responsible for a syndrome known as "blind staggers" in sheep and may be both neurotoxic and hepatoxic. Experimental studies show that a high level of exposure to selenate may be carcinogenic but that sodium selenide, an antioxidant, may reduce the carcinogenic potential of other known carcinogens. This aspect of selenium toxicology deserves considerably more study. The toxicology of tellurium is even less well understood than selenium. Experimental studies suggest it is primarily a neurotoxin and may produce demyelination of peripheral nerves.

The last two chapters in the book are concerned with conceptual aspects of metal toxicology that are presently emerging as highly important and relevant. These are interactions of metals with one another and with essential metals and the potential carcinogenicity of metals.

The chapter on nutritional interaction with toxic elements by Sandstead points out that the major exposure of humans to toxic elements occurs through the food supply. Although such exposures are usually many times below those that cause toxicity, the potential for toxicity may be influenced by variations, particularly deficiencies, in the diet content of essential metals. The review is restricted to dietary factors that influence toxicity of lead, cadmium, and mercury. Dietary deficiencies of calcium and iron enhance the toxicity of lead in experimental animals and may have a role in lead poisoning in young children with subclinical lead poisoning. Lead toxicity may also be influenced by vitamin D, zinc, protein, and other vitamins and essential nutrients. Many of these are thought to act at the level of absorption from the gastrointestinal tract, although some substances like calcium may influence mobilization of lead from storage sites. Cadmium toxicity is also antagonized by dietary content of essential minerals in the diet, particularly zinc and calcium and possibly iron, copper, and selenium.

The formation of a mercury-selenium complex appears to reduce the potential toxicity of a particular dose of methylmercury. This observation may be of great importance in protection of human populations with unavoidable exposure to mercury. The influence of other metals on mercury poisoning such as zinc and copper may also be of interest, but they need more study.

The chapter by Sunderman summarizes the epidemiological and experimental support for the carcinogenicity of certain metal compounds. Particular emphasis is given to nickel carcinogenesis because of the detailed knowledge presently available.

Occupational exposures of workers to arsenic, chromium, and nickel compounds have been associated with increased incidence of specific types of cancer. Compounds of eight metals (beryllium, cadmium, chromium, cobalt, lead, nickel, zinc, and iron-carbohydrate complexes) have been shown to induce cancers in experimental animals. Carcinogenic nickel and beryllium compounds administered to rats become localized in nuclei and inhibit mitosis. Beryllium blocks the induction of certain enzymes needed for DNA synthesis without affecting other enzyme systems, whereas nickel may inhibit RNA polymerase activity. It appears that study of the mechanism of carcinogenesis of these compounds may be helpful in providing an understanding of critical steps in the pathogenesis of metal-induced malignancies. It is interesting that several of the metals suspected of inducing cancer in experimental animals have not been identified as carcinogenic in humans. On the other hand, arsenic compounds have been associated with certain cancers in humans, but the induction of tumors by arsenic in experimental animals has not yet been accomplished.

Finally, it is intended that the information in these chapters provide a convenient and useful resource of detailed current information for toxicologists and related scientists. It is also hoped that the unanswered questions raised in this volume may serve as stimulus for further studies.

Robert A. Goyer

TOXICOLOGY OF TRACE ELEMENTS

Chapter 1

METABOLISM OF
MERCURIAL COMPOUNDS

Tsuguyoshi Suzuki
Department of Public Health
Tohoku University School of Medicine
Sendai, Japan

INTRODUCTION

Mercury is found in the environment in various chemical forms, and the different forms have different pharmacokinetic properties as regards absorption, bodily distribution, accumulation, and excretion. Elemental mercury, inorganic mercury compcunds, short-chain alkylmercurials (methyl and ethyl), and other organomercury compounds can be distinguished by their toxicological properties.

Mercury was known in prehistoric times, and as an occupational hazard it has a long history of causing disease (Bidstrup, 1964). But the recent increasing concern with mercury stems from repeated outbreaks of epidemics of methylmercury poisoning (Tsubaki et al., 1967; Kutsuna, 1968; Curley et al., 1971; Pierce et al., 1972; Bakir et al., 1973), wide existence of mercury contamination in the environment (Miller and Berg, 1969; Nordiskt Symposium, 1969; Goldwater, 1971; Berglund et al., 1971; Nelson et al., 1971; Hartung and Dinman, 1972; D'Itri, 1972; Friberg and Vostal, 1972; Joselow et al., 1972), and the conversion of elemental mercury and mercury compounds into methylmercury in the natural environment (Jensen and Jernelöv, 1969; Jernelöv, 1969; Landner, 1971). Thus it has become urgent to know the actual degree of mercury contamination and to evaluate the risk. The danger of industrial mercury poisoning can be minimized by proper preventive

This review is restricted to the metabolism of mercurial compounds, partly because of space and time limitations and partly because mercurial effects and dose–response relationships are covered in the proceedings of the Task Group on Dose–Response Relationships (1976).

The author is grateful for the editorial support in preparing this manuscript. Dr. L. Magos kindly read the main part of the manuscript and gave much advice. The author also appreciates the assistance by Miss S. Shishido, Mrs. K. Honda, and Miss F. Takizawa.

measures, although it still exists in various industrial fields. In addition, mercury compounds other than methylmercury, for instance ethylmercury, have been reported to cause disease and death in populations without occupational exposure but with consumption of dressed seed or the intake via other routes (Jalili and Abbasi, 1961; Schmidt and Harzman, 1970; Suzuki et al., 1973; Derban, 1974).

We must not underestimate the risk due to mercury compounds other than methylmercury, but the most threatening mercury compound is methylmercury in relation to the health of human beings, as well as to the health of some animals in the ecosystem (Peakall and Lovett, 1972; Wood, 1972). Thus the facets of toxicology of mercury involve both industrial toxicology and ecotoxicology, that is, the environmental pollution by mercury and its implications for human health (Clarkson, 1971; FAO/WHO, 1972; Skerfving, 1972; Suzuki, 1976).

METABOLISM

The metabolism of mercury compounds has been repeatedly discussed in several reviews and proceedings of conferences on mercury toxicology (Clarkson, 1972a; Nordberg and Skerfving, 1972; Miller and Clarkson, 1973; Task Group on Metal Accumulation, 1973). This review will cover mainly the recent literature.

Absorption

Pulmonary absorption. Mercury vapor easily penetrates the alveolar membrane into the blood, but its retention depends on its oxidation. Nielsen-Kudsk (1965, 1969a,b) have reported that ethanol at blood concentrations less than 0.04% (w/v) depresses the pulmonary absorption of mercury, while the alcohol concentration of 0.2% (w/v) inhibits the *in vitro* oxidation of the metal by 60%. In rats oxidation of elemental mercury is also inhibited *in vivo* after alcohol treatment and is the probable cause of decreased retention (Magos et al., 1973).

There are no detailed data on respiratory uptake of inorganic mercury compounds and organomercury compounds. Östlund (1969) has reported the retention of dimethylmercury after a single inhalation exposure by mice, and Fang and Fallin (1973) have compared the uptake and distribution of ethylmercury chloride by inhalation with that by oral administration, but no quantitative estimations of pulmonary absorption are possible from these data. The Task Group on Metal Accumulation (1973) has stated that it is known from both animal experiments and human exposures that toxic amounts of several of the compounds could be absorbed by inhalation. Some of the organomercury compounds including methylmercury chloride vaporize easily, while inorganic mercury compounds exist mostly in particulate form in air. Volatility, solubility, and particle size are the main factors that influence the pulmonary absorption of mercury compounds.

Gastrointestinal absorption. As for inorganic mercury compounds, the Task Group on Metal Accumulation (1973) has concluded from the results of Rahola et al. (1971; 1973) in human volunteers that 1.4–15.6% (mean, 7%) of administered dose of inorganic mercury (Hg^{2+}) was absorbed in the gastrointestinal tract.

The absorption of methylmercury in human beings is about 95% of administered dose either given as an aqueous solution or bound to fish protein (Åberg et al., 1969; Miettinen, 1973).

In squirrel monkeys at least 95% of the administered dose of methylmercury was absorbed (Berlin et al., 1975a). In rats Takahashi (1974) has shown that the transfer of methylmercury chloride to the lymph in the thoracic duct is about 0.1% of administered dose, and the transfer rate is not influenced by simultaneous addition of salad oil to methylmercury chloride. Thus in rats the methylmercury after the intestinal absorption must reach the portal vein.

For other mercury compounds, no quantitative human data were available. In rats or mice phenylmercury compounds were more efficiently absorbed than inorganic mercury compounds (see Clarkson, 1972a). The absorption of ethylmercury was as much as that of methylmercury in rats and mice (Ogawa et al., 1973). After a single tracer oral dose of methylmercury, absorption was 59% in lactating cows (Neathery et al., 1974).

Regarding the efficiency of gastrointestinal absorption of mercury compounds, the dose, the form of mercury compound, and the influence of coexistent substances in food have to be taken into consideration. The dose dependence was suggested by Clarkson (1972a) for the case of inorganic mercury compounds, but no precise data have been obtained. The difference due to the form of mercury in intestinal reabsorption was noticed in the study of biliary excretion of mercury and its implication for enterohepatic circulation. This topic will be discussed in the section on biliary excretion. As to the influence of coexistent substances, the experiments by Sahagian et al. (1966) using strips of rat small intestine have shown that inorganic mercury enhances the uptake of zinc and cadmium, and the uptake of inorganic mercury is enhanced by zinc, cadmium, or manganese. No further progress has been observed in this matter. The effect of various thiol compounds on the absorption of methylmercury was tested by the *in vitro* experiment using isolated intestinal tract of rats (Takahashi, 1974). Transfer rates through the intestine to the Krebs-Ringer solution outside the intestine were about the same among methylmercury cysteine, methylmercury glutathione, and methylmercury acetylcysteine. Addition of excess cysteine to methylmercury cysteine markedly enhanced the transfer, but this finding was not supported by Sugiyama et al. (1975). Wide variations of molar ratio of methylmercury to cysteine will have to be studied.

Percutaneous absorption. Topical application of methylmercury compound preparations to the skin has resulted in serious and fatal poisonings in humans. Other mercury compounds also penetrate the skin (see Clarkson,

1972a; Nordberg and Skerfving, 1972). It is still unknown whether the transdermal or the follicular pathway route of skin absorption is more important. After topical application of mercuric chloride on human skin, Silberberg et al. (1969) using their histochemical method found abnormal electron-dense structures in the epidermis in intracellular and extracellular sites below the stratum corneum. These electron-dense structures may represent mercury or a mercury-cell component complex. Comparative studies on various forms of mercury are necessary.

Transport

In the case of mercury vapor exposure, the rate of oxidation is not fast enough to prevent some of the elemental mercury from reaching the brain in this highly diffusible form (Magos, 1967). Except for this particular case, the mercury of various chemical forms is transported in binding plasma protein or blood cells.

Partition between erythrocytes and plasma. The factors that determine partition of mercury between erythrocytes and plasma have not been adequately elucidated. The Task Group on Metal Accumulation (1973) has listed the form of the metal, dose, time after dosing, and animal species as relevant factors. The red blood cell-to-plasma ratio of methylmercury in human blood has been reported to be dependent on the level of methylmercury in blood but usually it is as high as 9:1 (Swensson et al., 1959; Suzuki et al., 1971a,b; Birke et al., 1972), whereas the ratio after oral absorption of inorganic mercury is 1:2.5 (Miettinen, 1972) and 1:1 after industrial exposure to mercury vapor (Lundgren et al., 1967). In industrial workers exposed to mercury vapor with simultaneous exposure to methylmercury from eating fish, the ratio is 1.8:1 for inorganic mercury and 2.4:1 for organic mercury (Suzuki et al., 1970). Both the dose and the form of mercury are responsible for the variation in the partition. In short-chain alkylmercury compounds, the mercurial penetrates the red cell membrane and binds to hemoglobin (Takeda et al., 1968; White and Rothstein, 1973). The binding of methylmercury to intracellular hemoglobin of humans, rats (White and Rothstein, 1973), and rainbow trout (Giblin and Massaro, 1974) is reversible by extracellular SH groups. The binding of methylmercury is tighter in rats than in humans; no obvious differences in binding were observed between human adult and fetal erythrocytes (White and Rothstein, 1973). Garrett and Garrett (1974) have reported by the incubation experiment on red blood cells of rabbits that the uptake of methylmercury is approximately the same for erythrocytes and reticulocytes, but reticulocytes accumulated more mercury from inorganic mercury than did mature erythrocytes.

As the determining factors of partition of mercury between blood cells and plasma, the permeability of erythrocytes to mercurial compounds and the balance of the number of binding sites between intracellular and extracellular components of blood are to be noted. Rothstein (1973) has demonstrated that the permeability of erythrocytes is inversely related to the degree of mercurial dissociation. Pericellular attachment of mercurial to erythrocytes or the binding to cellular membrane may play some role in cases of highly dissociable mercurial compounds.

In spironolactone-pretreated rats the removal of mercury (Hg^{2+}) from erythrocytes was decreased, despite the increased rate of plasma clearance (Haddow and Lester, 1973), and the uptake of mercury (Hg^{2+}) into human and rat erythrocytes suspended in saline was enhanced by pre-, simultaneous, and postincubation with spironolactone (Kushlan et al., 1975). The formation of a permeable complex of mercury–spironolactone is likely responsible for the increase of mercury uptake.

In experiments using intact red blood cells, various mercurial compounds of molecular weights up to 70,000 penetrate the cell membrane and inhibit both Na^+,K^+-ATPase and Na^+,K^+-insensitive ATPase, but a dextran derivative of molecular weight 250,000, to which p-chloromercuribenzoate is bound by the bridge of the aminoethyl group, does not penetrate the membrane and inhibits only Na^+,K^+-ATPase (Nakao et al., 1973). The mercurial compound of molecular weight 70,000, Hg anilinodextran, which penetrates the red cell membrane, does not penetrate the epithelium of frog bladder. Even though the effect of medium used for experiments is obscure, the permeability of erythrocytes to mercurial compounds is to be further studied in relation to the size of molecule and the dissociability.

Binding to plasma proteins. The Task Group on Metal Accumulation (1973) stated that plasma is generally considered as the main pathway for transport of metals, although the possibility of erythrocytes having a role in a direct exchange of metals between circulating metal in blood and metal in tissues cannot be excluded. The group emphasized that the low-molecular-weight or "diffusible" fraction of metal in plasma is of the greatest interest for transfer of the metal. This fraction had not yet been identified qualitatively or quantitatively for mercurial compounds. For methylmercury compounds the methylmercury halide (White and Rothstein, 1973; Giblin and Massaro, 1974) is nominated as a possible diffusible chemical form.

If this theoretical diffusible fraction in plasma plays a chief role in the transfer of metal to and from various organs, erythrocytes as well as plasma proteins should be just deposits from which the metal is exchanged to the diffusible fraction. However, most observations have revealed that all the mercury except for a very low percentage of the total amount in plasma is protein bound (see Clarkson, 1972a), although the kind of proteins to which mercury is bound changes according to the dose and the time after dosing in experiments on inorganic mercury in rats (Suzuki et al., 1967b; Cember et al., 1968; Jakubowski et al., 1970). Mercury (Hg^{2+}), which binds to the thiol and carboxy groups of human serum albumin (Perkins, 1961), was found by Katz and Samitz (1973) to bind at other sites in addition to those groups. Magos

(1974) has presented arguments against the idea that both valencies of mercury are bound to thiol groups.

The binding of mercury (Hg^{2+}) to plasma proteins can be changed by a simultaneous administration of selenite in rats (Burk et al., 1974). The meaning of this finding for the mechanism of selenite protection for mercury poisoning is not clear, but the change of binding to plasma protein involves both mercury and selenium and both elements are bound to the same protein in gel filtration chromatogram.

Placental Transfer

The placental barrier is less effective in preventing the passage of short-chain alkylmercury than of other mercurial compounds (see Nordberg and Skerfving, 1972). Comparing the fetal uptake of elemental mercury with inorganic mercury (Hg^{2+}), the former was over 10 times greater after injection of elemental mercury and 40 times greater after the inhalation of mercury vapor (Clarkson et al., 1972). At the equivalent body burden of mercury the amount of mercury in maternal blood was over 25 times greater after inhalation of mercury vapor than after injection of inorganic mercury in rats (Greenwood et al., 1972).

Garrett et al. (1972) injected inorganic mercury (Hg^{2+}) into the heart of a pregnant rat (day 16) and found after 30 min that although the chorio-allantois and the visceral yolk sac concentrated mercury, there was little transfer of the metal to the fetus. In addition, selenite induced a 100% increase in mercury accumulation in both these tissues. Methylmercury is more readily transferable than inorganic mercury in rats (Mansour et al., 1973) and mice (Suzuki et al., 1967a). Tuna meat was fed to mice, and the concentration of mercury in fetuses was found to be proportional to the maternal diet level of mercury (Childs, 1973).

Administration of methylmercury in a form of methyl methylmercuric sulfide (CH_3HgSCH_3) labeled with ^{203}Hg or ^{35}S to pregnant mice revealed that the Hg–S bond was likely to break easily in the maternal body and the metals to be transferred separately into fetuses (Fujita, 1969). By experiments in female rats that received injections of methylmercury or phenylmercury only before mating, the increase in mercury content was significantly higher in the newborn from mothers given methylmercury than in those from mothers administered phenylmercury (Yamaguchi and Nunotani, 1974). Clearance rates of mercury in the newborn were, however, more rapid in the case of methylmercury than of phenylmercury. Casterline and Williams (1972) found that the clearance rates of methylmercury from blood and brain were more rapid in the offspring than in the mother, who was given oral doses of methylmercury from day 6 through day 15 of gestation.

Detailed kinetics of transplacental passage of methylmercury have been reported, Olson and Massaro (1975) noticed peak fetal concentration attained 3 days after injection of pregnant mice on day 13 of gestation, and Reynolds and Pitkin (1975) revealed in the rhesus monkey, which has a placenta similar in structure and function to the human, that transfer of methylmercury from mother to fetus occurred more readily than movement in the opposite direction. The administration of penicillamine after 4–5 hr of oral dosing of methylmercury on pregnant rats prevented the morphological change and reduced the methylmercury content of the fetal brains (Matsumoto et al., 1967). In rats receiving methylmercury the mercury concentration in the fetal brain was about twice that in the maternal brain (Yang et al., 1972; Null et al., 1973). This relatively elevated concentration of methylmercury in the fetal brain is much greater if the blood volume of fetal and maternal brain is taken into consideration and each brain mercury concentration is corrected by respective blood volumes and the blood mercury concentrations (Wannag, 1976).

There are species differences in placental structure and function that are likely to influence the passage of mercurials, but no systematic comparison has yet been undertaken. The relative permeability of short-chain alkyl-mercury, phenylmercury, and inorganic mercury seems to be invariable in all the placenta studied. The influence of materials that modify the metabolism and toxicity of mercurial compounds such as penicillamine or selenite has not been studied fully in terms of transplacental passage.

The human placenta is permeable to methylmercury, as has been shown by several cases of intrauterine poisoning (see Task Group on Metal Accumulation, 1973). In humans, fetal erythrocytes at birth have been found to contain an amount of mercury about 30% higher than maternal erythrocytes (Bakir et al., 1973; Suzuki et al., 1971a; Tejning, 1970). The hypothesis by Löfroth (1969) that an increasingly selective concentration of mercury in the fetus occurs at high methylmercury exposure levels has not been supported by human experiences and animal experiments. The relative excess in fetal blood was postulated as being due to differences in hematocrit and binding characteristics of hemoglobins between the fetus and the adult (Suzuki et al., 1971a; Task Group on Metal Accumulation, 1973). White and Rothstein (1973) did not find differences in binding of fetal and adult hemoglobins to methylmercury and suggested that plasma be studied more carefully in this connection. Ishihara (1974) examined this postulation from a theoretical point of view of hemoglobin structure and pointed out that the number of thiol groups that are capable of binding mercurials in adult hemoglobin is greater than that in fetal hemoglobin.

Biotransformation and Other Metabolic Change

The term "biotransformation" is used here to refer to changes in oxidation state of mercury and the formation and cleavage of covalent organomercurial bonds. Side-by-side biotransformation, the metabolic modification of the organic moiety of organomercurial compound, is one more important concern in the metabolism of mercurial compounds in the biological system.

Elemental mercury and inorganic mercury. Mercury vapor (Hg^0) is oxidized to mercuric ion (Hg^{2+}) not only in blood but also in other tissue (see Clarkson, 1972a). The inhibition or acceleration of mercury vapor uptake in blood was observed using a variety of metabolic inhibitors and substrates (Nielsen-Kudsk, 1969a,b). Increasing concentrations of crystalline beef liver catalase caused a pronounced acceleration in the rate of uptake of mercury vapor in 3 mM glutathione solutions (Nielsen-Kudsk, 1973). This catalase-stimulated rate of uptake was inhibited by 0.2% w/v ethyl alcohol and by 3-amino-1,2,4-triazole. The oxidation of mercury vapor by erythrocytes was not inhibited by 3-amino-1,2,4-triazole (Nielsen-Kudsk, 1969a,b); later, however, Magos et al. (1974b) proved that in aminotriazole-pretreated erythrocytes with decreased catalase activity, the mercury uptake decreased and the pretreatment of rats with 1 g/kg aminotriazole had the same effect as ethanol (i.e., decreased the lung content and increased the liver content of mercury after mercury vapor or metallic mercury exposure). From these results catalase is considered the enzyme responsible for the oxidation of elemental mercury (Hg^0), a hypothesis supported by experiments using acatalasemic mice (Sugata, 1975).

The reduction of ionic mercury to elemental mercury, which results in volatilization of mercury, is caused by certain commonly existing microorganisms (Magos et al., 1964), some of which acquire the mercury-resistant capability by this mechanism (Tonomura et al., 1968).

After the injection of divalent mercury into animals, mercury volatilizes through the lung and skin. The chemical form of this volatile mercury has not been identified. It may be mercury vapor or some Hg complex like $HgCl_4$.

Methylmercury compounds. Studies of methylmercury compounds by *in vivo* experiments revealed that the cleavage of covalent bond occurs in rats (Gage, 1964; Norseth and Clarkson, 1970a,b; Magos, 1971), mice (Östlund, 1969; Norseth, 1971), ferrets (Hanko et al., 1970), guinea pigs (Iverson et al., 1973; Iverson and Hierlihy, 1974), and squirrel monkeys (Berlin et al., 1975a), although the extent of cleavage is the least among organomercurial compounds (Clarkson, 1972b). The cleavage was also observed in human cases of methylmercury poisoning in Iraq (Bakir et al., 1973).

The inverse dose dependence of methylmercury covalent bond cleavage was observed in guinea pigs (Iverson and Hierlihy, 1974), rats (Syverson, 1974a), and squirrel monkeys (Berlin et al., 1975a). These results are of considerable interest in view of dose dependence of brain accumulation. As possible mechanisms, inhibition or saturation of an enzymic mechanism responsible for the cleavage, an influence of overdosed mercury on a nonenzymic breakdown pathway, or the disturbed transport of inorganic mercury from the liver to the kidney have been discussed (Iverson and Hierlihy, 1974). As far as the knowledge of actual mechanisms on the cleavage is now known, we cannot explain the dose dependence of methylmercury metabolism.

The idea that the release of inorganic mercury from methylmercury is caused by microorganisms in the gastrointestinal tract of rat is rejected by Norseth's experiment (1971) using germfree rats. Fang and Fallin (1974) studied the transformation of phenylmercuric acetate, ethylmercury chloride, and methylmercury chloride by rat liver, kidney, and brain slices. In the kidney and liver, there was no cleavage of C—Hg bond from methylmercury, while a slight cleavage occurred in the brain. In contrast, the cleavage of the C—Hg bond from phenylmercury and ethylmercury was the greatest in the kidney, less in the liver, and even less in the brain. The homogenate of rat liver and kidney showed the activity to break the C—Hg bond of phenyl-mercury and ethylmercury, but not of methylmercury (Fang, 1974a,b). Sodium selenite supplemented in drinking water with 0.5 ppm or 5 ppm increased the cleavage activity of homogenate for phenylmercury. Ishihara and Suzuki (1976) studied the release of inorganic mercury from methylmercury in chopped tissue of mice liver and kidney in Eagle minimum essential medium and found an increase of inorganic mercury formation. Homogenates did not show the increase of cleavage, and selenite added to chopped tissue had no effect.

Other alkylmercury compounds. Ethylmercury compounds undergo conversion to inorganic mercury much more than methylmercury compounds. The cleavage of covalent bonds has been shown by *in vivo* experiments in rats (Takeda and Ukita, 1970; Fang and Fallin, 1973), mice (Suzuki et al., 1973), squirrel monkeys (Blair et al., 1975), and guppies and coontails (Fang, 1974a,b), as well as in patients administered human plasma containing ethylmercurithiosalicylate (Suzuki et al., 1973). As mentioned earlier, in the *in vitro* experiments using homogenates or slices of rat liver and kidney, ethylmercury underwent cleavage of the C—Hg bond, although less than did phenylmercury (Fang, 1974a; Fang and Fallin, 1974).

Other than methyl- or ethylmercury, the study of alkylmercurial compound metabolism has been scanty. Recently, Kitamura et al. (1975) studied the elimination rates of four alkylmercury compounds (methyl-, ethyl-, *n*-propyl-, and *n*-butylmercury) from rat brain, kidney, liver, and muscle after peroral administration and found the elimination of *n*-propyl- and *n*-butyl-mercury (intact alkylmercury) from the organs was more rapid than methyl-mercury and slower than ethylmercury. The proportion of intact alkylmercury to total mercury was measured on the blood, liver, kidney, and muscle (S. Kitamura and K. Hayakawa, private communication). In the blood no significant difference was found among four alkylmercurials. Conversely, in the liver, kidney, and muscle, the proportion of intact alkylmercury to total mercury was fairly constant after administration: methyl- > *n*-propyl- > *n*-butyl- > ethylmercury.

Other organomercury compounds. To explain rapid and complete breakdown of phenylmercury, Daniel and Gage (1971) postulated that phenyl-mercury is hydroxylated in the liver microsome and then undergoes a nonenzymic breakdown to phenol and inorganic mercury. Later, Daniel et al. (1972) rejected their own hypothesis because the presence of *o*- or *p*-hydroxy-phenylmercury in urine and bile had proved negative, and the *in vitro* release of inorganic mercury by liver homogenates occurred in the absence of the microsomal fraction. Fang and Fallin's (1974) finding also proved that the split to benzene and inorganic mercury is the primary pathway. The cleavage activity concentrated in the nuclear fraction and some enzymological exami-nations suggested strongly that an enzyme capable of breaking down phenyl-mercury may exist in rat tissue.

The intracellular site responsible for the breakdown of phenylmercury is disputed. Daniel et al. (1972) found the breakdown was effected by the soluble, but not the microsomal, fraction of rat liver homogenates and requires neither NADPH nor NADH. In the mercury-degrading enzyme of cell-free extracts of a mercury-resistant pseudomonas in which cytochrome c_1 is involved, there is an absolute requirement for NADPH, and elemental mercury (Hg^0) is produced (Furukawa and Tonomura, 1972, 1973). The system degrading phenylmercury and methoxyethylmercury, which was examined in the soluble fraction of the liver, brain, and kidney of several mammals and the chicken, demonstrated some species difference in activity (Lefevre and Daniel, 1973). The cleavage activity in the liver-soluble fraction was stimulated by the addition of such thiol compounds as dithiothreitol or 2-mercaptoethanol.

Methoxyethylmercury compounds, which like phenylmercury undergo easily the cleavage of the C–Hg bond, were shown by *in vivo* experiments to be rapidly broken down to yield ethylene and inorganic mercury (Daniel et al., 1971). It is possible that the cleavage is not catalyzed by an enzyme system (Daniel et al., 1971); rather cleavage actually occurs by the degrada-tion system of the liver-soluble fraction like phenylmercury (Lefevre and Daniel, 1973).

Whether the cleavage of the C–Hg bond in organomercurials is due to enzymic or nonenzymic mechanisms in the mammalian body has not been conclusively elucidated. Though recent studies mentioned above suggest that enzymic mechanisms play the main role, the possibility of nonenzymic mechanisms cannot be abandoned; for instance, ascorbate attacks the C–Hg bond of phenyl, methyl, and methoxyethylmercury in the presence of cupric ion, and γ-globulin and albumin can degradate phenylmercury to yield inorganic mercury (Hg^{2+}) (Gage, 1975). To prove that these nonenzymic mechanisms work in the body, more careful studies simulating actual condi-tions are needed.

Implications of toxicity and C–Hg bond breakage. The idea that inorganic mercury is a mediator of the toxic effects of organomercurials is a recurring one (see Clarkson, 1972a). This is also the case for the toxic effect of short-chain alkylmercury on the central nervous system. Syverson (1974a) injected rats with [203]Hg-labeled methylmercury chloride or mercuric chloride and found that more inorganic mercury in the brain was retained after the injection of methylmercury than after an equal dose of mercuric chloride. However, he also noticed the inverse dose dependence of the amount of inorganic mercury in the brain. At present, the significance of the inorganic mercury released from alkylmercury as a mediator of the toxic effect is still obscure. In the case of ethylmercury, which produces damage in the central nervous system similar to methylmercury, the proportions of inorganic mercury in the brain of experimental animals reach around 50% in mice and rats (see Suzuki et al., 1973) and 90% in squirrel monkeys (Blair et al., 1975). Levels of inorganic mercury were slightly over 6 μg/g tissue; the proportion to total mercury was about 35% in the brain of a patient who died of ethylmercury poisoning (Suzuki et al., 1973). This discrepancy between methyl- and ethylmercury compounds will not be clarified until more refined mechanisms better define the toxicity of alkylmercurials to the central nervous system. The finding by Kitamura and Hayakawa (private communication) that *n*-propylmercury is less than methyl- and ethylmercury and *n*-butylmercury is not able to induce the central nervous system damage suggests that the inorganic mercury release is not the only factor that explains the toxicity.

   Biotransformation of mercurials to methylmercury. This area was recently reviewed by Neville and Berlin (1974). Since Jensen and Jernelöv (1969) were able to demonstrate microbial methylation of mercuric chloride to both mono- and dimethylmercury using lake bottom sediments and the homogenate of decaying fish, the possibility of biotransformation of inorganic mercury to methylmercury in the mammalian body has been considered. Westöö (1968) suggested possible methylmercury formation in cattle liver. Kivimäe et al. (1969) found methylmercury in organs and eggs of hens fed wheat containing mercuric nitrate, methoxyethyl-, or phenylmercury. The reason why methylmercury was found was not clear. Various microorganisms have been detected that possess the ability to convert inorganic mercury to methylmercury, for example, *Clostridium cochlearium* (Yamada and Tonomura, 1972), *Metanobacterium omelianskii* (Wood et al., 1968), and *Neurospora* sp. (Kitamura et al., 1969). In the intestinal bacterial flora of animals, some microorganisms capable of converting inorganic mercury to methylmercury may exist. It is reported that inorganic mercury can be converted to methylmercury in a surgically arranged "blind loop" of the rat's intestine (Abdulla et al., 1973). No bacteriological investigation of the blind loop was conducted, but the conversion by the intestinal flora would be possible under certain conditions that facilitate the growth of microorganisms that convert inorganic mercury to methylmercury. Methylation *in vitro* of inorganic mercury was reported when methylcobalamin, a vitamin B_{12} analog, and inorganic mercury were mixed (Bertilsson and Neujahr, 1971; Imura et al., 1971). The kinetics and mechanism of cobalamin-dependent methyl and ethyl transfer of mercuric ion were studied (DeSimone et al., 1973). Another methylated complex of cobalt, methylcobaloxim, also showed the same characteristic, though the conversion was slower than methylcobalamin (Kim et al., 1971).

Inorganic mercury was converted to methylmercury by incubation with the liver homogenate of a certain kind of tuna, but not of other fishes (Ukita and Imura, 1971). Species differences were observed in nontreated or pronase-treated fish livers. The liver's ability to make this conversion after pronase treatment can be related to the content of vitamin B_{12} in the liver (Imura, 1974).

Excretion

Biliary excretion. General aspects of gastrointestinal excretion are well covered by the report of the Task Group on Metal Accumulation (1973). Biliary excretion of mercurial compounds has been one of the most explored themes in recent years. Biliary excretion of inorganic and organomercurial compounds has been demonstrated directly in rats (Norseth and Clarkson, 1971; Cikrt, 1972) and indirectly in mice (Magos and Clarkson, 1973). No direct data on humans have been reported on the role of biliary excretion and reabsorption (Task Group on Metal Accumulation, 1973), but the effectiveness of the peroral administration of polythiol resin in decreasing the body burden of mercury in methylmercury-poisoned patients has indirectly shown the biliary excretion of methylmercury (Bakir et al., 1973).

In rats 10 μg methylmercury cysteine, 2.5 μg protein-bound methylmercury, and 0.5 μg protein-bound inorganic mercury were found in the bile after intravenous injection of 170 μg Hg per rat as methylmercury chloride; by other routes of gastrointestinal excretion, including intestinal cell shedding and pancreatic excretion, less than 2 μg Hg was excreted (Norseth and Clarkson, 1971). Also in rats 24 hr after intravenous injection of mercuric chloride, the amount of mercury excreted via bile predominated over other routes of excretion (Cikrt, 1972). The mercury in rat bile after intravenous injection of mercuric chloride was predominantly found on the gel zone of high-molecular, bile-specific protein by disc electrophoresis (Cikrt and Tichý, 1972) and on the high-molecular-weight fraction by Sephadex G-100 chromatography (Havrdová et al., 1974). Because biliary excretion of methylmercury was clearly higher than the fecal excretion, Norseth and Clarkson (1971) considered that methylmercury cysteine excreted in bile was reabsorbed by the intestine. Reabsorption of methylmercury was inhibited by an oral administration of polythiol resin that is insoluble and not absorbable in the intestine of mice (Clarkson et al., 1973). The administration of reduced human hair decreased the body burden of methylmercury in mice (Takahashi and Hirayama, 1971), that is understood by inhibition of reabsorption. Thus attention has been focused on the enterohepatic circulation of methylmercury and its implication for biological half-life and accumulation (Norseth, 1973c).

Norseth and Clarkson (1971) postulated that the dominant small molecule complex of methylmercury in rat bile is methylmercury cysteine, but Norseth (1973c) found that the dominant compound in mouse bile may be glutathione and that in rat bile it is not cysteine but peptides other than glutathione. Refsvik and Norseth (1975) reexamined the rat bile after methylmercury administration and found that the predominant complex was methylmercury glutathione, although a small amount was methylmercury cysteine.

According to Ohsawa and Magos (1974), the methylmercury in rat bile is mainly complexed with a compound that has a molecular weight between cysteine and glutathione. According to Takahashi (1974), the methylmercury in rat bile is found mainly on the part corresponding to methylmercury cysteine and slightly on the part corresponding to methylmercury acetyl-cysteine by ion-exchange chromatography. Later methylmercury glutathione was nominated the more likely compound, not methylmercury cysteine (Hirata and Takahashi, 1975). The experiments—by which the effect of mercaptodextran, a dextran thiolated by using N-acetylhomocysteine, and N-acetylhomocysteine was tested in mice exposed to methylmercury—revealed that the N-acetylhomocysteine–methylmercury complex is easily absorbed when formed in the gastrointestinal tract and easily excreted in the urine (Aaseth and Norseth, 1974). N-acetylhomocysteine is released from mercaptodextran in the gastrointestinal tract.

After a single intravenous injection of methylmercury chloride the rat bile was collected and perorally administered to the duodenum of other rats (Norseth, 1973b). In these experiments about 7% of the originally injected dose was excreted via the bile of the injected rat, and 3.9 or 4.8% was excreted via the bile of second rats receiving the bile of injected rats. Because the intestinal absorption of methylmercury excreted via bile is nearly 90% (Norseth and Clarkson, 1971), the above figures of reexcretion via bile do not show the existence of perfect enterohepatic circulation.

Reabsorbed methylmercury complex may undergo some modification in the liver and may change a chemical form that is not excreted via bile. Takahashi (1974) postulated that methylmercury cysteine is acetylated into methylmercury acetylcysteine in the liver after reabsorption from the intestinal tract because the predominant chemical form in the urine of rat injected with methylmercury was methylmercury acetylcysteine. But as already mentioned, the chemical form of methylmercury complex in bile is still a debatable question, and the difficulty is certainly due to the intricacy of chemical analysis.

There are differences in the quantity of reabsorption of inorganic mercury, phenylmercury, and methylmercury. After intraduodenal administration of biliary excreted mercury, inorganic mercury was absorbed in $21.1 \pm 10.7\%$ (Cikrt, 1973), and phenylmercury absorption was 17.7–25.6% (Cikrt and Tichý, 1974). Norseth and Clarkson (1971) indicate 9% in inorganic mercury after 24 hr, 40% in methylmercury after 2 hr, and 70% 6 hr after administration. The similar rate of absorption of phenylmercury to that of inorganic mercury indirectly shows the cleavage of C—Hg bond before biliary excretion.

Change of biliary excretion of mercurial compounds. Cherian and Vostal (1971) demonstrated in rats a 3- to 5-fold increase and a more than 50-fold increase in the rate of biliary excretion of methylmercury and mercuric mercury, respectively, by 2,3-dimercaptopropanol (BAL). Both mercurial compounds were found in the low-molecular-weight fraction of bile.

Pretreatment with spironolactone before a high dose of mercuric chloride induced a more than 10-fold increase of fecal excretion (Haddow and Marshall, 1972) and a 3-fold increase of biliary excretion of mercury in rats (Haddow et al., 1972), although Garg et al. (1971) failed to demonstrate an increase in biliary excretion of mercuric mercury (Hg^{2+}) by spironolactone administration.

D,L-penicillamine and BAL administered after the injection of methylmercury chloride increased the biliary excretion of mercury in rat. This effect was sex dependent and dependent on the time interval between mercury injection and treatment (Norseth, 1973a). Pretreatment with phenobarbitone caused an 1.4-fold increase of mercury excretion accompanied by an increase in bile flow of methylmercury-injected rats; mice pretreated with phenobarbitone showed a 1.3-fold increase of fecal excretion of mercury, which was trapped by polythiol resin, after methylmercury injection (Magos and Clarkson, 1973). This effect by barbiturates was confirmed again in rats; infusion of sodium dehydrocholate increased bile flow 2.6-fold, but did not affect methylmercury excretion (Magos et al., 1974a). Increased bile flow is thus considered as not having a direct influence for biliary excretion of methylmercury (Norseth, 1973a; Magos et al., 1974a). Diethyldithiocarbamate (DDC) and tetraethylthiuram disulfide (disulfiram) decreased the biliary excretion of mercury after the injection of methylmercury chloride in rats (Norseth, 1974). Klaassen (1975) compared the effects of pretreatment with phenobarbital, spironolactone, pregnenolone-16-carbonitrile (PCN), or 3-methylcholanthrene on biliary excretion of mercuric chloride and methylmercury and found PCN to be the most effective.

Renal excretion. Glomerular filtration, tubular reabsorption, uptake via peritubular membrane, and tubular secretion are all concerned with the renal excretion of mercurial compounds. In these processes the chemical form of mercurial compounds is of primary importance. The rate of filtration is very much related to the molecular weight or the size of molecule; about 1×10^4 has been determined as the threshold molecular weight, over which decrease of filtration rates begins. The diffusible fraction of mercury in plasma consists of two kinds of mercury—that bound to small molecule compounds and to proteins of filterable molecular weight.

The reabsorbability of the mercury complexes has not been elucidated adequately. In the review by Vostál (1968), inorganic mercury and other organomercurials were considered to be present in the tubular fluid as cysteine complexes. But this thought is likely to be modified in light of recent results on the chemical form of mercury in plasma (see the section on transport) and in urine. The form of small molecules complexed with mercury in urine was reported as cysteine in the case of diuretic mersalyl in dogs (see Clarkson, 1972a), as a compound of molecular weight between cysteine and glutathione in the case of inorganic mercury in rats (Jakubowski et al., 1970), and as acetylcysteine in the case of methylmercury in rats (Takahashi, 1974). Elemental mercury, ionic mercury, and mercurials of other forms were detected (Henderson et al., 1974) in the urine of workers exposed to mercury vapor.

The inorganic mercury and organomercurial diuretics filtered through glomerulus were considered to be reabsorbed in the tubule, and the reabsorption could occur with small molecule complexing compounds like cysteine (Vostál, 1968). But the mercury was complexed with proteins of moderate molecular weight (about 11,000) in urine of rats (Jakubowski et al., 1970). The reabsorption and the tubular secretion of the complex of such molecular weights need further study.

The difference in amount of urinary excretion between two routes of injection (iv and sc) of inorganic mercury was demonstrated in rabbits and found to be dose dependent; at a single low dose (0.5 mg/kg), the rabbits injected sc excreted a much greater amount; at higher doses the rabbits injected iv excreted greater amounts of mercury (Miyama et al., 1968). Subcutaneous injection produced higher mercury concentrations in brain and kidney and lower levels in liver and blood than intravenous injection without showing a dose dependence. The difference may be explained by the possibly variant chemical forms of transported mercury or by the difference in blood flow after injections. In the case of intravenous injection of inorganic mercury, the reaction with vascular tissue must be considered (Perry et al., 1970).

The removal of inorganic mercury from subcutaneous injection sites and the uptake in kidney tissues was facilitated by cysteine given with mercury and delayed by ascorbic acid pretreatment (Magos, 1973). Foulkes (1974) demonstrated that in the presence of mercaptoethanol, essentially all inorganic mercury, which was injected into the renal artery, disappeared from the blood of renal vein. Probenecid failed to prevent this clearance of mercury, and urinary excretion of mercury was absent. Thus a small molecule thiol compound like cysteine is considered to accelerate the glomerular filtration, and to increase the peritubular accumulation of mercury.

In a review of renal excretory mechanisms of mercury compounds, Vostál (1968) stated that the glomerular transport of mercury was supposed to play only a secondary, if any, role in the excretion of mercury. Another way mercurial compounds can reach the tubular lumen is by transport across the tubular wall. Recent reviews (Clarkson, 1972a; Nordberg and Skerfving, 1972) also are of the same opinion as that by Vostál (1968). Tubular secretion of mercury was demonstrated in avian kidneys (Vostál and Heller, 1968). Berlin and Vostál (cited in Vostál, 1968) found by micropuncture measurements that the amount accumulated in tubular cells and not the plasma concentration proved more decisive for the transport of mercury into the nephron than did filtration through the glomerulus. Peritubular transfer of inorganic mercury from the peritubular capillaries was suggested by the experiment on the effect of various metabolic inhibitors to the uptake and excretion of mercury in rats (Clarkson and Magos, 1970). The proportion of inorganic mercury to total mercury in urine followed the proportion of inorganic mercury in kidney tissue and was not related to the proportion of inorganic mercury in plasma in rats injected with a single dose of methyl-mercury chloride (Norseth and Clarkson, 1970a). All these results are compatible with the idea that tubular secretion is the main pathway in renal excretion of mercurial compounds.

The mechanisms of tubular secretion of mercurial compounds are, however, unknown. Exocytosis and exfoliation of epithelial cells are possible mechanisms relevant to tubular secretion.

High doses of inorganic mercury (0.4–1.2 mg Hg/kg, iv) caused the increase of urinary mercury excretion without the change in fecal excretion that accompanied renal damage (Piotrowski et al., 1969). In relation to the effect of renal damage on urinary mercury excretion, a significant role of normal turnover and abnormal desquamation of tubular cells was pointed out by Cember (cited in Magos, 1973). Tubular cell damage with the increase of urinary mercury excretion was reported after the administration of thio-acetamide in rats (Trojanowska et al., 1971) and sodium fluoride (Magos, 1973) in rats. In rats injected with mercuric chloride, thioacetamide administration caused an increased urinary excretion of high-molecular-weight mercury-containing protein (Piotrowski et al., 1973). The release of alkali and acid phosphatase in urine without change of the enzymes in serum at the earliest stage after inorganic mercury injection in dogs (Ellis et al., 1973) also suggests the contribution of tubular cell damage to urinary mercury excretion. In human urine a considerable proportion of the excreted amount of mercury was found in the precipitate after a centrifugation of $10^5 \times g$ for 60 min (Suzuki and Shishido, 1975b). The difference between the total mercury and the measurable mercury without any oxidation in urine of workers exposed to mercury vapor (Henderson et al., 1974) may also have resulted from the excretion of mercury by exocytosis or epithelial cell desquamation. Fowler (1972a,b) observed that a long-term feeding of methylmercury chloride (2 ppm Hg in ration) to female rats for 12 wk caused extrusion of numerous cytoplasmic masses in tubule lumens. These masses were characterized by the presence of a smooth endoplasmic reticulum. This finding is postulated to be related to the cytotoxic effect of methylmercury in the proximal tubule and the possible route of urinary excretion of mercury as methylmercury or inorganic mercury released from methylmercury.

If the amount of mercury accumulated in the tubule is decisive in excretion, the storage and release of mercurials to and from kidney tissues should not be neglected in the discussion of renal excretion.

As factors affecting the kidney level of mercury in rats Magos (1973) identified the following: (1) absorption from the injection site; (2) chemical form of mercury; (3) complexing agents; (4) kidney metabolism; and (5) desquamation of tubular cells. Chemical form differences of mercurials (elemental, inorganic mercury, or organomercury) serve to regulate uptake in kidneys, the brain, or other tissue (see Clarkson, 1972a; Nordberg and Skerfving, 1972; Magos, 1973). For instance, kidneys take up a greater amount of mercury after metallic mercury is injected than after the same dose of inorganic mercury is injected (Magos, 1973).

The mercury taken up in kidney tissues is bound to two different classes of binding sites (Clarkson and Magos, 1966). Proteins of high and moderate molecular weights and a trace amount of dialyzable compounds in an aqueous homogenate of kidney were found to bind mercury after inorganic mercury administration (Jakubowski et al., 1970). The protein of moderate molecular weight (11,000) in kidney tissues was claimed to be identical with metallothionein (Wiśniewska et al., 1970). It was demonstrated that mercury was able to induce synthesis of this metallothionein in kidney tissues but not in the liver in contrast to cadmium, which induces the synthesis in both liver and kidney tissues (Piotrowski et al., 1973, 1974). Mercury present in the kidney was bound predominantly to this metallothionein. A very large portion of mercury after inorganic mercury, phenylmercury, or ethylmercury administration was bound to a soluble protein of 11,000 molecular weight in aqueous extracts of kidney of rats (Ellis and Fang, 1971; Fang, 1973). In contrast to the above-mentioned mercurials, after feeding methylmercury at toxic levels to rats and Japanese quail, a few percent of the total mercury was found in the peak of gel filtration chromatography corresponding to metallothionein in the supernatant fraction of kidney and liver homogenates. Methylmercury had a low affinity for the metal-free form of metallothionein isolated from horse kidney (Chen et al., 1973). Based on these results, the induction of metallothionein synthesis and the binding of mercury to this protein is considered to be mainly due to the inorganic form of mercury, which is released at a slower rate in methylmercury than in ethyl- or phenylmercury.

The chemical properties of inorganic mercury-induced metallothionein were not quite identical with cadmium-induced metallothionein in the kidney of rabbits (Nordberg et al., 1974). For this reason, the Task Group on Dose–Response Relationships of Toxic Metals (1975) adopted the term "metallothionein-like protein" instead of metallothionein as far as mercurial compounds are concerned. Both metallothionein induced by cadmium and metallothionein-like protein induced by mercury might help combat the toxic effects of inorganic mercury (Magos et al., 1974c; Yoshikawa, 1974).

Change of urinary mercury excretion due to various chemical compounds. Various chemical compounds have been tested for their effect as antidotes against toxicity of mercury compounds, and some have been used in human cases. The change of urinary mercury excretion using these chemical compounds has been described frequently. Swensson and Ulfvarson (1967) extensively tested the effect of such compounds as penicillamine, BAL, EDTA, ascorbic acid, thioctic acid, Rongalite-c, thiomalic acid, thioacetamide, and PAS in rats. In the case of mercuric nitrate, no compounds showed the enhancement of fecal excretion, and only BAL and thiomalic acid increased urinary excretion. For phenylmercury, several compounds (e.g., BAL, thioctic acid, thioacetamide, and PAS) enhanced both urinary and fecal excretion, and ascorbic and thiomalic acids enhanced only urinary excretion. In the case of methylmercury, no compounds increased fecal excretion, and penicillamine, BAL, thioctic acid, thiomalic acid, and thioacetamide enhanced urinary excretion. In rats spironolactone prevented renal damage caused by mercuric chloride and reduced urinary mercury excretion and mercury content in organs, in contrast to increased fecal mercury excretion (Garg et al., 1971). Haddow and Marshall (1972) confirmed decreased renal retention and increased fecal excretion of mercury by spironolactone in rats injected with mercuric chloride, but they found that urinary mercury excretion was only delayed by spironolactone; the amount excreted within 24 hr after injection did not show a significant difference.

Fowler et al. (1975) found that phenobarbital-injected rats given a low oral dose level of methylmercury for 2 or 4 wk had increased urinary excretion of inorganic mercury and increased blood concentrations of methylmercury.

Effects of several thiol compounds such as 2-mercaptopropionylglycine, mercaptoacetic acid (MAA), D,L-penicillamine, glutathione (GSH), BAL, and penthanil (DTPA) were compared in terms of distribution and excretion of mercury in rats or mice given an oral dose of methylmercury chloride (Ogawa et al., 1972a,b). Significant increases of urinary, but not fecal, mercury excretion were observed after monothiol compounds were administered (e.g., 2-mercaptopropionylglycine, MAA, D,L-penicillamine, and GSH) in rats and mice. Increase of brain mercury level due to redistribution by thiol compounds was observed in the case of dithiol compounds as in the experiment using BAL and organomercurials in mice by Berlin and Rylander (1964) and Berlin et al. (1965). The oral administration of cysteine during and after oral doses of methylmercury demonstrated no significant increase of urinary and fecal mercury excretion in rats (Sugiyama et al., 1975). Mercaptodextran, *N*-acetylhomocysteine thiolactone, or *N*-acetylhomocysteine given orally significantly increased urinary, but not fecal, mercury excretion in mice injected with methylmercury chloride (Aaseth and Norseth, 1974). Both cysteine and selenium, plus a combination of cysteine and selenium, prevented and delayed the onset of toxicity from methylmercury in rats, but did not increase the elimination of mercury from the body via urine and feces; on the contrary, they retained slightly more mercury (Stillings et al., 1974).

In a factory worker poisoned by mercury vapor, a markedly large increase of urinary mercury (total mercury) excretion and a slight increase of fecal mercury excretion were observed during treatment with *N*-acetyl-D,L-penicillamine (Kark et al., 1971). Yamaguchi et al. (1974) reported that 2-mercaptopropionyl enhanced urinary mercury excretion in a patient with mercury vapor poisoning; Hamada et al. (1974) and Kaku et al. (1975) also reported a positive effect of 2-mercaptopropionyl among heavy fish-eating populations. Urinary mercury excretion in a patient severely poisoned with mercury vapor was remarkably elevated by D-penicillamine (Mikawa et al., 1975).

In human cases of methylmercury poisoning, D-penicillamine enhanced urinary mercury excretion (Tsubaki and Shirakawa, 1966; Suzuki and Yoshino, 1969). Peroral administration of D-penicillamine to a patient with mercury vapor poisoning enhanced the urinary excretion of organic but not inorganic mercury (Ishihara et al., 1974). This selective effect of D-penicillamine for organic mercury excretion was also observed in five normal persons (Suzuki et al., 1976), when a significant correlation between plasma levels and 24-hr urinary output was found for inorganic but not organic mercury.

Based on these results, it is postulated that the effect of antidotes, except selenium, on urinary mercury excretion is related to the character of mercury complexes and small molecule compounds formed from the administration of antidotes, in addition to the difference of mecurial compounds in binding to the tissue binding sites. The level of body burden of mercurial compounds and the time lapse between exposure and the administration of antidotes also affect the results achieved with the antidotes. To get an enhanced urinary mercury excretion, the antidote has to be able to compete with the tissue binding of mercury, and the formed complex of mercury and a small molecule compound has to be filterable in the glomerulus and not reabsorbable in the tubulus. The peritubular uptake of the formed complex and the effect of antidotes on tubular cells modify the urinary mercury excretion. In the sense of decreasing reabsorbability of the complex, the acetylated thiol compound has been claimed to enhance the urinary excretion of methylmercury more than the corresponding nonacetylated thiol compound. *N*-acetyl cysteine (Takahashi, 1974) and *N*-acetylhomocysteine (Aaseth, 1975) were nominated as such compounds. As a filterable and less reabsorbable candidate, a protein of molecular weight about 10,000 was of particular interest. In case mercury can form a complex with such a protein and the formed complex is filtered, the urinary excretion of mercury should be enhanced. The intravenous injection of the mercaptodextran of molecular weight about 10,000 may suggest a possible role of a protein of such molecular weight. In contrast to oral administration, the intravenously administered mercaptodextran did not undergo a release of small molecule thiols (Aaseth, 1973). In mice administered a dose of mercuric chloride, the mercaptodextran of molecular weight about 10,000 (SH-10) eliminated the whole-body retention of mercury remarkably without increasing brain mercury level; conversely, other thiol compounds such as BAL, sodium diethyl dithiocarbamate, D-penicillamine, and 2-mercaptopropionyl glycine failed to eliminate the whole-body retention, and dithiol compounds induced an increased retention of mercury in the brain (Aaseth, 1973). This effect of SH-10 is interpreted as due to the increased urinary excretion of mercury.

Secretion in milk. The Task Group on Metal Accumulation (1973) pointed out that newborns can be exposed to mercurial compounds through milk, based on the report on epidemics of methylmercury poisoning in Iraq (Bakir et al., 1973). Several animal experiments revealed the secretion of mercury in milk after administering various mercurial compounds in different species, for example, inorganic mercury in goats (Howe et al., 1972), phenylmercury in cows (Miyamoto, 1974a; Neathery et al., 1974), goats (Miyamoto, 1974a), and guinea pigs (Miyamoto, 1974b), and methylmercury in guinea pigs (Trenholm et al., 1971).

The total mercury concentrations in human milk from lactating women who had been exposed to methylmercury in fish was about 5% of the total mercury concentrations in blood (Skerfving and Westöö, cited by Nordberg and Skerfving, 1972). In experiences with methylmercury exposure in Iraq, transmission via milk contributed significantly to infant blood levels (Amin-Zaki et al., 1974).

Because the percentage in milk of the total administered dose was as low as 0.17% in 14 days after a single oral dose of methylmercury in cows, Neathery et al. (1974) concluded that milk was protected from methylmercury. However, as already mentioned, human experiences in Iraq (Bakir et al., 1973; Amin-Zaki et al., 1974) do not agree with this conclusion.

In the milk of women exposed to methylmercury, 40% of the mercury was an inorganic form (Bakir et al., 1973). The uptake of mercury in a baby's brain via milk should be smaller than that in the mother's brain because of different kinetic properties of methylmercury and inorganic mercury. The more rapid clearance rate of mercury from the brain of offspring than of mothers, which was found in rats that had received an oral dose of methylmercury during gestation (Casterline and Williams, 1972), does not necessarily support this contention. The clearance rate of methylmercury once retained in the brain is more rapid than phenylmercury (Suzuki et al., 1969).

Observations by Yang et al. (1973) in rats dosed orally with methyl-mercury at day 16 of gestation revealed that the brain of offspring contained greater concentrations of mercury than did their mothers' brain at the end of 5 days of nursing. From these results, it seems that the incomplete blood-brain barrier in the fetus and the newborn has to be taken into consideration.

Elimination via other routes. Mercury is eliminated via routes other than biliary and renal excretion; for example mercury concentrations ranging from 6 to 31 ng/ml were found in the pooled parotid fluid (Windeler, 1973). Sweating, pancreatic secretion, and intestinal wall secretion are other routes. Hair and nails also contribute to the elimination of mercurial compounds. In normal Japanese excretion by these means was estimated at 2-3 µg Hg/day (Kondo, 1971). The different concentrations in hair of the various chemical forms of mercury have been seen in the available data, but the adsorption of mercury from outside the body makes the quantitative interpretation difficult in humans.

In methylmercury-exposed people, a linear relationship was observed between total mercury levels in blood and hair (Berglund et al., 1971; Birke et al., 1972). In light of selective accumulation of methylmercury in red blood cells, the relationship was reexamined of total mercury levels in red blood cells and in hair (Skerfving, 1974) and of total mercury level in red blood cells and organic mercury level in hair (Suzuki and Miyama, 1975) in populations eating a lot of fish. A linear relationship was confirmed by these studies. The equation obtained by Skerfving (1974) is Y (total Hg in hair, µg/g) $= 230X$ (total Hg in blood cells, µg/g) $- 3.6$; that of Suzuki and Miyama (1975) is $Y = 59X + 4.4$ (in males) and $Y = 48X + 2.2$ (in females). As the noticeable difference in regression coefficients shows, we need much more data from various conditions of exposure to mercurial compounds. Even in populations whose exposure to mercurials is mainly limited to the consumption of methylmercury-containing fish, there is an additive intake of other forms of mercury.

METABOLIC PROPERTIES AND TOXICITY

Metabolic Properties and Distribution

The study of the differing patterns of distribution of the various chemical forms of mercurials with time after administration would be included, in a broad sense, in the study of the metabolic properties of mercurial compounds. Even though the distribution is very closely related to the metabolic properties, a single generalized rule determining the pattern of distribution does not exist. Physicochemical properties, for example, lipid or water solubility or dissociability of mercurials, are of primary importance as factors influencing distribution. As already mentioned in the section on metabolism, however, these properties are modified during metabolism. From the data accumulated to date, mercurials seem to be metabolized decreasing the lipid solubility and increasing the water solubility ($Hg^0 \to Hg^{2+}$; $RHg^+ \to R + Hg^{2+}$). The metabolism in the opposite direction ($Hg^{2+} \to Hg^0$; $Hg^{2+} \to RHg^+$) seems not to be the main pathway in mammals, but it has been found in microorganisms.

The subcellular distribution is also related to metabolic properties of mercurials. Mercury in any chemical form is usually found in all the centrifugal fractions (see Clarkson, 1972a), but the pattern differs with time after dosing and chemical forms. In rat liver mercury after inorganic mercury (Hg^{2+}) injection distributes similarly to that after methoxyethylmercury injection; mercury after methylmercury injection shows less content in the lysosome/peroxicome fraction and more in the microsome fraction (Norseth, 1969). The lysosomal content of mercury is mainly inorganic mercury after methylmercury injection (Norseth and Brendeford, 1971).

Morphological and functional changes in liver subcellular components have been reported after administration of inorganic mercury or methylmercury (Cornish et al., 1970; Brubaker et al., 1971; Pekkanen, 1971; Alvares et al., 1972; Lucier et al., 1972; Pekkanen and Lindberg, 1972; Pekkanen and Pekkarinen, 1972; Pekkanen and Salmine, 1973; Wagstaff, 1973; Chang and Yamaguchi, 1974; Carlson, 1975), and some results are interpreted by relatively selective accumulation of mercury in a certain subcellular component.

Lysosomal uptake of mercury in rat renal proximal tubule cells after inorganic mercury administration (Timm et al., 1966) and after a long-term feeding of methylmercury (Fowler et al., 1974) has been demonstrated by histochemical or electron microscope methods. The mercury found in lysosomes after methylmercury feeding is likely to be the inorganic form. The fact that the damage of tubular epithelial cells in rat kidney was histologically the same in both inorganic and methylmercury administrations (Klein et al., 1973) supports this interpretation.

In subcellular fractions of nervous tissue, a discrepancy between the results of histological (Chang and Hartmann, 1972) and biochemical examinations (Yoshino et al., 1966, Syversen, 1974b) has been noticed after methylmercury administration. By histological methods no mercury was detected in the nucleus, but all the intracellular components included mercury by biochemical methods. The subcellular distribution of mercury was correlated with the protein content, but the myelin fraction contained four times as much inorganic mercury relative to total mercury, compared with other fractions after methylmercury injections (Syversen, 1974b).

Metabolic Properties and Accumulation

From the theoretical point of view, at continuous exposure to a certain dose level, a dynamic equilibrium between absorption and excretion will be attained after a period of time, providing the dose level does not affect body functions and does not change absorption and excretion. The length of time necessary to attain dynamic equilibrium depends on the rate of turnover of the materials. The rate of turnover differs with both the dose level and the phase of exposure. The phases of accumulation, equilibrium, and elimination may have different rates of turnover. The turnover rate usually has been estimated from the elimination phase, but an attempt to determine turnover rate from accumulation data for humans has been made (Tsuchiya et al., 1975).

The rate of elimination can be approximated with reasonable accuracy from a single exponential first-order function (Task Group on Metal Accumulation, 1973). By this approximation the decay kinetics of mercury after methylmercury uptake has been examined on the whole body or each organ and tissue of humans and animals (Berglund and Berlin, 1969; Suzuki, 1969; Berglund et al., 1971). The biological half-time was shorter in a greater dose level than in a smaller dose level in every organ and tissue studied in mice receiving 10 injections of methylmercury; this half-time varied from 4.1 days for blood to 6.2 days for cerebellum and liver at the larger dose level (100 μg Hg, 10 times) (Suzuki, 1969). In methylmercury the smaller variation of biological half-times in organs and tissues makes the approximation possible on the one-compartment model on a whole-body level. However, inorganic mercury (Hg^{2+}) and phenylmercury have been excreted in feces or urine according to two-compartment models (Cember and Donagi, 1964). A four-compartment model has been postulated for elimination of mercury from the whole body of rats injected with inorganic mercury (Hg^{2+}) (Cember, 1969). In most organs of mouse injected subcutaneously with inorganic mercury (Hg^{2+}), patterns of diminution fitted the one-compartment model, but the brain showed that decay followed the two-compartment model; the elimination pattern of total and organic mercury in liver and kidneys after intravenous injection of ethylmercury in mice was one-compartmental and that in brain was two-compartmental (Matsubara-Khan, 1974). These data show the difficulty of application of a simple one-compartment model to the accumulation kinetics of mercurials, except methylmercury. Even for methylmercury, a careful treatment is necessary. The dose seems to be especially important in this regard. After repeated intraperitoneal injections of methylmercury in mice, all organs and tissues except hair reach a point of mercury accumulation saturation, that is, the mercury concentration in each organ and tissue becomes similar to the whole-body concentration (accumulated dose/body weight) (Salvaterra et al., 1975). The time required to reach saturation differs among the organs and tissues.

Accumulation of mercury in humans with or without particular exposure to mercurials has been recently examined. In the brain of workers exposed to mercury vapor, very high concentrations remained after the cessation of exposure (Takahata et al., 1970; Watanabe, 1971). In postmortem samples from mercury mine workers, inhabitants of a mercury mining town (Idrija,

Yugoslavia) and from a control group, the accumulation of mercury in organs is demonstrated in mercury workers and, to a lesser extent, in inhabitants of the town. Among organs of workers, the highest accumulation of mercury was found in the thyroid (35.2 μg Hg/g fresh weight) and pituitary (27.1), both markedly higher than the level in kidneys (8.4). Selenium was found at a 1:1 molar ratio to mercury in these workers' organs (Kosta et al., 1975). In these Yugoslavian people, methylmercury was not remarkable, but in postmortem samples from normal Japanese, the proportion of methylmercury to total mercury was 2.8% in the kidney, 10-15% in other organs including parts of the brain, and 58% in hair. A higher concentration was found among males than females (Kitamura et al., 1974). In normal Japanese, mercury, which is taken into the body in daily life, consists of methylmercury from fish and other foodstuffs and inorganic mercury from various sources. The proportion of methylmercury to total mercury depends on the fish consumption, but is over 50% in people consuming the national average amount of fish (Shishido and Suzuki, 1974). Therefore, the values found in postmortem samples from Japanese have to be interpreted as having resulted from a relatively longer biological half-time of inorganic mercury than methylmercury in organs and tissues except hair.

Individual variations of the biological half-time of methylmercury are large enough to cause a significant difference in susceptibility to this mercurial, according to Iraqi data (Al-Shahristani and Shihab, 1974; Nordberg and Strangert, 1974). The cause of this individual variation is not known. Sex differences and the effect of age on the metabolism of mercurials should be considered (Kitamura et al., 1974; Lin et al., 1974; Suzuki and Miyama, 1975).

Distribution, Accumulation, and Toxicity

The earliest toxic effect of mercurial compounds is expected to appear when the retention of mercury in a critical organ reaches critical organ concentration. As defined in the report of the Task Group on Dose–Response Relationships of Toxic Metals (1975), critical organ concentration is the mean concentration in the organ at the time any of its cells reach critical concentration at which undesirable reversible or irreversible functional changes occur in the cell, and critical organ is the particular organ that first attains the critical concentration of a metal. As these definitions show, the concept is rather operational in its character. There is, in general, no fixed value of critical organ concentration because it varies depending on many factors. Nevertheless, mercury has to reach the site of action in the cell in a quantity large enough to induce changes. Therefore, distribution and accumulation are basic data in studying the toxicity.

Whether or not all mercury in the organ contributes to the toxic phenomenon is debatable. In other words, the proportion of biologically active mercury to total mercury deposited in the organ must be known. With regard to inactivation of mercury, metallothionein may play an important role in some organs.

In terms of the relationship between distribution and toxicity, methylmercury has become controversial in recent years based on the report of human cases in the Minamata district. According to Takeuchi (1972), the original reports of Minamata human autopsy cases minimized the findings of peripheral nerve changes. A specific reexamination, however, confirmed the presence of peripheral sensory neuropathy. The earliest clinical sign is paresthesia in human short-chain alkylmercury poisonings, which has been reconfirmed in epidemics of methylmercury poisoning in Iraq (Bakir et al., 1973). Thus the problem centers on whether the paresthesia has resulted from the peripheral neural damage.

In rats dosed with methylmercury, there are both morphological and functional evidences that the primary site of the damage is in the cell bodies in dorsal root ganglia; secondary damage is found in the deterioration of their fibers (Cavanagh and Chen, 1971; Herman et al., 1973 Somjen et al., 1973a). Different evidence, however, has indicated that the primary site of damage is the peripheral sensory fibers, not the dorsal root ganglia (Miyakawa et al., 1970; Chang and Hartmann, 1972). The clinically unapparent but morphologically evident nervous tissue damage was found in peripheral nerves, as well as dorsal roots, of rats fed methylmercury (Grant, 1973). Nonetheless, spinal dorsal root ganglia contained the highest mercury concentration, followed closely by the cerebral cortex and the cerebellum in rats injected with methylmercury (Somjen et al., 1973b). Spinal cord, spinal roots, and peripheral nerves contained significantly less mercury than the sensory ganglia. It is not equivocal in rats that the earliest damage induced by methylmercury is in the peripheral nervous system, but the precise site of damage is controversial.

In experimentally poisoned monkeys fed methylmercury (Berlin et al., 1973, 1975b; Grant, 1973) and methylmercury poisoned humans in Iraq (Rustam and Hamdi, 1974; Von Burg and Rustam, 1974a,b), no evidence that proves the peripheral nervous system as the primary site of damage has been obtained, although a neuromuscular disorder was noticed in Iraqi cases with moderate to severe poisoning (Rustam et al., 1975). However, the motor conduction velocity was significantly slower in persons with total mercury levels in hair above 101 μg/g than in normal controls in the investigation of a Niigata epidemic of methylmercury poisonings (Kanbayashi et al., 1974).

The dose level of methylmercury in feeding experiments changes the onset and the course of intoxication in monkeys (Berlin et al., 1975b) and mice (Suzuki and Miyama, 1971; Suzuki and Shishido, 1975a). From these experimental and human experiences, the mode of dosing has to be carefully evaluated for the distribution and accumulation of mercury. The dose dependence of methylmercury metabolism may be associated with the site of the earliest damage in the nervous system. The critical organ problem in methylmercury toxicology is just an example of the complicated character of metabolism and toxicity of mercurial compounds. Methylmercury and other mercurial compounds must be studied more extensively and intensively to know all the features of toxicity.

REFERENCES

Aaseth, J. 1973. The effect of mercaptodextran on distribution and toxicity of mercury in mice. *Acta Pharmacol. Toxicol.* 32:430–441.

Aaseth, J. 1975. The effect of *N*-acetylhomocysteine and its thiolactone on the distribution and excretion of mercury in methyl mercuric chloride injected mice. *Acta Pharmacol. Toxicol.* 36:193–202.

Aaseth, J. and Norseth, T. 1974. The effect of mercaptodextran and *N*-acetylhomocysteine on the excretion of mercury in mice after exposure to methyl mercury chloride. *Acta Pharmacol. Toxicol.* 35:23–32.

Abdulla, M., Arnesjö, B. and Ihse, I. 1973. Methylation of inorganic mercury in experimental jejunal blind-loop. *Scand. J. Gastroentol.* 8:565–567.

Åberg, B., Ekman, L., Falk, R., Greitz, U., Person, G. and Snihs, J. O. 1969. Metabolism of methyl mercury (^{203}Hg) compounds in man (excretion and distribution). *Arch. Environ. Health* 19:478–484.

Al-Shahristani, H. and Shihab, K. M. 1974. Variation of biological half-life of methylmercury in man. *Arch. Environ. Health* 28:342–344.

Alvares, A. P., Leigh, S., Cohn, J. and Kappas, A. 1972. Lead and methyl mercury: Effects of acute exposure on cytochrome P-450 and the mixed function oxidase system in the liver. *J. Exp. Med.* 135:1406–1409.

Amin-Zaki, L., Elhassani, S., Majeed, M. A., Clarkson, T. W., Doherty, R. A. and Greenwood, M. R. 1974. Studies of infants postnatally exposed to methylmercury. *J. Pediatr.* 85:81–84.

Bakir, F., Damluji, S. F., Amin-Zaki, L., Murtadha, M., Khalidi, A., Al-Rawi, N. Y., Tikriti, S., Dhahir, H. I., Clarkson, T. W., Smith, J. C. and Doherty, R. A. 1973. Methylmercury poisoning in Iraq. *Science* 181:230–241.

Berglund, F. and Berlin, M. 1969. Risk of methylmercury cumulation in man and mammals and the relation between body burden of methylmercury and toxic effects. In *Chemical fallout*, ed. M. W. Miller and G. G. Berg, pp. 258–273. Springfield: Charles C Thomas.

Berglund, F., Berlin, M., Birke, G., Cederlöf, R., Euler, U., Friberg, L., Holmstedt, B., Jonsson, E., Lüng, K. G., Ramel, C., Skerfving, S., Swensson, Å. and Tejning, S. 1971. Methyl mercury in fish. A toxicologic–epidemiologic evaluation of risks. Report from an expert group. *Nord. Hyg. T. Suppl.* 4.

Berlin, M. and Rylander, R. 1964. Increased brain uptake of mercury induced by 2,3-dimercaptopropanol (BAL) in mice exposed to phenylmercuric acetate. *J. Pharmacol. Exp. Ther.* 146:236–240.

Berlin, M., Jerksell, L. G. and Nordberg, G. 1965. Accelerated uptake of mercury by brain caused by 2,3-dimercaptopropanol (BAL) after injection into the mouse of a methylmercuric compound. *Acta Pharmacol. Toxicol.* 23:312–320.

Berlin, M., Nordberg, G. and Hellberg, J. 1973. The uptake and distribution of methylmercury in the brain of *Saimiri sciureus* in relation to behavioral and morphological changes. In *Mercury, mercurials and mercaptans*, ed. M. W. Miller and T. W. Clarkson, pp. 187–208. Springfield: Charles C Thomas.

Berlin, M., Carlson, J. and Norseth, T. 1975a. Dose-dependence of methyl-mercury metabolism. A study of distribution: Biotransformation and excretion in the squirrel monkey. *Arch. Environ. Health* 30:307–313.

Berlin, M., Grant, C. A., Hellberg, J., Hellström, J. and Schütz, A. 1975b. Neurotoxicity of methylmercury in squirrel monkeys. Cerebral cortical pathology, interference with scotopic vision, and changes in operant behavior. *Arch. Environ. Health* 30:340–348.

Bertilsson, L. and Neujahr, H. Y. 1971. Methylation of mercury compounds by methylcobalamin. *Biochemistry* 10:2805–1808.

Bidstrup, P. L. 1964. *Toxicity of mercury and its compounds.* Amsterdam: Elsevier.

Birke, G., Johnels, A. G., Plantin, L. O. Sjöstrand, B., Skerfving, S. and Westermark, T. 1972. Studies on humans exposed to methyl mercury through fish consumption. *Arch. Environ. Health* 25:77–91.

Blair, A. M. J. N., Clark, B., Clarke, A. J. and Wood, P. 1975. Tissue concentrations of mercury after chronic dosing of squirrel monkeys with thiomersal. *Toxicology* 3:171–176.

Brubaker, P. E., Luicier, G. W. Klein, R. 1971. The effects of methylmercury on protein synthesis in rat liver. *Biochem. Biophys. Res. Commun.* 44:1552–1558.

Burk, R. F., Foster, K. A., Greenfield, P. M. and Kiker, K. W. 1974. Binding of simultaneously administered inorganic selenium and mercury to a rat plasma protein. *Proc. Exp. Biol. Med.* 145:782–785.

Carlson, G. P. 1975. Protection against carbon tetrachloride-induced hepato-toxicity by pretreatment with methylmercury hydroxide. *Toxicology* 4:83–89.

Casterline, J. C., Jr., and Williams, C. H. 1972. Elimination pattern of methyl mercury from blood and brain of rats (dams and offspring) after delivery, following oral administration of its chloride salt during gestation. *Bull. Environ. Contam. Toxicol.* 7:292–295.

Cavanagh, J. B. and Chen, F. C. K. 1971. The effects of methyl-mercury-dicyandiamide on the peripheral nerves and spinal cord of rats. *Acta Neuropathol. (Berlin)* 19:208–215.

Cember, H. 1969. A model for the kinetics of mercury elimination. *Am. Ind. Hyg. Assoc. J.* 30:367–371.

Cember, H. and Donagi, A. 1964. The influence of dose level and chemical form on the dynamics of mercury elimination. *Excerpta Med. Int. Congr. Ser.* 62:440–442.

Cember, H., Gallagher, P. and Faulkner, A. 1968. Distribution of mercury among blood fractions and serum proteins. *Am. Ind. Hyg. Assoc. J.* 29:233–237.

Chang, L. W. and Hartmann, H. A. 1972. Ultrastructural studies of the nervous system after mercury intoxication II. Pathological changes in the nerve fibers. *Acta Neuropathol. (Berlin)* 20:316–334.

Chang, L. W. and Yamaguchi, S. 1974. Ultrastructural changes in the liver after long-term diet of mercury-contaminated tuna. *Environ. Res.* 7:133–148.

Chen, R. W., Ganther, H. E. and Hoekstra, W. G. 1973. Studies on the binding of methylmercury by thionein. *Biochem. Biophys. Res. Commun.* 51:383–390.

Cherian, M. G. and Vostal, J. J. 1971. BAL-induced biliary excretion of mercuric and methyl mercuric ions in the rat. *Fed. Proc.* 32:261.

Childs, E. A. 1973. Kinetics of transplacental movement of mercury fed in a tuna matrix to mice. *Arch. Environ. Health* 27:50–52.

Cikrt, M. 1972. Biliary excretion of ^{203}Hg, ^{64}Cu, ^{52}Mn, and ^{210}Pb in the rat. *Br. J. Ind. Med.* 29:74–80.

Cikrt, M. 1973. Enterohepatic circulation of ^{64}Cu, ^{52}Mn and ^{203}Hg in rats. *Arch. Toxikol.* 31:51–59.

Cikrt, M., and Tichý, M. 1972. Polyacrylamide gel disc electrophoresis of rat bile after intravenous administration of ^{52}MnCl$_2$, ^{64}CuCl$_2$, ^{203}HgCl$_2$, and ^{210}Pb(NO$_3$)$_2$. *Experientia* 28:383–384.

Cikrt, M. and Tichy, M. 1974. Biliary excretion of phenyl- and methyl mercury chlorides and their enterohepatic circulation in rats. *Environ. Res.* 8:71–81.

Clarkson, T. W. 1971. Epidemiological and experimental aspects of lead and mercury contamination of food. *Food Cosmet. Toxicol.* 9:229–243.

Clarkson, T. W. 1972a. The pharmacology of mercury compounds. *Annu. Rev. Pharmacol.* 12:375–406.

Clarkson, T. W. 1972b. Biotransformation of organo-mercurials in mammals. In *Environmental mercury contamination*, ed. R. Hartung and B. D. Dinman, pp. 229–238. Ann Arbor: Ann Arbor Science.

Clarkson, T. W. and Magos, L. 1966. Studies on the binding of mercury in tissue homogenates. *Biochem. J.* 99:62–70.

Clarkson, T. W. and Magos, L. 1970. Effect of 2,4-dinitrophenol and other metabolic inhibitors on the renal deposition and excretion of mercury. *Biochem. Pharmacol.* 19:3029–3037.

Clarkson, T. W., Magos, L. and Greenwood, M. R. 1972. The transport of elemental mercury into fetal tissues. *Biol. Neonate* 21:239–244.

Clarkson, T. W., Small, H., Mich, M. and Norseth, T. 1973. Excretion and absorption of methyl mercury after polythiol resin treatment. *Arch. Environ. Health* 26:173–176.

Cornish, H. H., Wilson, G. E. and Abar, E. L. 1970. Effect of foreign compounds on liver microsomal enzymes. *Am. Ind. Hyg. Assoc. J.* 31:605–608.

Curley, A., Sedlak, V. A., Girling, E. F., Hawk, R. E., Barthel, W. F., Pierce, P. E. and Likosky, W. H. 1971. Organic mercury identified as the cause of poisoning in humans and hogs. *Science* 172:65–67.

Daniel, J. W. and Gage, J. C. 1971. The metabolism by rats of phenylmercury acetate. *Biochem. J.* 122:24p.

Daniel, J. W., Gage, J. C. and Lefevre, P. A. 1971. The metabolism of methoxyethylmercury salts. *Biochem. J.* 121:411–415.

Daniel, J. W., Gage, J. C. and Lefevre, P. A. 1972. The metabolism of phenylmercury by the rat. *Biochem. J.* 129:961–967.

Derban, L. K. A. 1974. Outbreak of food poisoning due to alkyl-mercury fungicide on southern Ghana State farm. *Arch. Environ. Health* 28:49–52.

DeSimone, R. E., Penley, M. W., Charbonneau, L., Smith, S. G., Wood, J. M., Hill, H. A. O., Pratt, J. M., Ridsdale, S. and Williams, R. J. P. 1973. The kinetics and mechanism of cobalamin-dependent methyl and ethyl transfer to mercuric ion. *Biochim. Biophys. Acta* 304:851–863.

D'Itri, F. M. 1972. *The environmental mercury problem.* Cleveland: Chemical Rubber Company.

Ellis, B. G., Price, R. G. and Topham, J. C. 1973. The effect of tubular damage by mercuric chloride on kidney function and some urinary enzymes in the dog. *Chem.-Biol. Interact.* 7:101–113.

Ellis, R. W. and Fang, S. C. 1971. The *in vivo* binding of mercury to soluble proteins of the kidney. *Toxicol. Appl. Pharmacol.* 20:14–21.

Fang, S. C. 1973. Uptake and biotransformation of phenylmercuric acetate by aquatic organism. *Arch. Environ. Contam. Toxicol.* 1:18–26.

Fang, S. C. 1974a. Induction of C–Hg cleavage enzymes in rat liver by dietary selenite. *Res. Commun. Chem. Pathol. Pharmacol.* 9:579–582.

Fang, S. C. 1974b. Uptake, distribution, and fate of ^{203}Hg-ethylmercuric chloride in the guppy and the coontail. *Environ. Res.* 8:112–118.

Fang, S. C. and Fallin, E. 1973. Uptake, distribution, and metabolism of inhaled ethylmercuric chloride in the rat. *Arch. Environ. Contam. Toxicol.* 1:347–361.

Fang, S. C. and Fallin, E. 1974. Uptake and subcellular cleavage of organomercury compounds by rat liver and kidney. *Chem.-Biol. Interact.* 9:57–64.

FAO/WHO 1972. Evaluation of certain food additives and the contaminants mercury, lead, and cadmium. *16th Rep. FAO Nutr. Mtg. Rep. Ser., No. 51A; World Health Organ. Tech. Rept. Ser. No. 505.*

Foulkes, E. C. 1974. Excretion and retention of cadmium, zinc, and mercury by rabbit kidney. *Am. J. Physiol.* 227:1356–1360.

Fowler, B. A. 1972a. The morphologic effects of dieldrin and methyl mercuric chloride on pars recta segments of rat kidney proximal tubules. *Am. J. Pathol.* 69:163–178.

Fowler, B. A. 1972b. Ultrastructural evidence for nephropathy induced by long-term exposure to small amounts of methyl mercury. *Science* 175:780–781.

Fowler, B. A., Brown, H. W., Lucier, G. W. and Beard, M. E. 1974. Mercury uptake by renal lysosomes of rats ingesting methyl mercury hydroxide. *Arch. Pathol.* 98:297–301.

Fowler, B. A., Lucier, G. W. and Mushak, P. 1975. Phenobarbital protection against methyl mercury nephrotoxicity. *Proc. Soc. Exp. Biol. Med.* 149:75–79.

Friberg, L. and Vostál, J. 1972. *Mercury in the environment.* Cleveland: Chemical Rubber Company.

Fujita, E. 1969. Experimental studies of organic mercury poisoning: on the behaviors of Minamata disease causal agent in the maternal bodies and its transference to their infants via either placenta or breast milk. *Kumamoto Med. J.* 43:47–62.

Furukawa, K. and Tonomura, K. 1972. Metallic mercury-releasing enzyme in mercury-resistant *Pseudomonas. Agric. Biol. Chem.* 36:217–226.

Furukawa, K. and Tonomura, K. 1973. Cytochrome *c* involved in the reductive decomposition of organic mercurials: Purification of cytochrome *c*-I from mercury-resistant *Pseudomonas* and reactivity of cytochrome *c* from various kinds of bacteria. *Biochim. Biophys. Acta* 325:413–423.

Gage, J. C. 1964. Distribution and excretion of methyl and phenyl mercury salts. *Br. J. Ind. Med.* 21:197–202.

Gage, J. C. 1975. Mechanisms for the biodegradation of organic mercury compounds: The actions of ascorbater and soluble proteins. *Toxicol. Appl. Pharmacol.* 32:225–238.

Garg, B. D., Solymoss, B. and Tuchweber, B. 1971. Effect of spironolactone on the distribution and excretion of ^{203}HgCl$_2$ in the rat. *Arzneimittelforsch.* 21:815–816.

Garrett, N. E., Garrett, R. J. B. and Archdeacon, J. W. 1972. Placental transmission of mercury to the fetal rat. *Toxicol. Appl. Pharmacol.* 22:649–654.

Garrett, R. J. B. and Garrett, N. E. 1974. Mercury incorporation by mature and immature red blood cells. *Life Sci.* 15:733–740.

Giblin, F. J. and Massaro, E. J. 1974. The mechanism of methylmercury transport and transfer to the tissues of the rainbow trout (*Salmo gairsneri*). In *Trace substances in environmental health,* ed. D. D. Hemphill, vol. VIII, pp. 349–355. Columbia: University of Missouri.

Goldwater, L. J. 1971. Mercury in the environment. *Sci. Am.* 224:15–21.

Grant, C. A. 1973. Pathology of experimental methylmercury intoxication: Some problems of exposure and response. In *Mercury, mercurials and mercaptans,* ed. M. W. Miller and T. W. Clarkson, pp. 294–312. Springfield: Charles C Thomas.

Greenwood, M. R., Clarkson, T. W. and Magos, L. 1972. Transfer of metallic mercury into the foetus. *Experientia* 28:1455–1456.

Haddow, J. E. and Lester, R. 1973. Mechanism of action of spironolactone in preventing mercury toxicity. *Gastroenterology* 64:159. Abstr.

Haddow, J. E. and Marshall, P. 1972. Increased stool mercury excretion in the rat: The effect of spironolactone. *Proc. Soc. Exp. Biol. Med.* 140: 707–709.

Haddow, J. E., Fish, C. A., Marshall, P. C. and Lester, R. 1972. Biliary excretion of mercury enhanced by spironolactone. *Gastroenterology* 63:1053–1058.

Hamada, R., Nakamura, N. and Igata, A. 1974. An experience of tiopronin administration. *Igaku no Ayumi* 91:285–286 (in Japanese).

Hanko, E., Erne, K., Wanntorp, H. and Borg, K. 1970. Poisoning in ferrets by tissues of alkyl mercury-fed chickens. *Acta Vet. Scand.* 11:268–282.

Hartung, R. and Dinman, B. D. 1972. *Environmental mercury contamination.* Ann Arbor: Ann Arbor Science.

Havrdová, J., Cikrt, M. and Tichý, M. 1974. Binding of cadmium and mercury in the rat bile: Studies using gel filtration. *Acta Pharmacol. Toxicol.* 34:246–253.

Henderson, R., Shotwell, H. P. and Krause, L. A. 1974. Analyses for total, ionic, and elemental mercury in urine as a basis for a biologic standard. *Am. Ind. Hyg. Assoc. J.* 35:576–580.

Herman, S. P., Klein, R., Talley, A. and Krigman, M. R. 1973. An ultra-structural study of methylmercury-induced primary sensory neuropathy in the rat. *Lab. Invest.* 28:104–118.

Hirata, E. and Takahashi, H. 1975. On chemical forms of methylmercury in bile. *Seikagaku* 47:524 (in Japanese).

Howe, S. M., Mcgee, J. and Lengemann, F. W. 1972. Transfer of inorganic mercury to milk of goats. *Nature* 237:516–518.

Imura, N. 1974. On the formation of organometallic compounds with special reference to mercurials. *Sogo Rinsho* 23:65–75 (in Japanese).

Imura, N., Sukegawa, E., Pan, S.-K., Nagano, K., Kim, J.-Y., Kwan, T. and Ukita, T. 1971. Chemical methylation of inorganic mercury with methylcobalamin, a vitamin B_{12} analog. *Science* 172:1248–1249.

Ishihara, N. 1974. Placental transfer and accumulation in the fetus on mercurial compounds. *Igaku no Ayumi* 90:127–132 (in Japanese).

Ishihara, N. and Suzuki, T. 1976. Biotransformation of methylmercury *in vitro. Tohoku J. Exp. Med.* (in press).

Ishihara, N., Shiojima, S. and Suzuki, T. 1974. Selective enhancement of urinary organic mercury excretion by D-penicillamine. *Br. J. Ind. Med.* 31:245–249.

Iverson, F. and Hierlihy, S. L. 1974. Biotransformation of methyl mercury in the guinea pig. *Bull. Environ. Contam. Toxicol.* 11:85–91.

Iverson, F., Downie, R. H., Paul, C. and Trenholm, H. L. 1973. Methyl mercury: Acute toxicity, tissue distribution and decay profiles in the guinea pig. *Toxicol. Appl. Pharmacol.* 24:545–554.

Jakubowski, M., Piotrowski, J. and Trojanowska, B. 1970. Binding of mercury in the rat: Studies using $^{203}HgCl_2$ and gel filtration. *Toxicol. Appl. Pharmacol.* 16:743–753.

Jalili, M. A. and Abbasi, A. H. 1961. Poisoning by ethyl mercury toluene sulphonanilide. *Br. J. Ind. Med.* 18:303–308.

Jensen, S. and Jernelöv, A. 1969. Biological methylation of mercury in aquatic organisms: *Nature* 223:753–754.

Jernelöv, A. 1969. Conversion of mercury compounds. In *Chemical fallout*, ed. M. W. Miller and G. G. Berg, pp. 68–74. Springfield: Charles C Thomas.

Joselow, M. M., Louria, D. B. and Browder, A. A. 1972. Mercurialism: Environmental and occupational aspects. *Ann. Int. Med.* 76:119–130.

Kaku, S., Yamaguchi, S., Shiramizu, M., Shimojo, N., Hirota, Y., Ishida, N., Kuwahara, Y. and Kurata, S. 1975. A provocation test of incorpolated mercury in a biological milieu. *Jap. J. Ind. Health* 17:110–111 (in Japanese).

Kanbayashi, K., Shirakawa, K., Hirota, K., Tsubaki, T. and Tanaka, H. 1974. Relationship between methylmercury contaminated and the maximal motor conductive velocity. *Ikagu no Ayumi* 91:339–340 (in Japanese).

Kark, R. A. P., Poskanzer, D. C., Bullock, J. D. and Boylen, G. 1971. Mercury poisoning and its treatment with *N*-acetyl-D,L-penicillamine. *New Engl. J. Med.* 285:10–16.

Katz, S. A. and Samitz, M. H. 1973. The binding of mercury to bovine serum albumin. *Environ. Res.* 6:144–146.

Kim, J. Y., Imura, N., Ukita, T. and Kwan, T. 1971. The methylation of the mercuric ion by methylcobaloximes. *Bull. Chem. Soc. Jap.* 44:300.

Kitamura, S., Sumino, K. and Taira, M. 1969. Synthesis and decomposition of organomercury compounds by microorganism. *Jap. J. Hyg.* 24:76 (in Japanese).

Kitamura, S., Sumino, K., Hayakawa, K. and Shibata, T. 1974. Mercury content in "normal" human tissues. Working paper for the Subcommittee on the Toxicology of Metals, Tokyo, November 1974.

Kitamura, S., Hayakawa, K. and Shibata, T. 1975. Studies on biological half life of alkyl mercury compounds in rats and mice. *Jap. J. Hyg.* 30:56 (in Japanese).

Kivimäe, A., Swensson, Å., Ulfvarson, U. and Westöö, G. 1969. Methylmercury compounds in eggs from hens after oral administration of mercury compounds. *Agric. Food Chem.* 17:1014–1016.

Klaassen, C. D. 1975. Biliary excretion of mercury compounds. *Fed. Proc.* 34:266. Abstr.

Klein, R., Herman, S. P., Bullock, B. C. and Talley, F. A. 1973. Methyl mercury intoxication in rat kidneys. *Arch. Pathol.* 96:83–90.

Kondo, M. 1971. Studies on the intake and excretion of heavy metals in human being. *Kankyo Hoken Rep.* 5:47–64 (in Japanese).

Kosta, L., Byrne, A. R. and Zelenko, V. 1975. Correlation between selenium and mercury in man following exposure to inorganic mercury. *Nature* 254:238–239.

Kushlan, M. C., Haddow, J. E. and Lester, R. 1975. *In vitro* enhancement of erythrocyte mercury uptake by spironolactone. *Toxicol. Appl. Pharmacol.* 31:527–533.

Kutsuna, S., ed. 1968. *Minamata disease.* Kumamoto University, Japan: Study Group of Minamata Disease.

Landner, L. 1971. Biochemical model for the biological methylation of mercury suggested from methylation studies *in vivo* with *Neurospora crassa. Nature* 230:452–454.

Lefevre, P. A. and Daniel, J. W. 1973. Some properties of the organomercury-degrading system in mammalian liver. *FEBS Lett.* 35:121–123.

Lin, F. M., Romero-Sierra, C. and Malaiyanda, M. 1974. The age factor in susceptibility to methylmercury intoxication. In *Trace substances in environmental health,* ed. D. D. Hemphill, vol. VIII, pp. 399–401. Columbia: University of Missouri.

Löfroth, G. 1969. Methylmercury. A review of health hazards and side effects associated with the emission of mercury compounds into natural systems. *Ecol. Res. Commun. Bull.* 4:1 (stencils).

Lucier, G. W., McDaniel, O. S., Williams, C. and Klein, R. 1972. Effects of chlordane and methylmercury on the metabolism of carbaryl and carbofuran in rats. *Pestic. Biochem. Physiol.* 2:244–255.

Lundgren, K. D., Swensson, Å. and Ulfvarson, U. 1967. Studies in humans on the distribution of mercury in the blood and the excretion in urine after exposure to different mercury compounds. *Scand. J. Clin. Lab. Invest.* 20:164–166.

Magos, L. 1967. Mercury-blood interaction and mercury uptake by the brain after vapour exposure. *Environ. Res.* 1:323–337.

Magos, L. 1971. Selective atomic-absorption determination of inorganic mercury and methylmercury in undigested biological samples. *Analyst* 96:847–853.

Magos, L. 1973. Factors affecting the uptake and retention of mercury by kidneys in rats. In *Mercury, mercurials and mercaptans,* ed. M. W. Miller and T. W. Clarkson, pp. 167–186. Springfield: Charles C Thomas.

Magos, L. 1974. Binding sites for mercury and the synergism between penicillamine and maleate. *Postgrad. Med. J. Suppl.* 50:22–23.

Magos, L. and Clarkson, T. W. 1973. Effect of phenobarbitone on the biliary excretion of methylmercury in rats and mice. *Nature New Biol.* 246:123.

Magos, L., Tuffery, A. A. and Clarkson, T. W. 1964. Volatilization of mercury by bacteria. *Br. J. Ind. Med.* 21:294–298.

Magos, L., Clarkson, T. W. and Greenwood, M. R. 1973. The depression of pulmonary retention of mercury vapor by ethanol: Identification of the site of action. *Toxicol. Appl. Pharmacol.* 26:180–183.

Magos, L., MacGregor, T. and Clarkson, T. W. 1974a. The effect of phenobarbital and sodium dehydrocholate on the biliary excretion on methylmercury in the rat. *Toxicol. Appl. Pharmacol.* 30:1–6.

Magos, L., Sugata, Y. and Clarkson, T. W. 1974b. Effects of 3-amino-1,2,4-triazole on mercury uptake by *in vitro* human blood samples and by whole rats. *Toxicol. Appl. Pharmacol.* 28:367–373.

Magos, L., Webb, M. and Butler, W. H. 1974c. The effect of cadmium pretreatment on the nephrotoxic action and kidney uptake of mercury in male and female rats. *Br. J. Exp. Pathol.* 55:589–594.

Mansour, M. M., Dyer, N. C., Hoffman, L. H., Schulert, A. R. and Brill, A. B. 1973. Maternal–fetal transfer of organic and inorganic mercury via placenta and milk. *Environ. Res.* 6:479–484.

Matsubara-Khan, J. 1974. Compartmental analysis for the evaluation of biological half-lives of cadmium and mercury in mouse organs. *Environ. Res.* 7:54–67.

Matsumoto, H., Suzuki, A., Morita, C., Nakamura, K. and Saeki, S. 1967. Preventive effect of penicillamine on the brain defect of fetal rat poisoned transplacentally with methyl mercury. *Life Sci.* 6:2321–2326.

Miettinen, J. K. 1972. Gastrointestinal absorption and whole-body retention of toxic heavy metal compounds (methyl mercury, ionic mercury, cadmium) in man. *Proc. XVII. Int. Congr. Occup. Health,* Buenos Aires, 1972.

Miettinen, J. K. 1973. Absorption and elimination of dietary mercury (Hg^{2+}) and methylmercury in man. In *Mercury, mercurials and mercaptans,* ed. M. W. Miller and T. W. Clarkson, pp. 233–243. Springfield: Charles C Thomas.

Miller, M. W. and Berg, G. G. 1969. *Chemical fallout.* Springfield: Charles C Thomas.

Miller, M. W. and Clarkson, T. W. 1973. *Mercury, mercurials and mercaptans.* Springfield: Charles C Thomas.

Miyakawa, T., Deshimaru, M., Sumiyoshi, S., Teraoka, A., Udo, N., Hattori, E. and Tatetsu, S. 1970. Experimental organic poisoning—Pathological changes in peripheral nerves. *Acta Neuropathol.* 15:45–55.

Mikawa, T., Monma, M., Hirayama, F. and Yamamura, Y. 1975. Clinical observation of a mercury poisoning. *Proc. 48th Jap. Ind. Health,* Sapporo, pp. 414–415 (in Japanese).

Miyama, T., Murakami, M., Suzuki, T. and Katsunuma, H. 1968. Retention of mercury in the brain of rabbit after intravenous or subcutaneous injection of sublimate. *Ind. Health* 6:107–115.

Miyamoto, S. 1974a. Transfer into milk and metabolism of mercury (Hg) administered as phenyl mercuric acetate (PMA) in domestic animals. *J. Jap. Soc. Food Nutr.* 27:47–53 (in Japanese).

Miyamoto, S. 1974b. Distribution and excretion of mercury (Hg) administered as phenyl mercuric acetate (PMA) in lactating guinea pigs and its transfer into the sucklings. *J. Jap. Soc. Food Nutr.* 27:109–115 (in Japanese).

Nakao, M., Nakao, T., Ohta, H., Nagai, F., Kawai, K., Fujihira, Y. and Nagano, K. 1973. The Na^+, K^+-ATPase molecule. In *Organization of energy-transducing membranes,* ed. M. Nakao and L. Packer, pp. 23–34. Tokyo: Tokyo University Press.

Neathery, M. W., Miller, W. J., Gentry, R. P., Stake, P. E. and Blackmon, D. M. 1974. Cadmium-109 and methyl mercury-203 metabolism, tissue distribution, and secretion into milk of cows. *J. Dairy Sci.* 57: 1177–1183.

Nelson, N., Byerly, T. C., Kolbye, Jr., A. C., Kurland, L. T., Shapiro, R. E., Shibuko, R. E., Stickel, W. H., Thompson, J. E., Berg, L. A. and Weissler, L. A. 1971. Hazards of mercury. *Environ. Res.* 4:1–69.

Neville, G. A. and Berlin, M. 1974. Identification and biotransformation of organomercurial compounds in living systems. *Environ. Res.* 7:75–82.

Nielsen-Kudsk, F. 1965. The influence of ethyl alcohol on the absorption of mercury vapour from the lungs in man. *Acta Pharmacol. Toxicol.* 23:263–274.

Nielsen-Kudsk, F. 1969a. Uptake of mercury vapour in blood *in vivo* and *in vitro* from Hg-containing air. *Acta Pharmacol. Toxicol.* 27:149–160.

Nielsen-Kudsk, F. 1969b. Factors influencing the *in vitro* uptake of mercury vapour in blood. *Acta Pharmacol. Toxicol.* 27:161–172.

Nielsen-Kudsk, F. 1973. Biological oxidation of elemental mercury. In *Mercury, mercurials and mercaptans,* ed. M. W. Miller and T. W. Clarkson, pp. 355–371. Springfield: Charles C Thomas.

Nordberg, G. F. and Skerfving, S. 1972. Metabolism. In *Mercury in the environment*, ed. L. Friberg and J. Vostál, pp. 29–91. Cleveland: Chemical Rubber Company.

Nordberg, G. F. and Strangert, P. 1974. Estimation of a dose-response curve for long-term methyl-mercury exposure in human beings taking into account variability of critical organ concentration and biological half-time—A preliminary communication. Working paper for the Subcommittee on the Toxicology of Metals, Tokyo, November 1974.

Nordberg, M., Trojanowska, B. and Nordberg, G. F. 1974. Studies on metal-binding proteins of low molecular weight from renal tissue of rabbits exposed to cadmium or mercury. *Environ. Physiol. Biochem.* 4:149–158.

Nordiskt Symposium. 1969. Nordiskt symposium kring kvicksilverproblematiken., 10–11 Oktober 1968, på Handelns Gård, Lidingö, Sverige.

Norseth, T. 1969. Studies of intracellular distribution of mercury. In *Chemical fallout*, ed. M. W. Miller and G. G. Berg, pp. 408–419. Springfield: Charles C Thomas.

Norseth, T. 1971. Biotransformation of methyl mercuric salts in germfree rats. *Acta Pharmacol. Toxicol.* 30:172–176.

Norseth, T. 1973a. The effect of chelating agents on biliary excretion of methyl mercuric salts in the rat. *Acta Pharmacol. Toxicol.* 32:1–10.

Norseth, T. 1973b. Biliary excretion and intestinal reabsorption of mercury in the rat after injection of methyl-mercuric chloride. *Acta Pharmacol. Toxicol.* 33:280–288.

Norseth, T. 1973c. Biliary complexes of methylmercury: A possible role in organ distribution. In *Mercury, mercurials and mercaptans*, ed. M. W. Miller and T. W. Clarkson, pp. 264–276. Springfield: Charles C Thomas.

Norseth, T. 1974. The effect of diethyldithiocarbamate on biliary transport, excretion and organ distribution of mercury in the rat after exposure to methyl mercuric chloride. *Acta Pharmacol. Toxicol.* 34:76–87.

Norseth, T. and Brendeford, M. 1971. Intracellular distribution of inorganic and organic mercury in rat liver after exposure to methylmercury salts. *Biochem. Pharmacol.* 20:1101–1107.

Norseth, T. and Clarkson, T. W. 1970a. Studies on the biotransformation of [203]Hg-labeled methyl mercury chloride in rats. *Arch. Environ. Health* 21:717–727.

Norseth, T. and Clarkson, T. W. 1970b. Biotransformation of methylmercury salts in the rat studied by specific determination of inorganic mercury. *Biochem. Pharmacol.* 19:2775–2783.

Norseth, T. and Clarkson, T. W. 1971. Intestinal transport of [203]Hg-labeled methyl mercury chloride. Role of biotransformation in rats. *Arch. Environ. Health* 22:568–577.

Null, D. H., Gartside, P. S. and Wei, E. 1973. Methylmercury accumulation in brain of pregnant, non-pregnant and fetal rats. *Life Sci.* 12:65–72.

Ogawa, E., Suzuki, S., Tsuzuki, H., Tobe, M., Kobayashi, K. and Hojo, M. 1972a. Experimental studies on body distribution and excretion of methylmercuric chloride (1). *Traumatology* 15:222–228 (in Japanese).

Ogawa, E., Suzuki, S., Tsuzuki, H., Tobe, M., Kobayashi, K. and Hojo, M. 1972b. Experimental studies on body distribution and excretion of methylmercuric chloride (2). *Traumatology* 15:371–376 (in Japanese).

Ogawa, E., Suzuki, S., Tsuzuki, H. and Kawajiri, M. 1973. Experimental studies on body distribution and excretion of ethylmercuric chloride. *Folia Pharmacol. Jap.* 69:271p (in Japanese).

Ohsawa, M. and Magos, L. 1974. The chemical form of the methylmercury complex in the bile of the rat. *Biochem. Pharmacol.* 23:1903–1905.

Olson, F. C. and Massaro, E. J. 1975. Pharmacodynamics of placental transfer of methyl mercury and induced teratology in mice. *Fed. Proc.* 34:773. Abstr.

Östlund, K. 1969. Studies on the metabolism of methyl mercury and dimethyl mercury in mice. *Acta Pharmacol. Toxicol. Suppl.* 27:1.

Peakall, D. B. and Lovett, R. J. 1972. Mercury: Its occurrence and effects in the ecosystem. *BioScience* 22:20–25.

Pekkanen, T. J. 1971. The effect of experimental methyl mercury poisoning on the distribution of acid phosphatase during autolysis in cat liver. *Acta Vet. Scand.* 12:523–535.

Pekkanen, T. J. and Lindberg, L.-A. 1972. Ultrastructure and microsomal protein content of mouse liver treated with methyl mercury. *Acta Pharmacol. Toxicol.* 31:337–340.

Pekkanen, T. J. and Pekkarinen, K. 1972. The effect of methyl mercury pretreatment on the duration of hexobarbital hypnosis on rat. *Acta Vet. Scand.* 13:149–150.

Pekkanen, T. J. and Salminen, K. 1973. Inductive effects of methyl mercury on the hepatic microsomes of mice. *Acta Pharmacol. Toxicol.* 32:289–293.

Perkins, D. J. 1961. Studies on the interaction of zinc, cadmium and mercury ions with native and chemically modified human serum albumin. *Biochem. J.* 80:668–672.

Perry, H. M., Erlanger, M., Yunice, A. Schoepfle, E. and Perry, E. E. 1970. Hypertension and tissue metal levels following intravenous cadmium, mercury and zinc. *Am. J. Physiol.* 219:755–761.

Pierce, P. E., Thompson, J. F., Likosky, W. H., Nickey, L. N., Barthel, W. F. and Hinman, A. R. 1972. Alkyl mercury poisoning in humans. Report of an outbreak. *J. Am. Med. Assoc.* 220:1439–1442.

Piotrowski, J. K., Bolanowska, W., Trojanowska, B. and Szendzikowski, S. 1969. The relation between the urinary excretion of mercury and the damage of the kidney in rats given different doses of mercuric chloride. *Med. Pracy* 20:589–599.

Piotrowski, J. K., Trojanowska, B., Wiśniewska-Knypl, J. M. and Bolanowska, W. 1973. Further investigations on binding and release of mercury in the rat. In *Mercury, mercurials and mercaptans*, ed. M. W. Miller and T. W. Clarkson, pp. 247–263. Springfield: Charles C Thomas.

Piotrowski, J. K., Trojanowska, B., Wiśniewska-Knypl, J. M. and Bolanowska, W. 1974. Mercury binding in the kidney and liver of rats repeatedly exposed to mercuric chloride: Induction of metallothionein by mercury and cadmium. *Toxicol. Appl. Pharmacol.* 27:11–19.

Rahola, T., Aaran, R. K. and Miettinen, J. K. 1971. Half-time studies of mercury and cadmium by whole body counting. Paper presented at IAEA/WHO Symposium on the Assessment of Radioactive Organ and Body Burdens, Stockholm, November 22–26.

Rahola, T., Hattula, T., Korolainen, A. and Miettinen, J. K. 1973. Elimination of free and protein-bound ionic mercury ($^{203}Hg^{2+}$) in man. *Ann. Clin. Res.* 5:214–219.

Refsvik, T. and Norseth, T. 1975. Methyl mercuric compounds in rat bile. *Acta Pharmacol. Toxicol.* 36:67–78.

Reynolds, W. A. and Pitkin, R. M. 1975. Transplacental passage of methylmercury and its uptake by primate fetal tissues. *Proc. Soc. Exp. Biol. Med.* 148:523–526.

Rothstein, A. 1973. Mercaptans, the biological targets for mercurials. In *Mercury, mercurials and mercaptans*, ed. M. W. Miller and T. W. Clarkson, pp. 68–95. Springfield: Charles C Thomas.

Rustam, H. and Hamdi, T. 1974. Methyl mercury poisoning in Iraq, a neurological study. *Brain* 97:499–510.

Rustam, H., Burg, R. V., Amin-Zaki, L. and Hassani, S. E. 1975. Evidence for a neuromuscular disorder in methylmercury poisoning. *Arch. Environ. Health* 30:190–195.

Sahagian, B. M., Harding-Barlow, I. and Perry, H. M., Jr. 1966. Uptakes of zinc, manganese, cadmium and mercury by intact strips of rat intestine. *J. Nutr.* 90:259–267.

Salvaterra, P., Massaro, E. J., Morganti, J. B. and Lown, B. A. 1975. Time-dependent tissue/organ uptake and distribution of ^{203}Hg in mice exposed to multiple sublethal doses of methyl mercury. *Toxicol. Appl. Pharmacol.* 32:432–442.

Schmidt, H., and Harzman, R. 1970. Humanpathologische und tierexperimentelle Beobachtungen nach Intoxikation mit einer organischen Quecksilberverbindung ("Fusariol"). *Int. Arch. Arbeitsmed.* 26:71–83.

Shishido, S. and Suzuki, T. 1974. Estimation of daily intake of inorganic or organic mercury via diet. *Tohoku J. Exp. Med.* 114:369–377.

Silberberg, I., Prutkin, L. and Leider, M. 1969. Electron microscopic studies of transepidermal absorption of mercury. *Arch. Environ. Health* 19: 7–15.

Skerfving, S. 1972. Mercury in fish—Some toxicological considerations. *Food Cosmet. Toxicol.* 10:545–556.

Skerfving, S. 1974. Methylmercury exposure, mercury levels in blood and hair, and health status in swedes consuming contaminated fish. *Toxicology* 2:3–23.

Skerfving, S. and Westöö, G. cited by Nordberg, F. and Skerfving, S. 1972. Metabolism. In *Mercury in the environment*, ed. L. Friberg and J. Vostál, pp. 29–91. Cleveland: Chemical Rubber Company.

Somjen, G. G., Herman, S. P. and Klein, R. 1973a. Electrophysiology of methyl mercury poisoning. *J. Pharmacol. Exp. Ther.* 186:579–592.

Somjen, G. G., Herman, S. P., Klein, R., Brubaker, P. E., Briner, W. H., Goodrich, J. K., Krigman, M. R. and Haseman, J. K. 1973b. The uptake of methyl mercury (^{203}Hg) in different tissues related to its neurotoxic effects. *J. Pharmacol. Exp. Ther.* 187:602–611.

Stillings, B. R., Lagally, H., Bauersfeld, P. and Soares, J. 1974. Effect of cystine, selenium, and fish protein on the toxicity and metabolism of methylmercury in rats. *Toxicol. Appl. Pharmacol.* 30:243–254.

Sugata, Y. 1975. Oxidation mechanism of metallic mercury-distribution of metallic mercury in the brain and the fetus of acatalasemic mice. *Jap. J. Hyg.* 30:52 (in Japanese).

Sugiyama, S., Hashimoto, M., Chiba, N., Hayase, Y., Yokoyama, T. Aikawa, H., Odaka, Y. and Miyamoto, S. 1975. The influence of L-cystein on toxicity of methyl mercuric chloride. *J. Med. Soc. Toho Univ.* 22:78–85 (in Japanese).

Suzuki, T. 1969. Neurological symptoms from concentration of mercury in the brain. In *Chemical fallout*, ed. M. W. Miller and G. G. Berg, pp. 245–257. Springfield: Charles C Thomas.

Suzuki, T. 1976. Human ecology in understanding environmental health problems. *Rev. Environ. Health* (in press).

Suzuki, T. and Miyama, T. 1971. Neurological symptoms and mercury concentration in the brain of mice fed with methylmercury salt. *Ind. Health* 9:51–58.

Suzuki, T. and Miyama, T. 1975. Mercury in red cells in relation to organic mercury in hair. *Tohoku J. Exp. Med.* 116:379–384.

Suzuki, T. and Shishido, S. 1975a. A possible change of the critical concentration of methylmercury in the brain of mice: Observations by a quantitative assessment of early neurological signs. In *Effects and dose–response relationships of toxic metals*, ed. G. F. Nordberg, pp. 283–289. Amsterdam: Elsevier.

Suzuki, T. and Shishido, S. 1975b. Desquamation of renal epithelial cells as a route of mercury excretion in man.–A preliminary study. *Tohoku J. Exp. Med.* 117:397–398.

Suzuki, T. and Yoshino, Y. 1969. Effects of D-penicillamine on urinary excretion of mercury in two cases of methylmercury poisoning. *Jap. J. Ind. Health* 11:487–488.

Suzuki, T., Matsumoto, N., Miyama, T. and Katsunuma, H. 1967a. Placental transfer of mercuric chloride, phenyl mercury acetate and methyl mercury acetate in mice. *Ind. Health* 5:149–155.

Suzuki, T., Miyama, T. and Katsunuma, H. 1967b. Mercury in the plasma after subcutaneous injection of sublimate of mercuric nitrate in rat. *Ind. Health* 5:290–292.

Suzuki, T., Miyama, T. and Katsunuma, H. 1969. Differences in mercury content in the brain after injection of methylphenyl or inorganic mercury. *Proc. XVI. Int. Congr. Occup. Health*, Tokyo, pp. 563–565.

Suzuki, T., Miyama, T. and Katsunuma, H. 1970. Mercury contents in the red cells, plasma, urine and hair from workers exposed to mercury vapour. *Ind. Health* 8:39–47.

Suzuki, T., Miyama, T. and Katsunuma, H. 1971a. Comparison of mercury contents in maternal blood, umbilical cord blood and placental tissues. *Bull. Environ. Contam. Toxicol.* 5:502–507.

Suzuki, T., Takemoto, T., Shimano, H., Miyama, T., Katsunuma, H. and Kagawa, Y. 1971b. Mercury content in the blood in relation to dietary habit of the women without any occupational exposure to mercury. *Ind. Health* 9:1–8.

Suzuki, T., Takemoto, T., Kashiwazaki, H. and Miyama, T. 1973. Metabolic fate of ethylmercury salts in man and animals. In *Mercury, mercurials and mercaptans*, ed. M. W. Miller and T. W. Clarkson, pp. 209–232. Springfield: Charles C Thomas.

Suzuki, T., Shishido, S. and Ishihara, N. 1976. Different behaviour of inorganic and organic mercury in renal excretion with reference to effects of D-penicillamine. *Br. J. Ind. Med.* 33:88–91.

Swensson, Å. and Ulfvarson, U. 1967. Experiments with different antidotes in acte poisoning by different mercury compounds. *Int. Arch. Gewerbepathol. Gewerbehyg.* 24:12–50.

Swensson, Å., Lundgren, K. D. and Lindstrom, O. 1959. Distribution and excretion of mercury compounds after single injections. *AMA Arch. Ind. Health* 9:1–8.

Syversen, T. L. M. 1974a. Biotransformation of Hg-203 labelled methyl mercuric chloride in rat brain measured by specific determination of Hg^{2+}. *Acta Pharmacol. Toxicol.* 35:277–283.

Syversen, T. L. M. 1974b. Distribution of mercury in enzymatically characterized subcellular fractions from the developing rat brain after injections of methylmercuric chloride and diethylmercury. *Biochem. Pharmacol.* 23:2999–3007.

Takahashi, H. 1974. Absorption, excretion and body distribution of mercury. *Sogo Rinsho* 23:37–45 (in Japanese).

Takahashi, H. and Hirayama, K. 1971. Accelerated elimination of methyl mercury from animals. *Nature* 232:201–202.

Takahata, N., Hayashi, H., Watanabe, B. and Anso, T. 1970. Accumulation of mercury in the brains of two autopsy cases with chronic inorganic mercury poisoning. *Folia Psychiatr. Neurol. Jap.* 24:59–69.

Takeda, Y. and Ukita, T. 1970. Metabolism of ethylmercuric chloride-[203]Hg in rats. *Toxicol. Appl. Pharmacol.* 17:181–188.

Takeda, Y., Kunugi, T., Terao, T. and Ukita, T. 1968. Mercury compounds in the blood of rats treated with ethyl mercuric chloride. *Toxicol. Appl. Pharmacol.* 13:165–173.

Takeuchi, T. 1972. Biological reactions and pathological changes in human beings and animals caused by organic mercury contamination. In *Environmental mercury contamination*, ed. R. Hartung and B. D. Dinman, pp. 247–289. Ann Arbor: Ann Arbor Science.

Task Group on Metal Accumulation. 1973. Accumulation of toxic metals with special reference to their absorption, excretion, and biological half-times. *Environ. Physiol. Biochem.* 3:65–107.

Task Group on Dose–Response Relationships of Toxic Metals. 1975. *Effects and dose-response relationships of toxic metals*, ed. G. Nordberg. Amsterdam: Elsevier (in press).

Tejning, S. 1970. The mercury contents in blood corpuscles and in blood plasma in mothers and their newborn children. Rept. 70-05-20 Dep. Occup. Med., Univ. Hosp. Lund (stencils).

Timm, F., Naudorf, Ch. and Kraft, M. 1966. Zur Histochemie und Genese der chronischen Quecksilbervergiftung. *Int. Arch. Gewerbepathol.* 22:236–245.

Tonomura, K., Maeda, K. and Futai, F. 1968. Studies on the action of mercury-resistant microorganism on mercurials. II. The vaporization of mercurials stimulated by mercury-resistant bacterium. *J. Ferment. Technol.* 46:685–692.

Trenholm, H. L., Paul, C. L., Baer, H. and Iverson, F. 1971. Methyl mercury [203]Hg excretion by lactation in guinea pigs. *Toxicol. Appl. Pharmacol.* 19:409. Abstr.

Trojanowska, B., Piotrowski, J. K. and Szendzikowski, S. 1971. The influence of thioacetamide on the excretion of mercury in rats. *Toxicol. Appl. Pharmacol.* 18:374–386.

Tsubaki, T. and Shirakawa, K. 1966. Therapeutics of organic mercury poisoning with special reference to α-mercaptopropionyl-glycine. *Shinyaku to Rinsho* 15:1471–1475 (in Japanese).

Tsubaki, T., Sato, T., Kondo, K., Shirakawa, K., Kanbayashi, K., Hirota, K., Yamada, K. and Murone, I. 1967. Outbreak of inotoxication by organic mercury compound in Niigata prefecture: An epidemiological and clinical study. *Jap. J. Med.* 6:132–133.

Tsuchiya, K., Uchiyama, G., Sugita, M. and Yasuda, K. 1975. Total mercury concentrations in cerebrum, cerebellum, hair and urine from "normal" Japanese people—Interrelationships and biological half times. *Jap. J. Hyg.* 30:50 (in Japanese).

Ukita, T. and Imura, N. 1971. Experimental investigation of mercury poisoning. II. Formation of methylmercury from inorganic mercury. *Kagaku* 41:586–592 (in Japanese).

Von Burg, R. and Rustam, H. 1974a. Conduction velocities in methylmercury poisoned patients. *Bull. Environ. Contam. Toxicol.* 12:81–84.

Von Burg, R. and Rustam, H. 1974b. Electrophysiological investigations of methylmercury intoxication in humans. Evaluation of peripheral nerve by conduction velocity and electromyography. *Electroencephalogr. Clin. Neurophysiol.* 37:381–392.

Vostál, J. 1968. Renal excretory mechanisms of mercury compounds. Working paper for the Symposium on MAC values, Stockholm.

Vostál, J. and Heller, J. 1968. Renal excretory mechanisms of heavy metals. *Environ. Res.* 2:1–10.

Wagstaff, D. J. 1973. Enhancement of hepatic detoxification enzyme activity by dietary mercuric acetate. *Bull. Environ. Cont. Toxicol.* 9:10–14.

Wannag, A. 1976. The importance of organ blood mercury when comparing fetal and maternal rat organ distribution of mercury after methylmercury exposure. *Acta Pharmacol. Toxicol.* 38:289–298.

Watanabe, S. 1971. Mercury in the body 10 years after long term exposure to mercury. *Proc. 16th Int. Congr. Occup. Health*, Tokyo, September, pp. 553–555.

Westöö, G. 1968. Determination of methylmercury salts in various kinds of biological material. *Acta Chem. Scand.* 22:2277–2280.

White, J. F. and Rothstein, A. 1973. The interaction of methyl mercury with erythrocytes. *Toxicol. Appl. Pharmacol.* 26:370–384.

Windeler, A. S., Jr. 1973. Determination of mercury in parotid fluid. *J. Dent. Res.* 52:19–22.

Wiśniewska, J. M., Trojanowska, B., Piotrowski, J. and Jakubowski, M. 1970. Binding of mercury in the rat kidney by metallothionein. *Toxicol. Appl. Pharmacol.* 16:754–763.

Wood, J. M. 1972. A progress report on mercury. *Environment* 14:33–39.

Wood, J. M., Kennedy, F. S. and Rosen, C. G. 1968. Synthesis of methylmercury compounds by extracts of a methanogenic bacterium. *Nature* 220:173–174.

Yamada, M. and Tonomura, K. 1972. Microbial methylation of mercury in hydrogen sulfide-envolving environments. *J. Ferment. Technol.* 50: 901–909.

Yamaguchi, S., and Nunotani, H. 1974. Transplacental transport of mercurials in rats at the subclinical dose levels. *Environ. Physiol. Biochem.* 4:7–15.

Yamaguchi, S., Kaku, T., Hirota, Y. and Inoguchi, T. 1974. A consideration on the therapy for the remaining symptom and sign after acute phase of metallic mercury poisoning. *Nippon Iji Shinpo* 2618:12–15 (in Japanese).

Yang, M. G., Krawford, K. S., Garcia, J. D., Wang, J. H. C. and Lei, K. Y. 1972. Deposition of mercury in fetal and maternal brain. *Proc. Soc. Exp. Biol.* 141:1004–1007.

Yang, M. G., Wang, J. H. C., Garcia, J. D., Post, E. and Lei, K. Y. 1973. Mammary transfer of 203Hg from mothers to brains of nursing rats. *Proc. Soc. Exp. Biol.* 142:723–726.

Yoshikawa, H. 1974. Tolerance to acute metal toxicity in mice having received a daily injection of its low dose. *Ind. Health* 12:175–177.

Yoshino, Y., Mozai, T. and Nakano, K. 1966. Distribution of mercury in the brain and its subcellular units in experimental organic mercury poisoning. *J. Neurochem.* 13:397–406.

Chapter 2

LEAD TOXICITY LABORATORY ASPECTS

Robert A. Goyer
Department of Pathology
University of Western Ontario
London, Ontario

Paul Mushak
Department of Pathology
University of North Carolina
Chapel Hill, North Carolina

INTRODUCTION

Lead poisoning in the United States is largely a pediatric problem with major interest shifting from a few individuals with symptomatic intoxication to masses of urban children suffering from asymptomatic or subclinical lead poisoning. Among adults the lead problem is confined to a few accidental poisonings from such sources as continued use of pottery with improperly fired lead glazes or renovation of an old residence heavily painted with lead-containing paint. Occupational lead exposure is under continued surveillance by local regulatory agencies and the National Institute of Occupational Safety and Health.

The purpose of this review is to briefly define the lead problem in children in the United States and to review the metabolism and pathological effects of lead. These introductory comments should serve as a basis for considering the significance of blood lead levels and other biological parameters helpful in the recognition of lead intoxication. Details of the performance of these tests is also presented.

BACKGROUND OF PRESENT STATUS OF COMMUNITY LEAD POISONING

Lead poisoning in young children living in deteriorating inner city housing emerged as an urban problem in the United States as early as the

This work was supported in part by grants from the U.S. Public Health Service and Medical Research Council of Canada.

second and third decades of this century (Williams et al., 1952). Thomas and Blackfan (1914) of the Johns Hopkins Hospital of Baltimore were the first to describe in American literature the clinical and pathological effects of lead-induced encephalopathy in children. Williams and co-workers (1952) point out that prior to 1951 multiple cases of lead poisoning had already been reported from 19 separate communities, and it was decided early that this problem was related to pica and the habits of chewing cribs, toys, furniture, and woodwork such as window sills and the eating of painted plaster and fallen paint flakes. Sixty percent of the affected children were over 5 yr of age. During the 20-yr period 1931–1951 in Baltimore, 293 cases were identified as having clinical lead poisoning identified by a blood lead greater than 50 μg/100 ml coupled with a clinical manifestation of lead toxicity (anemia or central nervous system symptoms). Eighty-five cases were fatal.

Following the early Baltimore experience public health departments of other major American cities began efforts to determine the extent to which lead poisoning did exist in the urban community. Generally, two steps were taken. First, lead poisoning was made a reportable disease, and second, separate programs for control of lead poisoning were established. A report of the results of these measures in Philadelphia from 1955 to 1960 (Ingalls et al., 1961) has noted that about 50 cases of lead poisoning were reported each year. This is comparable to the average of 41 per year reported by the Baltimore City Health Department for the period 1948–1951 (Williams et al., 1952). From 10 to 20% of diagnosed patients died in that period, and neurological disorders including mental retardation persist as sequelae in those who survive therapy (Perlstein and Attala, 1966).

Also, methods of measurement of blood lead have been inconsistent, but dithizone methods and atomic absorption spectroscopy were employed most often. Anodic stripping voltammetry has been introduced recently (DHEW, 1972).

The studies do indicate, however, that 9.1–45.5% of children surveyed have blood lead levels above 40 μg/100 ml, and up to 12.5% are above 60 μg/100 ml. The Bureau of Community Environmental Management and National Bureau of Standards (Gilsinn, 1972) reports estimate that about 23 or 24% of children at risk have elevated blood lead levels (above 40 μg/100 ml), and about 5% of these children have clinical symptoms.

The National Bureau of Standards has published the results of mathematical models together with the assumptions and data used to formulate the models (Gilsinn, 1972). Estimates are given of the number of children who have elevated blood lead levels (>40 μg/100 ml) in 241 standard metropolitan statistical areas in the United States. Present estimate based on these models suggests that approximately 600,000 children would show elevated blood lead levels if tested. At the present time the models and assumptions are only partially validated.

RESULTS OF RECENT SCREENING PROGRAMS

The city public health programs of the past 20 yr have been expanded to include large-scale screening of children at risk. The results of some of the most significant of these screening programs are summarized in Table 1. A more complete summary of screening program results is found in a National Bureau of Standards report (Gilsinn, 1972).

The children included in these surveys are usually between 1 to 6 yr of age and said to be *at risk for undue absorption of lead* because they reside in deteriorated, pre-World War II housing. The surveys, for the most part, involve at risk areas of very large metropolitan communities, but the study from Portland, Maine (Clark and Hallett, 1971) suggests that the risk factor is not necessarily the size of the city but rather the age of the housing. This notion is supported by the study of children at risk in 14 intermediate Illinois communities (Fine et al., 1972).

Another characteristic of the problem is that there is a definite seasonal variation in blood lead levels, and the peak incidence of lead poisoning in young children occurs during the summer months (Chisolm and Harrison, 1956; Blanksma et al., 1969; Guinee, 1972). However, the larger programs have collected blood samples all months of the year; some smaller studies have been restricted to summer months.

SOURCES OF LEAD EXPOSURE

The principal source of lead intake by members of the general population is from diet. Lead intake by adults without undue lead exposure is usually between 100 and 300 mg/day. If blood lead levels are correlated with daily dietary lead intake as given in recent published studies (Coulston et al., 1972; Tepper and Levin, 1972; Rabinowitz et al., 1974), it is estimated that every 100 mg of dietary lead contributes about 10 mg (6.4–15.0) of blood lead per 100 ml of blood. With blood lead levels in the normal range, 20–30 mg/100 ml, air lead present in the ambient environment contributes relatively little to blood lead levels (about 0.3–1.75 mg lead/100 ml blood per mg Pb/m^3 of air).

For children these estimates are more difficult and less well established. Dietary lead intake for U.S. children 1-3 yr of age has been estimated to be about 100 mg lead/day (Kolbye et al., 1974), but absorption may be as high as 50% (Alexander, 1974) so that diet lead may contribute to a greater degree to blood lead than occurs in adults. The contribution of air lead to blood lead in children is presently not known.

The difference in lead exposure between persons in the general population and children with elevated blood lead levels has been a matter of considerable debate. Control of excessive lead exposure cannot be effectively regulated until two aspects of this problem became clearly established. There

TABLE 1 Results of Screening Programs for Increased Lead Absorption in 1–6-yr-old Children

City	Year	Number of children	Blood lead levels			
			40 µg/100 ml	50 µg/100 ml	60 µg/100 ml	70 µg/100 ml
Baltimore	1968	665	25.3[a]		5.6	2.1
	1969	746	28.9		7.2	2.8
	1970	939	31.5		7.4	1.4
Chicago	1967–1970	120,000	20.0			
New Haven	1969–1970	1,897	29.8	4	9.5	
Newark	1970	594	38.9		7.4	
Washington, D.C.	1970	1,158	22.0	12		
New York City	1969	2,648	45.5	24.5	12.5	2.2
	1970	84,493	28.7	12.6	5.9	
	1971	87,559	23.7	10.3	4.6	
BCEM survey 27 cities	1971–1972	2,309	9.1–41.6			
Portland, Me.	1970	905			1.5	
Boston	1972	705	16.3	5.8	5.4	1.0

[a]Values are expressed as percent of children.

is the question of, first, what level of blood lead is harmful to health at any particular age of life including the fetus *in utero*, and second, what the relative significance of particular metabolic factors and environmental sources of the excessive lead is.

The incidence of acute lead poisoning in childhood with clinically evident central nervous system symptoms is, in fact, declining. The excessive lead in the majority of these children has been attributed to ingestion of lead-containing paint from walls of deteriorating urban housing. Ingestion of leaded paint was observed or demonstrated by X-ray in 90% of 2,200 patients treated in a lead clinic in Chicago (Sachs, 1975). In addition, there are occasional cases of acute lead poisoning with nervous system symptoms and even fatalities due to uncommon sources of lead exposure such as improperly glazed ceramics (Klein et al., 1970).

Apart from the direct ingestion of lead paint the relative importance of lead intake from other sources is not really known. Direct inhalation of automobile exhausts may be of some importance as a source of lead to children living within 200 ft of a major highway (Caprio et al., 1974).

The major problem, however, for the public health scientist is the matter of small elevations of blood lead (>40 mg/100 ml) without obvious clinical symptoms but perhaps with testable differences in central nervous system function when compared with children with lower blood lead levels. Increase in lead in soil or street dirt and urban dust has been attributed to both automotive exhausts (Lepow et al., 1974) and house paints (Ter Haar and Aronow, 1974). Recent studies indicate that soil lead, probably because of its large particulate size, may be a relatively minor contributor to lead in children (Barltrop et al., 1974). On the other hand, dust on interior surfaces of urban houses has been identified as a significant source of lead in children with elevated blood lead levels, but, again, whether the lead in house dust is derived from lead paint or automobile exhausts is unclear (Sayre et al., 1974). Lockeretz (1975) utilized lead content of deciduous teeth as an index of body burden and past exposure to lead in children in five different environments in order to determine the relative contribution of different sources of lead. The importance of lead paint in children living in dilapidated housing was confirmed, but no effect attributable to automobile exhaust or to industrial emissions could be identified.

Finally, the problem of undue exposure to lead in childhood may be enhanced by increased absorption of lead secondary to dietary deficiencies of metabolically related essential metals, iron and calcium.

METABOLISM OF LEAD

Lead intake in the general population is usually between 150 and 300 μg/day. Most of this is from food, but as much as 20 μg may be added from inhalation of lead vapor and particulate matter in polluted urban environments. In addition, children may take in lead from ingestion or inhalation of dust contaminated with lead or from pica or the chewing of lead-containing nonfood items, particularly paint and plaster. Only about 10% of ingested lead is absorbed, although recent studies suggest a higher rate of absorption for young children particularly when fasting (Alexander, 1974).

A schematic view of lead metabolism is shown in Fig. 1. Absorbed lead equilibrates with plasma and red blood cell lead very quickly. Most of the blood lead is bound to red blood cells and is generally regarded as non-diffusible but is in equilibrium with plasma lead, which contains a small diffusible fraction. The nature of this diffusible fraction is unknown, but it must be bound to biomolecules of sufficiently small size to permit passage across cell membranes. Diffusible plasma lead, in turn, must be in equilibrium with other body pools of lead, which may be divided into two types of tissues: hard tissues such as bone, hair, and teeth and soft tissues such as brain, kidney, liver, and bone marrow. Lead content of soft tissues is directly related to the adverse health effects of lead, whereas lead in bone, nails, etc. reflects tightly bound stores of lead, which are not regarded as immediately harmful but potentially toxic in that this pool may serve as a source of soft tissue lead.

Lead is excreted by the kidneys, but a small fraction may be lost in sweat or by way of gastrointestinal excretion. This includes lead returned to the gastrointestinal tract in bile and in secretions from the salivary glands and

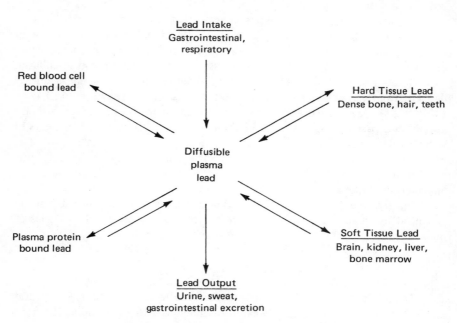

FIGURE 1   Schematic view of lead metabolism.

other intestinal glands. The quantitative extent of net gastrointestinal excretion has not been definitively measured.

In terms of lead balance, nearly all lead taken in by individuals in the general population is excreted, and there is nearly a steady state between bone lead and blood lead. However, lead content of most soft tissues and bone, in particular, tends to increase with age resulting in a total body burden of lead of about 240 mg for the standard 70 μg man (Barry and Mossman, 1970). Thompson (1971) estimates that this level of lead accumulation must be about 4% of the average daily intake or about 10 μg per day.

Blood lead levels are regarded as the most reliable single measure of the body burden of lead particularly in terms of reflecting potential adverse health effects of lead. Blood lead levels in terms of recognition of lead poisoning are discussed in more detail later. Conceptually, it would seem to be more important to measure only the diffusible fraction of blood lead, that is, that portion of plasma lead bound to transportable ligands. However, it has not been possible to measure the small amount of lead present in plasma accurately in the average laboratory.

In recently reported kinetic and metabolic balance studies in a healthy man fed a diet normal in lead content and labelled with the stable isotope ^{204}Pb, it was learned that the dynamics of the body burden of lead can be interpreted by a three-compartment model (Rabinowitz et al., 1973). Compartment one contains the most freely diffusible lead pool and probably contains blood lead and easily exchangeable soft tissue lead. The second compartment must include soft tissue lead and perhaps the more actively exchanging portions of skeletal lead. The most readily exchangeable fraction of blood lead has a half-life of 27 days, whereas the most stable compartment of bone lead is thought to have a half-life of greater than 20 yr.

Lead content of individual organs in control persons without excessive exposure to lead and persons with lead intoxication is shown in Table 2. The largest concentration of lead is usually in the bone, where lead is bound in a nondiffusible form. Only the diffusible lead fraction from plasma passes in and out of capillaries, permeates cell membranes, and enters parenchymal cells of the central nervous system, liver, kidneys, and other organs. The relatively large content of lead in liver and kidney may be related to the excretory function of these two organs, whereas only trace amounts of lead are present in muscle and brain. The major symptoms of lead intoxication are related to the content of lead of soft tissue, particularly the hematopoietic system, the liver, and the kidneys, and factors that enhance lead content of these organs may influence susceptibility to lead toxicity.

Exposure to an organic form of lead is likely to result in rapid accumulation of lead in the tissues most sensitive to the toxic effects of lead, particularly the central nervous system. Experimental studies have shown that tetraethyl lead is converted to triethyl lead and inorganic lead. Triethyl lead is relatively stable and becomes rapidly distributed between brain, liver, kidney,

TABLE 2 Lead Content in Tissue from 15 Persons with No
Abnormal Exposure to Lead (Controls) and Persons Dying
from Inorganic and Organic (Tetraethyl) Lead Intoxication[a,b]

| Tissue | Controls | Lead intoxication | |
		Inorganic	Organic
Bone	0.67–3.59	5.6–17.6	2.9
Liver	0.04–0.28	1.8–8.0	2.35–3.4
Kidney	0.02–0.16	0.6–5.5	0.79
Spleen	0.01–0.07	1.13	0.29
Heart	0.04	0.2–0.8	
Brain	0.01–0.09	0.24–1.2	0.74–1.9

[a]Values are milligrams per 100 g wet tissue; range or single value given
in original reference.
[b]From Goyer and Rhyne (1973).

and blood (Bolanowska, 1968). Intoxication by tetraethyl lead, therefore, is
likely to result in more acute onset of intoxication than exposure to inorganic
lead compounds.

PATHOLOGICAL EFFECTS

The pathological effects of lead are of particular importance in three
organ systems: the nervous system, kidney, and hematopoietic system. Other
effects may occur, particularly endocrine and reproductive abnormalities, but
these are less well understood and have been summarized elsewhere (Goyer
and Rhyne, 1973).

Nervous System Effects

Children with overt lead poisoning usually have central nervous system
symptoms. These vary from ataxia to stupor, coma, and convulsions. The
most prominent pathological changes noted in the brain are cerebral edema
associated with an increase in cerebrospinal fluid (CSF) pressure, proliferation
and swelling of endothelial cells accompanied by dilatation of capillaries and
arterioles, proliferation of glial cells, focal necrosis, and neuronal degeneration.
The changes in capillaries resemble the angioblastic response seen in other
toxic and inflammatory processes and are sometimes associated with perivascular hemorrhages. The vascular alterations are thought to be the most
constant occurrence in lead encephalopathy (Pentschew, 1965). There may
also be a diffuse astrocytic proliferation in gray and white matter. The glial
response may be focal with astrocytic aggregates and microglial nodules.
Neuronal loss in cortical gray matter is variable and may also be seen in the
thalamus and basal ganglia.

Adults may have encephalopathy from lead intoxication, but usually they present with muscle weakness and wrist or foot drop. Schlaepfer (1969) has demonstrated Wallerian degeneration of posterior nerve roots of sciatic and tibial nerves, which suggests a cellular basis for lead-induced paresthesia and sensory nerve loss. Ganglion cells showed no consistent pathological alterations, but surrounding capsular cells contained increased numbers of organelles and dense bodies.

An effect of lead on anterior horn cells of rats, consistent with lead-induced motor neuron disease in humans, has not been found experimentally.

Hematological Effects of Lead

Anemia is an early manifestation of acute or chronic lead intoxication. It is nearly always present when other symptoms of lead toxicity occur, and it may be the only clinical feature of chronic exposure to low levels of lead.

In lead-induced anemia red blood cells are microcytic and hypochromic, as in iron deficiency, and usually reticulocytosis and basophilic stippling are also observed. Iron deficiency may be a coincident factor, but microcytic hypochromic anemia is seen in children with lead poisoning even when serum iron is normal or elevated.

Basophilic stippling of red blood cells has long been recognized as a feature of lead-induced anemia and has been employed as a method of monitoring workers in lead industry, but this test has the disadvantages of being nonspecific and probably does not correlate well with levels of lead exposure (Griggs, 1964). Stippling is more common in erythroblastic cells in bone marrow than in cells in peripheral blood (Waldron, 1966). The nature of the basophilic stippling has received considerable attention and is thought to represent clustered ribosomes (Jensen et al., 1965).

Large concentrations of lead in blood *in vivo* (25–100 mg Pb given intravenously) or added *in vitro* (20 μg/ml blood) produce red cell shrinkage, distortion, and wrinkling of the membrane (Waldron, 1966).

The anemia that occurs in lead poisoning results from two basic defects: shortened erythrocyte life span and impairment of heme synthesis. Whether shortened erythrocyte survival is due to an effect on the developing red cells in bone marrow or on the membrane of the circulating cell has been debated (Waldron, 1966). A more recent study of heme and porphyrin metabolism in an adult with lead poisoning (Berk et al., 1970) has shown that there is a direct hemolytic effect of lead on mature red blood cells, which is independent of effects on heme biosynthesis. Previous workers who postulated decreased erythrocyte survival on the basis of shortened half-life of ^{51}Cr-labeled erythrocytes in lead poisoning were misled by the fact that lead increases the rate of ^{51}Cr elution from tagged cells (Berk et al., 1970). This conclusion was based on the demonstration of an early excretion of ^{14}C-labeled stercobilin and on erythrokinetic studies with multiple tracers.

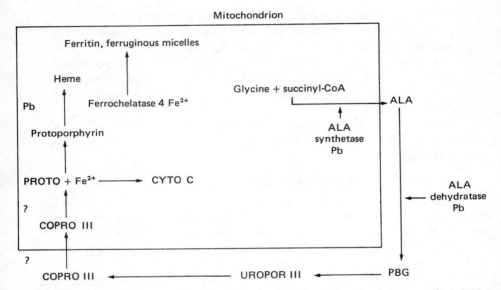

FIGURE 2 Scheme of heme synthesis showing sites where lead has an effect. PBG, porphobilinogen; UROPOR III, uroporphyrinogen III; COPRO III, coproporphyrinogen III; PROTO, protoporphyrinogen; CoA, coenzyme A; ALA, δ-aminolevulinic acid; CYTO C, cytochrome c.

A schematic presentation of the effect of lead on heme synthesis is shown in Fig. 2. At least three steps in heme synthesis may be affected by lead. δ-Aminolevulinic acid dehydratase (ALA-D) is probably the enzyme in the heme pathway that is most sensitive to lead. Inhibition of this enzyme results in a block in utilization of δ-aminolevulinic acid (δ-ALA) and in subsequent decline in heme synthesis. Second, in the scheme of negative feedback control of heme synthesis proposed by Granick and Levere (1964), δ-ALA synthetase activity is depressed, which results in increased activity of the enzyme and increased synthesis of δ-ALA.

A third abnormality of heme synthesis in lead intoxication is inhibition of the enzyme ferrochelatase (Ortzonsek, 1967). Ferrochelatase catalyzes the incorporation of the ferrous ion into the porphyrin ring structure. Bessis and Jensen (1965) have shown that iron in the form of apoferritin and ferruginous micelles may accumulate in mitochondria of bone marrow reticulocytes from lead-poisoned rats.

Other steps in the biosynthetic pathway of heme may also be abnormal in lead toxicity, but the evidence for this is incomplete. Increase in urinary excretion of coproporphyrin (CP), the degradative product of coproporphyrinogen III, is a sensitive reflection of lead toxicity. Metabolism of porphobilinogen to coproporphyrinogen proceeds unimpaired. However, since ALA-D is an extramitochondrial enzyme and later steps in heme synthesis are intramitochondrial, coproporphyrinogen must reenter the mitochondrion to be metabolized further. It has been suggested that transport of metabolites like coproporphyrinogen into the mitochondrial matrix might be impaired in the presence of altered inner membrane permeability and reduction in oxidation and phosphorylation (Haeger-Aronsen et al., 1968). Whether the increased urinary coproporphyrinogen occurring in lead poisoning is a reflection of a nonspecific alteration in mitochondrial membranes or reflects a more specific effect of lead on the intramitochondrial enzyme coproporphyrinogenase is not certain.

Renal Effects of Lead

Acute or early effects on the kidney. Acute renal effects of lead are seen in persons dying of acute lead poisoning or suffering from lead-induced anemia and/or encephalopathy and are usually restricted to nonspecific degenerative changes in renal tubular lining cells, usually causing swelling and some degree of cellular necrosis. Cells of the proximal convoluted tubules are most severely affected. There is little evidence that the glomerulus is affected in acute lead poisoning, although a recent report suggests that ultrastructural changes in the glomerular basement membrane may occur, including complete fusion of the epithelial cell foot processes and increased cytoplasmic density of epithelial cells adjacent to a normal-looking basement membrane. Changes in proximal tubules include the formation of inclusion bodies in cell nuclei cells and the development of functional as well as ultrastructural changes in renal tubular mitochondria.

Nuclear inclusion bodies were initially observed some 35 yr ago in hepatic parenchymal cells and renal tubular lining cells of children dying of acute lead encephalopathy.

The composition of lead-induced intranuclear inclusion bodies has been studied by histochemistry, autoradiography, and direct analysis after isolation. Histochemical studies have been summarized by Richter et al. (1968). Inclusion bodies do not stain with the Feulgen reaction, although they are sometimes surrounded by Feulgen-positive material. They do not stain with fast green after treatment with trichloroacetic acid but do stain strongly with mercuric bromophenol blue and with basic fuchsin, which suggests that they contain protein but probably not histones. The basis of the acid-fast reaction is uncertain; it may simply be a physical phenomenon. Landing and Nakai (1959) have shown that the inclusions contain at most small amounts of lipid, and they have suggested that the acid-fast properties of these bodies are related to the presence of sulfhydryl groups in the protein.

Further studies of the protein from inclusion bodies isolated by differential centrifugation confirms that they are composed of a lead–protein complex containing approximately 50 μg lead/mg protein (Moore et al., 1973). Lead within the inclusion bodies is 60–100 times more concentrated than in whole kidney. Amino acid composition and solubility characteristics of the inclusion body protein resemble those in the residual acidic fraction of proteins in normal nuclei.

Mitochondria in proximal renal tubular lining cells of workers (Cramer et al., 1974) and of experimental animals (Goyer et al., 1968) with lead poisoning show swelling and dilution of matrical granules probably secondary to increased membrane permeability. Such changes are similar to the non-specific swelling that occurs in the early stages of other forms of cellular injury and is probably reversible. The fate of such mitochondria is uncertain, but increased numbers of myelin figures and autophagocytosis suggest an increased rate of turnover. Mitochondria isolated from liver and kidneys of lead-intoxicated experimental animals have impaired respiratory and phosphorylative abilities.

Dysfunction of proximal renal tubules (Fanconi's syndrome) is manifested by aminoaciduria, glycosuria, and hyperphosphaturia and was first noted in acute lead poisoning by Wilson and co-workers in 1953. Plasma amino acids were normal, which suggested that the aminoaciduria and other functional abnormalities were of renal origin. Subsequently, aminoaciduria in children with acute lead poisoning has been observed by Marsden and Wilson (1955) in England, and Chisolm (1962) has found that 9 of 23 children with lead encephalopathy had aminoaciduria, glycosuria, and hypophosphatemia. The aminoaciduria was generalized in that the amino acids excreted in greatest amounts were those normally present in urine, and it was related to severity of clinical toxicity, most marked in children with encephalopathy. The aminoaciduria disappears after treatment with chelating agents and clinical remission of other symptoms of lead toxicity (Chisolm, 1962, 1968). This is an important observation relative to the long-term or chronic effects of lead on the kidney. Restoration of the functional integrity of renal tubular lining cells following treatment of acute lead poisoning implies restoration of normal morphology, but this has not been confirmed experimentally.

Chronic lead nephropathy. The occurrence of a chronic form of renal disease in humans is controversial. There are numerous reports in the medical literature of the past century that describe a form of end-stage renal disease and renal failure in humans said to follow many years of excessive exposure to lead. The morphology, however, of the kidney from such cases has no distinctive characteristics. There is considerable interstitial fibrosis with atrophy of tubular lining cells and tubular dilatation. Glomeruli are sclerotic. The number of inflammatory cells present is quite variable.

Recent studies of renal biopsies from lead workers suggest that typical lead inclusion bodies are not a constant feature of lead-induced nephropathy (Cramer et al., 1974). There has been a strong association between chronic lead nephropathy and gout since the 19th century, but the way in which lead interferes with uric acid excretion is not clear. The problem of chronic lead nephropathy has been discussed in Goyer (1971) and Emmerson (1973).

RECOGNITION OF LEAD POISONING

There seems to be general agreement that the single most informative index of exposure to lead is the blood lead level, particularly when correlated with biochemical and clinical parameters of lead effect. Blood lead levels have been used extensively for the control of workers with industrial exposure to lead as well as for the recognition of lead poisoning in young children with environmental or accidental exposure to lead. Nevertheless, from all of this experience it has been difficult to establish a level of blood lead above which a person can be regarded as having lead poisoning and a blood lead level below which one can confidently predict no adverse health effect. It is becoming recognized that with increasing blood lead levels there is a continuum of probable effects extending from a no-effect level to levels of almost certain manifestations of lead poisoning. It is also recognized that there may be an in-between range of blood lead level, which is accompanied by responses reflecting changes in biochemical activity invoked by lead. Such considerations have resulted in division of lead effect into three to five categories (Editorial, 1968; Goyer and Chisolm, 1972). Regardless of the number of levels of toxicity one wishes to define, conceptually it is important to recognize a transitional, adaptive or asymptomatic, and subclinical phase of lead poisoning.

A simple method for correlating blood lead levels with clinical manifestations of lead toxicity is presented in Table 3. Diagnostic criteria are divided into three phases: no-effect phase, an adaptive or subclinical phase,

TABLE 3 Relationship of Various Parameters of Lead Toxicity
with Severity of Lead Effect

Parameter[a]	Phase		
	No effect	Adaptive or subclinical	Clinical toxicity
Blood lead (μg/100 ml)	< 40	40–80	> 80
ALA-D	Decreased	Decreased	Decreased
Urinary ALA CP	Normal	Slight increase	5-fold
FEP	Normal	Increase	Increase
Anemia	None	Reticulocytosis	Usually
Nervous system	None	Nerve conduction?	Ataxia, coma, convulsions
Renal effects	None	None, inclusion bodies?	Fanconi syndrome, chronic nephropathy

[a]ALA-D, aminolevulinic acid dehydratase; ALA, δ-aminolevulinic acid; CP, coproporphyrin; FEP, free erythrocyte protoporphyrin.

and a phase of overt or clinical toxicity. The boundaries of these categories reflect present knowledge and are subject to modification as new information becomes available. Also, Table 3 does not reflect differences in susceptibility to lead among individuals. The compensated lead industry worker may have blood lead levels around 80 or 100 μg/100 ml without overt symptoms of lead toxicity (Subcommittee . . . , 1968). On the other hand, children with blood lead levels greater than 80 μg/100 ml are usually regarded as experiencing clinical lead intoxication because of the frequency of associated adverse effects due to lead (Chisolm, 1971). Blood lead concentration, however, is not in itself a measure of pathological effect of lead but must be correlated with other biochemical, functional, or clinical manifestations of lead toxicity to establish the diagnosis of lead poisoning in a particular individual.

Clinical Toxicity

The U.S. Public Health Service (Steinfeld, 1971) has suggested that blood lead level over 80 μg/100 ml is unequivocal evidence of clinical lead poisoning, whereas blood level over 40 μg/100 ml is evidence of undue lead absorption or subclinical lead poisoning. Blood lead levels over 80 μg/100 ml are usually associated with fivefold or more increase in urinary ALA and CP as well as an increase in free erythrocyte protoporphyrin (FEP). Each of these parameters reflects altered heme synthesis, and changes are usually detectable at lower blood lead level than anemia (Baloh, 1974).

Nervous system and renal effects of lead are usually recognizable only during the phase of overt lead toxicity. Children with blood lead concentrations above 120 μg/100 ml nearly always have lead encephalopathy, and adults with blood lead levels in this range are regarded as having dangerous levels of exposure with accompanying acute symptoms and long-term sequelae. Nerve conduction as expressed by nerve impulse amplitude may be decreased in the adaptive phase of lead toxicity as shown by Fullerton and Harrison (1969) in workers with excessive exposure to lead.

Clinical renal effects of lead toxicity, namely renal tubular dysfunction or the Fanconi syndrome, occur only during clinical lead toxicity. Industrial workers with excessive lead exposure and increased urinary excretion of lead but without signs of clinical lead toxicity have only minor increases in urinary amino acid excretion (Clarkson and Kench, 1956; Goyer et al., 1972). It is unlikely, therefore, that renal tubular function is impaired to a measurable degree during the adaptive phase. Nevertheless, intranuclear inclusion bodies (lead–protein complexes) occur in renal tubular lining cells of rats at a lower dosage of lead than do any functional signs of lead toxicity. It is suggested, therefore, that inclusion body formation involving the binding of renal lead in a nondiffusible lead–protein complex probably occurs during the adaptive phase of increased exposure to lead.

Overt clinical toxicity may be further subdivided into an early, reversible phase and irreversible effects. Reference to reversibility is more meaningful than use of the terms acute or chronic since these terms imply time as a factor. A child with acute lead intoxication may have more significant nonreversible, central nervous system sequelae than an individual with a more prolonged but less severe exposure to lead. Also, children with neurological sequelae of lead toxicity may have normal blood lead levels (Perlstein and Attala, 1966).

Subclinical Lead Poisoning

Although blood lead levels in the range of 40 or 60 μg/100 ml or in the early adaptive phase are not usually associated with clinical evidence of toxicity, it is not necessarily true that there is no harmful effect of marginal increases in body lead burden or blood lead levels. Two paths of investigation have been directed toward demonstrating that adverse health effects of lead may occur with blood lead levels as low as 40 μg/100 ml. These are the study of subclinical effects on the central nervous system and, second, measurement of effects of lead on parameters of heme synthesis.

Evidence for subclinical central nervous system effects. Detection of subclinical central nervous system effects of lead are important for helping to establish at which level of biochemical effect of lead or blood lead level adverse health effects actually occur. Recent studies suggest that subclinical or difficult-to-detect effects on central nervous function, particularly intelligence and behavioral activity, are present in children with blood lead levels over 40 μg/100 ml. A comprehensive study of 58 asymptomatic children with increased body burden of lead as demonstrated by blood lead over 50 μg/100 ml or over 40 μg/100 ml with urine lead exceeding 500 μg/24 hr following an EDTA provocation was conducted by Pueschel and co-workers in Boston (1972). A history of mild central nervous system symptoms such as clumsiness and irritability was obtained in one-third of the 58 children. Minor neurological dysfunction and various forms of motor impairment were detected by more elaborate testing in 22-27% of the children tested. Chelation therapy was given, and measures were taken to improve their home environment. Eighteen months later a significant increase in some areas of intellectual performance was observed.

Similarly, David and co-workers (1972) found higher body lead, that is, blood lead levels and urine levels, after challenge with a single dose of a chelating agent in hyperactive children than in nonhyperactive children suggesting a relationship between raised lead levels and hyperactivity. More than half of the hyperactive children (28 of 54) had blood lead levels between 25 and 65 μg/100 ml. These workers argue that any blood lead elevation over 24.5 μg/100 ml is dangerous and may produce central nervous system effects. In addition, de la Burde and Choate (1972) showed that 70 asymptomatic 4-yr-old children with elevated blood lead levels (mean 58 μg/100 ml, range 40-100 μg/100 ml) or a lead level of at least 30 μg/100 ml with positive radiographic findings of lead lines in long bones and/or metallic densities in the intestines had deficits in fine motor function and behavior. The common conclusion of each of these studies is that blood lead levels over 40 μg/100 ml must be regarded as potentially hazardous to health.

Measurement of effects of lead on heme synthesis. The following laboratory tests are advocated for detecting a subclinical effect of lead on heme synthesis. Each test has some limitation to its usefulness, however, and selection of a particular test is influenced by the population under study. Each of these tests is briefly discussed.

Free Erythrocyte Protoporphyrins (FEP). FEP is becoming recognized as a very sensitive and reliable measure of a biochemical effect of lead and has recently been recommended by the Center for Disease Control of the U.S. Public Health Service (DHEW, 1975) as an initial screening procedure. It offers another advantage of being relatively simple to perform and is quite reproducible but must be confirmed by actual measurement of blood lead level. Also, FEP is a useful test to confirm lead effect or lead poisoning in a child with elevated blood lead level. Although FEP may be the simplest and most easily performed primary test for large-scale screening programs for childhood lead poisoning, this test does have limitations. Relationship between increased lead absorption as measured by blood lead concentration is logarithmic (Piomelli, 1973) so that it is least sensitive in detecting slight increases in blood lead concentration, that is, the 40-60 μg/100 ml range, where detection of children with the early stages of lead poisoning is so important. Falsely positive increases in FEP occur only in two conditions: erythropoietica protoporphyria, which is very rare, and iron deficiency anemia, which is commonly associated with undue lead absorption.

Normal FEP values vary with the method of analysis, but normal values by the conventional ethyl acetate-acetic acid-HCl procedure are usually less than 250 μg/100 ml of erythrocytes. Children with clinical lead poisoning have FEP levels 25-250 times normal. Iron deficiency is not likely to produce FEP levels greater than 10 times normal.

Urine Coproporphyrin (COPRO). Mention should be made of increase in urinary coproporphyrins in the diagnosis of lead poisoning. To a greater degree than FEP, COPRO is not consistently elevated in children with slight increases in blood lead levels (40-60 μg/100 ml). Normal values as performed by this method are 75 μg/24 hr in small children, 250 μg/24 hr in adults, or 0.2 μg/ml in both children and adults. COPRO is even less specific than FEP and may be elevated in a number of hepatic disorders, particularly in acute hepatitis.

Urine δ-Aminolevulinic Acid (δ-ALA). Urinary ALA excretion as a measure of lead effect suffers from the same limitations as does COPRO and FEP. The relationship between blood lead and urinary ALA is curvilinear, and the reliable upper limit of ALA excretion acceptable as normal is difficult to determine. An upper limit of 0.6 mg/100 ml indicative of a blood lead level of 70 μg/100 ml or greater has been recommended for screening industrial workers, but a study of children with undue lead exposure has shown that even at this level about one-quarter of the tests are falsely negative or falsely positive (Blumenthal, 1972). ALA elevations when blood lead is in the 40–60 μg/100 ml range are only slight and usually not recognizable in the individual. Urinary ALA, therefore, is not suitable for mass screening of children for childhood lead poisoning. However, it has been widely used for the monitoring of lead industry workers and is generally performed on spot collected urine samples and reported in mg/100 ml urine.

Aminolevulinic Acid Dehydratase (ALA-D). The ALA-D activity of peripheral red blood cells may be the most sensitive biological parameter of lead effect that is presently measurable. Studies on venous blood samples from both children and adults show a negative correlation between the log of ALA-D activity and blood lead concentration from 10 to about 90 μg/100 ml (Hernberg et al., 1970; Miller et al., 1970; Haeger-Aronson et al., 1971; Weissberg et al., 1971). This test, therefore, has the advantage of being very sensitive at the low range of blood lead concentrations and is, in fact, adversely affected by blood lead levels in the normal individual. ALA-D is directly related to blood lead in persons with positive lead balance as well as those in the steady state. There is virtually no lag between lead exposure or absorption and inhibition of ALA-D. It is thought, therefore, that ALA-D activity of peripheral blood is directly related to blood lead levels. As blood lead levels are lowered by chelation therapy, ALA-D activity varies proportionately.

Lead in Hair, Bone, and Teeth. Lead in accessible body stores has been advocated as a measure of body burden of lead. In contrast to blood, lead levels in these hard tissues (Fig. 1) is a better measure of long-term or chronic exposure to lead (body burden lead) that is bone lead and does not correlate with symptoms of toxicity. However, it may be clinically important to be aware of increased body burden of lead since such individuals may have a greater susceptibility to lead toxicity.

Hair Lead. Lead becomes incorporated into the keratin molecule of hair in the hair follicle. The lead concentration of hair as it emerges from the scalp is relatively constant if exposure is continuous. Episodic exposure to lead is reflected by uneven distribution of lead in hair.

In children without excessive exposure to lead, hair lead ranges from 1 to 92 μg/gm hair; the mean is about 20–30 μg lead/gm hair (Baloh, 1974). Hair lead in lead workers has been found to range from 24 to 1,880 μg/gm (Barry, 1975) and in children with excessive lead exposure, from 70 to 975 μg/gm (Kopito et al., 1969). There seems to be good correlation between hair lead and body stores of lead as measured by blood lead levels and/or urine lead following provocative chelation (Kopito et al., 1969). Cleansing of hair to remove contamination from atmospheric lead is essential prior to analysis (Hammer et al., 1971).

Bone Lead. Bone contains the greatest concentration of lead in the body and accumulates during the lifetime of an individual. It is therefore a better measure of cumulative lead exposure rather than recent exposure. Lead lines in the body matrix at the metaphysis long bone as seen by X-ray have long provided a simple method of detecting increased body stores of lead in children. It is thought that lead deposited along with calcium is responsible for the increased densities. It must be remembered that metaphyseal bone formation occurs only in growing children and is not seen in adolescents or adults.

Treatment of lead poisoning with chelating agents produces an immediate decrease in blood lead and soft tissue lead, but bone lead decreases very little. Equilibration of bone lead with blood and soft tissue lead then follows lowering of blood lead as would be expected from the scheme in Fig. 1.

Tooth Lead. It has been shown that lead in deciduous teeth is concentrated in the body of the dentive in areas adjacent to the pulp (Carroll et al., 1972). Measurement of tooth lead concentration offers a way to estimate early childhood exposure to lead.

Strehlow's (1972) studies have shown that tooth lead content is dose dependent and is not reduced by chelation. The distribution of lead in various components of the tooth including enamel, root dentine, and coronal dentine is similar, whereas secondary or circumpulpal dentine, the area of dentine adjacent to pulp and in immediate contact with blood, is higher than other areas of the tooth and is thought to reflect actual exposure to lead throughout the life of the tooth. This area of the tooth may be obtained for analysis by dissection of the extracted tooth. The tooth may be sectioned apically to expose the pulp chamber and the root canal. The residual pulpal tissue may be removed with a dental broach and the tooth canal filed with dental reamers. The powder obtained from the secondary dentine is collected, dried, and analyzed for lead (Shapiro et al., 1972). Needleman and Shapiro (1974) found substantially higher dentine lead levels in teeth from urban children living in deteriorated housing or attending school in proximity to a major manufacturer of paint and lead products than in teeth from children considered to be a low risk for exclusive lead exposure. Comparison of lead content of teeth from a contemporary population of nonindustrialized Indians of the Lacandon Forest in Mexico with a modern urban industrial population shows a 45-fold difference in median tooth lead level (Shapiro et al., 1975).

ANALYTICAL PROCEDURES FOR THE EVALUATION OF LEAD AND BIOCHEMICAL PARAMETERS IN LEAD EXPOSURE

Quantitative Evaluation of Lead in Biological Media

For purposes of discussion here, lead shall be taken to mean the divalent, inorganic form of the element, whether arising as such from biological samples or converted to the inorganic form via sample treatment.

Since lead in biological media such as blood and urine is being assessed at trace levels, even under conditions of exposure, a number of precautions for sample collection and handling must be followed, regardless of specific instrumentation employed. It is first necessary to elaborate on these prior to discussion of various measurement methods currently in use.

Collection and handling of biological samples for lead analysis. Two key problems in any trace analytical procedure are sample contamination from some external source with the element also being measured and loss of the element somewhere along the analytical route. Sample collection can be especially troublesome in screening programs where large numbers of samples must be collected in the shortest possible time with minimum stress to the participants in the screening. A number of recent reports in the literature dwell almost exclusively on details of sampling for blood lead or other blood index for lead exposure.

Marcus et al. (1975) employed a double-blind study of 207 specimens that had been gathered as part of a screening procedure evaluation. These workers evaluated the Becton-Dickinson Unopette and Capillector and found them unsatisfactory on the basis of ease of use or precision in results obtained thereby. Similarly, heparinized capillary tubes pose problems with complete mixing of the collected sample, while Caraway tubes, in particular, were found to yield clotted blood volumes. An additional problem with these tubes has to do with the lead levels in the glass matrix itself, which defeats efforts at prior chemical debridement by acid washing, etc., especially if the samples must remain in the tubes for any period of time. Satisfactory results in the hands of these investigators were achieved with lead-free 100-μl capillary tubes containing deleaded heparin and capped at both ends with lead-free plastic closures (Environmental Sciences Associates, Inc., Burlington, Vermont). Satisfactory results also appear to be obtained by a number of laboratories using low-lead Vacutainer tubes containing anticoagulant.

The difficulty of dealing with collection of discrete volumes of blood has prompted a number of workers to employ the alternative of blood sample uptake on filter-paper discs selected for uniformity of fabrication, low lead, and uniform diffusion of blood through the paper matrix. In this connection Cooke et al. (1974) have carried out a comparison study varying only in the method of blood collection in which capillary blood collected on an area of paper roughly 1.0 cm in diameter was analyzed and compared to a conventional venous puncture procedure using 10.0 ml. A correlation of 0.78 was obtained between the two procedures.

Preparation of the puncture site prior to sample collection also requires considerable care. In the procedure of Marcus et al. (1975), cleansing is carried out using an ethanol solution of citric acid followed by a rinse with 70% ethanol. Prior soap and water washing was not found to be necessary. The work of Cooke et al. (1974) entailed a vigorous scrubbing with low-lead soap solution, followed by rinsing with distilled water.

Collection of urinary samples for lead should preferably be done in acid-washed plastic containers (polypropylene or polyethylene) with a low-lead bacteriocide if samples are to be stored.

Analytical methods for lead analysis. Consideration of the analytical techniques for the assessment of lead in biological media of relevance to the clinical laboratory narrows the choice to spectrophotometry, atomic absorption spectrometry, and several electrochemical techniques.

Classically, the method of widest use up to the present time has been a colorimetric procedure involving the interaction of lead in suitably treated samples with a chelating chromophore, such as dithizone, to form a complex that has a discrete and concentration-dependent absorption when assayed via a spectrophotometer. The method is economical in equipment and analyst expertise but suffers from sensitivity limitations, is quite tedious, and can be plagued by a variety of interferences.

Presently, the procedure enjoying the widest use, in all its instrumental ramifications and having largely supplanted colorimetry for lead analysis, is atomic absorption spectrometry. While equipment cost initially can be somewhat high, the training level of the operator in practice can be modest under appropriate supervision. The technique is sensitive and specific, and where use of various microsampling techniques is made, time consumption and sampling are minimal. For this reason the major portion of this section will be devoted to atomic absorption spectrometric techniques.

A somewhat newer, very sensitive technique for lead analysis, particularly in physiological fluids, is anodic stripping voltammetry, an electrochemical procedure. As is the case with most electrochemical techniques, results are influenced by the chemical composition of the matrix, requiring complete degradation of any organic matter, with attending input of time and effort.

Measurements of Lead in Biological Media by Atomic Absorption Spectrometry. A simple definition of atomic absorption spectrometry is the absorption of radiant energy by atoms. Atom production from various elements and compounds requires energy input, usually supplied by a flame or heated graphite rod in the case of atomic absorption spectrometry.

Absorption of thermal energy from a flame with subsequent emission of energy as a spectral line is atomic emission, while atomic absorption corresponds to energy absorption from a source other than a flame with a concomitant decrease in signal from the source. The relationship between these spectroscopic phenomena is given by

$$\text{Atom} \atop \text{(unexcited)} \; + \; h\nu \; {\overset{\text{absorb}}{\underset{\text{emit}}{\rightleftharpoons}}} \; {\text{Atom}^* \atop \text{(excited)}}$$

Photon absorption by a ground-state atom is atomic absorption, while photon emission by an excited atom with return to the unexcited ground state is atomic emission. For most elements, the spectral resonance line is the characteristic wavelength used in both atomic absorption and emission, which is to say the transition to the lowest excited state from the ground state.

If the metal atom is generated in a flame having sufficient energy, the atom may become ionized, the absorption of energy by the ion being similar to that of the atom but occurring at different wavelengths. Such ionization will reduce the population of atoms and will decrease both absorption and emission. With the exception of the alkali and alkaline earth elements, ionization is relatively negligible.

Detailed theoretical treatment of the relationship of atomic absorption to atomic concentration may be found elsewhere (Christian and Feldman, 1970; L'vov, 1970), but suffice it to say that within a given range there is a linear quantitative correlation of a metal such as lead in solution and atomic absorption.

Instrumentation. The basic components of an atomic absorption spectroscopic system are a primary source of radiation, a means of producing neutral atoms, a monochromator to isolate the line(s) of interest, a detector, and an amplification-readout system.

Hollow cathode lamps, which produce sharp spectral lines, are the most suitable source for atomic absorption. The cathode in such a lamp usually consists of a hollow cup configuration made of the same material as the element to be analyzed, while the anode is a wire or disc of tungsten. The electrodes are contained in a glass housing having some inert gas at reduced pressure. On application of a voltage between anode and cathode, the inert gas is charged at the anode and impelled at high velocity to the cathode. This collision dislodges metal atoms, which then collide with the gas ions to generate excited metal atoms.

The crucial process in atomic absorption spectrometry is the generation of neutral atoms of the metal of analytical interest, from either a solution or a solid matrix. In the former case, the general atomization technique involves the use of a nebulizer, spray chamber, and burner sequence, nebulization serving to break down the sample solution to fine particles. While various nebulizer designs have been reported, the most common design is the pneumatic, where the support gas provides the force to break up the liquid particles. With the latter type of atomizing technique, flash atomization occurs by drying a sample in a suitably fabricated vessel followed by direct insertion into the flame, or in the case of a heated graphite furnace, the sample is dried *in situ* followed by bringing the graphite container to a high temperature via passage of a current through the assembly.

Subsequent to nebulizing the aqueous or organic analyte solution, atomization is carried out using a burner–flame combination, the sequence of events leading to atomization being

$$\underset{\text{(solvent)}}{MX} \xrightarrow[\text{solvent}]{\Delta H \text{(vap.)}} \underset{\text{(solid)}}{MX} \xrightarrow[\text{solid}]{\Delta H \text{(vap.)}} \underset{\text{(gas)}}{MX} \xrightarrow{\text{dissoc.}} M^0 + X^0$$

Two types of burners are currently in use: (1) the premix burner, where nebulization occurs in a chamber with mixing of oxidant and fuel gases, and (2) the total consumption burner, where a sample is aspirated into a capillary and burned at the tip of the burner. The relative merits and disadvantages of these two configurations have been treated in some detail (Christian and Feldman, 1970).

The major component of the optical system of an atomic absorption spectrophotometer is the monochromator, which separates the resonance line from other nearby spectral lines. The types of monochromator in use include gratings, prisms, and interference filters. Signal detection is achieved mainly by use of photo tubes and photo diodes although in some cases barrier layer cells are employed, as with the alkali elements. With most instruments the readout device is a meter; many laboratories also find it desirable to use further accessories such as strip-chart recorders, which furnish easier-to-handle, permanent records.

Sample collection and handling of samples for lead analysis by atomic absorption spectroscopy. As with any technique for evaluation of trace levels of elements in biological media, special precautions must be exercised in sample collection and work-up to preclude contamination as well as loss of lead. Vessels employed for sample collection should be low in lead as should any other materials making contact with the sample. Venous blood gathering also requires cleaning of the skin area in the vicinity of the puncture. All aqueous standard lead solutions should be prepared in lead-free water. Similarly all reagents such as concentrated acids and buffers should be obtained and used as lead free as possible.

Degradation of the biological matrix prior to atomic absorption analysis for lead, where this is necessary, is usually carried out by wet or dry ashing techniques. Dry ashing has the advantage of simplicity, minimum recourse to reagents, and use of large numbers of samples. Loss of lead by volatilization and lead retention in ash residues are disadvantages, however, as is the protracted ashing time. Wet ashing is more rapid and eschews loss of lead by volatilization but requires the use of various concentrated, corrosive acids as well as more of the analysts' time and attention. Furthermore, contamination via reagents is a vexing problem, although the present commercial availability of many ultrapure reagents minimizes this objectionable feature.

The somewhat new technique of low-temperature ashing, whereby dried samples are exposed to an energy-rich plasma of oxygen via radio-frequency generation, avoids the negative features of both of the above techniques. The requisite instrumentation constitutes an economic factor against the technique; if this is no problem and competent personnel are available, this technique is preferable.

A modified procedure for wet ashing using commercially available ware is the combustion bomb, whereby samples are rapidly degraded by acid at high temperature in a sealed inert vessel. For large numbers of samples, this technique would be initially expensive but would yield considerable savings in time.

Micro sampling procedures using the Delves nickel cup (Delves, 1970) or the tantalum sample boat (Kahn and Sebestyen, 1970) as well as variations of the electrically heated graphite furnace usually do not require extensive sample treatment, which accounts in part for the increasing widespread use of these newer techniques. With blood samples the boat or cup procedures involve a short partial digestion with hydrogen peroxide followed by drying and direct insertion in the flame. A more recent procedure (Ediger and Coleman, 1972) involves preignition of the organic matrix of blood, thereby bypassing the peroxide treatment step.

In order to minimize matrix effects arising from any remaining organic matter or inorganic components in digests containing lead, a number of procedures employ an extraction step with tandem use of a chelating agent such as ammonium pyrrolidinodithiocarbamate. A common organic extractant is methylisobutyl ketone. In addition to yielding a cleaner lead-containing medium, burning of the organic extracting agent enhances the signal relative to that from an aqueous medium.

Sample analysis for lead content. Using hollow cathode lamps, there are four lead absorption lines of varying sensitivity, with the 283.3 nm line being commonly employed. While a number of fuel–oxidant combinations for flame generation in lead analysis may be used, the air–acetylene flame is employed in most cases. The others are air–coal gas, air–propane, and oxyhydrogen.

Analytical interferences encountered in lead determinations are a function of the flame type employed. With the air–acetylene combination, there is no interference from a number of physiological anions and cations present even in excess (Iida et al., 1966) although some interference arises from high levels of aluminum.

With the use of an air–acetylene flame and the 283.3 nm line, flame procedures are sensitive down to about 0.2–0.5 ppm lead at which direct aspiration of aqueous solutions of analyte is done. Where chelation–extraction steps are interposed (see above), this range is dropped proportionately as a function of the ratio of aqueous to organic extractant volume, there usually being a much smaller volume of organic extractant employed. In the case of the Delves cup and tantalum boat procedures, lower detection limits are in the range of 0.001–0.010 ppm lead. Similarly, the heated graphite furnace, depending on analyte volume, can be used to assess parts per billion or less.

A number of comparative studies have recently been carried out contrasting the results generated in blood analysis using various types of atomic absorption spectroscopy procedures Marcus et al. (1975) have carried out a comparison of the Delves cup micromethod with a macroscale method using a group of 207 specimens. A mean value of 276.6 μg/liter with the Delves assembly compared favorably with a corresponding mean of 273.2 achieved using a conventional macrolevel method. Anderson and co-workers

(1974) evaluated eight atomic absorption systems for the quantitation of capillary blood lead, including four commercially available analytical units, the Delves cup vs. furnace tube methods of atomization, and liquid blood samples vs. filter paper absorbance. Data precision was better with the Delves cup than with the graphite crucible, while the filter paper disc technique gave lead levels that were higher and less precise than the discrete liquid sampling method.

Colorimetric Determinations of Lead in Biological Media. As noted earlier the first widely employed technique for measuring lead in biological media involved a colorimetric procedure based upon the complexation of lead in suitably treated samples with a binding agent, leading to a chromophoric product. The most commonly used reagent in this connection has been dithizone, 1,5-diphenyl thiocarbazone. The spectral intensity of the lead dithizonate, which relates directly to sample lead concentration, is determined spectrophotometrically at 510 nm.

The two procedures employing the dithizone method that appear to be most reliable and that have enjoyed the benefit of interlaboratory evaluation are the USPHS and the APHA techniques. The USPHS method, recently published in detail (National Academy of Sciences, 1972) is a double-extraction, mixed-color procedure, which is satisfactory in the absence of bismuth. As is the case with other trace metal methods, reagents and water sources to be employed must be of low-lead quality. In particular, only commercial chloroform that is indicated as suitable for use as an extracting agent should be employed in dithizone procedures. Glassware must be rendered lead free and protected from contamination until use.

Both blood and urine samples are wet ashed using concentrated lead-free nitric acid. Digests are then treated with hydroxylamine hydrochloride and sodium citrate and the pH adjusted to 9–10 followed by cyanide addition. Extraction of lead dithizonate is carried out using a chloroform solution of dithizone. Lead is then transferred to an acidic aqueous solution using 1:99 nitric acid and the aqueous layer treated with an ammonia–cyanide mixture and the dithizone–chloroform reagent. The chloroform extracts are then analyzed in a spectrophotometer at 510 nm.

This method is regarded as essentially specific for lead. Tin is oxidized during ashing to the tetravalent form, which is not extractable. Bismuth poses interference problems, but this element is not frequently encountered in biological media. Thallium, another interferent, is similarly not commonly encountered. Interferences from other metals such as silver and copper are minimized by the use of cyanide, which strongly binds a variety of metals. Calony et al. (1963) have raised the objection that the method does not avoid the interference by cadmium. Dithizone is destroyed by free halogen, strong oxidizing agents, light, and large ferric ion concentrations.

The APHA method (American Public Health Association, 1955) varies chiefly in providing for the elimination of bismuth as the dithizonate at pH

3.4. The colorimetric procedures are adequate in the range of lead values 1–10 ppm. Any reliable spectrophotometer may be employed for these procedures; the USPHS technique employs the Beckman manual spectrometers as well as the B&L Spectronic 20.

Anodic Stripping Voltammetry. Anodic stripping voltammetry is an electrochemical technique that is presently being employed to evaluate lead levels in physiological fluids, chiefly blood. The method as developed by Matson (1968) relies on concentrating an ion such as divalent lead on a negative electrode during a long plating time (5–60 min), followed by reversing and increasing the electrode polarity for short periods yielding a sharp current peak proportional to ion concentration.

Prior sample treatment is required and involves digestion of the samples with perchloric acid. Since different forms of elements as well as remaining organic matter that interacts with the metal will materially influence results, thorough degradation of the sample is required. The method is considerably more sensitive than the dithizone colorimetric procedure, but the vulnerability of the assay to matrix interference points to the atomic absorption procedure as perhaps being preferable.

Polarography. Polarography is a second electrochemical technique applied to the assessment of lead in biological materials (Calony et al., 1963). Blood and urine samples are dry ashed and taken up in dilute (1 M) KCl solution. After deaeration with nitrogen, polarograms are obtained using a polarograph, the height of the half wave being directly related to element concentration.

Measurements of δ-Aminolevulinic Acid Dehydratase

The erythrocyte enzyme δ-aminolevulinate dehydratase (5-aminolevulinate hydrolase: E.C. 4.2.1.24, ALA-D) mediates the dehydrative cyclization of δ-aminolevulinate to porphobilinogen:

The enzyme has the properties of euglobulin (Gibson et al., 1955) and an apparent molecular weight of 270,000 (Coleman, 1966). That ALA-D is a sulfhydryl enzyme is evident from inhibition by heavy metals such as copper, mercury, silver, lead, and other thiol agents (Gibson et al., 1955; Granick and Mauzerall, 1958; De Barreiro, 1967; Tomio et al., 1968).

The characteristics of enzyme inhibition by lead comprise the basis for use of this enzyme to assess lead exposure. Lead inhibition appears to be of the noncompetitive type, as manifested in a reduced V_{max}. Apparently lead becomes attached to a sulfhydryl group at or in the vicinity of the active site or occasions some structural perturbation in the enzyme's conformation.

Highest enzyme activity to date has been found in mammalian erythrocytes and is confined to the soluble portion of the cytoplasm. Full activity of the enzyme requires either glutathione or dithiothreitol (Granick and Mauzerall, 1958). Furthermore, the latter agent fully activates the enzyme in lead-poisoned preparations and blood, indicating that lead does not inhibit the formation of, or occasion the destruction of, ALA-D (Granick et al., 1973).

Blood collection and storage. Blood collection is carried out with lead-free heparinized tubes and preferably lead-free syringes and needles. For microassay blood is collected with a heparinized microhematocrit tube (Granick et al., 1973). Strong chelating agents such as EDTA must be avoided because of both inhibition of the enzyme and also activation of lead-poisoned enzyme by preferential binding of the lead.

Blood samples must be cooled as soon after collection as is feasible. Enzyme activity is lost at varying rates upon storage depending on storage temperature and whether hemolyzed or intact cells are taken for assay (Granick et al., 1973). The activity of ALA-D in heparinized blood at 4°C is constant for 24 hr.

Enzyme assay procedures. The most widely used assay procedure for ALA-D is based on measurement of the amount of porphobilinogen generated from substrate δ-ALA spectrophotometrically via use of the Ehrlich reagent with modification (Mauzerall and Granick, 1956). The resulting Ehrlich color salt, a condensation product of p-dimethylaminobenzaldehyde with porphobilinogen, is measured at 553 nm (molar absorption coefficient, 6×10^4).

It is necessary that sulfhydryl compounds be absent from the final medium used for spectrochemical analysis, because these agents will interact with the chromophore to furnish a putative colorless addition product (Granick and Mauzerall, 1958):

(colored)   (colorless)

All procedures avoid sulfhydryl compounds by including mercury (II) ion in the quenching trichloroacetic acid solution. In addition, the method of Granick and Mauzerall (1958) calls for use of mercuric chloride in the modified Ehrlich reagent.

Whole blood directly (Weissberg et al., 1971a,b; Granick et al., 1973;

Lauwerys et al., 1973) as well as washed erythrocytes (Bonsignore et al., 1965; Hernberg et al., 1972) have both been employed as enzyme source. The micro procedure of Granick et al. (1973) requires only 5 μl of heparinized whole blood and appears to be of value in a screening-type program.

The enzyme incubation step is carried out using δ-ALA of the highest possible purity in phosphate buffer as the substrate (and phosphate buffer containing Triton X-100 or deionized water or saponin in water where whole blood is used). Incubation is carried out for 30–60 min at 37°C. While the incubation step should employ an inert atmosphere such as nitrogen, the micro method (see above) avoids this step. The reaction is quenched using trichloroacetic acid–mercuric ion mixture, the latter reagent assisting in removal of sulfhydryl compounds, which interfere with Ehrlich color salt formation.

Several ways of expressing ALA-D activity have been reported, and for convenience the mathematical forms of two are given here:

$$\delta\text{-ALA-D activity} = \Delta OD_{553} \text{ (sample tissue control)}$$

$$\times \frac{138,000}{\text{Hct}} \text{ nmol porphobilinogen/ml red blood cells per hr}$$

(Granick et al., 1973)

$$\delta\text{-ALA-D activity} = \Delta OD \, (t_{60}' - t_0') \times \frac{100}{\text{Hct}}$$

$$\times 131.48 \; \mu\text{mol porphobilinogen/ml red blood cells per min}$$

(Weissberg et al., 1971)

Quantitative Measurements of FEP in Lead Poisoning

Van den Bergh and Hyman (1928) first established the existence of free porphyrin in erythrocytes, while Grotepass (1937) ascertained that this porphyrin is protoporphyrin IX. The main factors that appear to influence FEP levels in the direction of elevation are increased erythropoiesis and perturbation of iron utilization for hemoglobin biosynthesis, as occurs in lead poisoning, as well as iron deficiency.

Pertinent chemical properties of FEP with reference to methods for its evaluation include its lability to light and strong acids, its ability to coordinate metals, and its sharp but solvent-dependent absorption spectrum in the near ultraviolet (Soret band) region. In addition, porphyrins such as FEP are among the most intensely fluorescing compounds known in nature. The spectro-chemical properties of FEP are quite similar to those of copro- and uropor-phyrin, whose spectra must be taken into account in any scheme where separation is not carried out.

The various techniques of FEP measurement have been reviewed and/or evaluated by Wranne (1960) and Heilmeyer (1966). The first method reported for FEP determination was that of van den Bergh et al. (1932) consisting of extraction from red cells with a mixture of ethyl acetate and glacial acetic acid (4:1). Subsequent steps include back extraction into 5% HCl, neutralization, and ethyl ether reextraction followed by a second HCl extraction. Estimation of FEP was carried out fluorometrically in the ultraviolet. A number of the earlier modifications of this procedure vary as to extraction time, acid concentrations, and techniques of FEP measurement once isolated.

Currently, both spectrophotometric and fluorometric procedures are in use for FEP analysis; the former involves venous samples and is quite laborious but is apparently preferred for limited numbers of samples in detailed studies; the latter technique has been recently adapted for screening purposes.

Spectrophotometric determination of FEP. Porphyrins such as FEP dissolved in mineral acids display two spectral bands in the visible and one in the near ultraviolet (Soret) region. The Soret band is sharp and is suitable for spectral analyses, while its maximum is dependent on solvent (acid) concentration (Wranne, 1960) being at 411 nm in 25% HCl and 407 nm in 5% HCl.

Small amounts of impurities, such as hematin, markedly affect spectral results for FEP in this region, prompting Rimington and Sveinssen (1950) to introduce a correction factor involving determination of extinction at 380 and 430 nm as well as extinction maximum, which on further refinement by With (1955) leads to the correction equation

$$\mu\text{g FEP/ml } 3 \, N \text{ HCl} = 1.225 \times [2E_{max} - (E_{380} + E_{430})]$$

Wranne (1960) has carried out a detailed study of spectrophotometric and fluorometric assays of FEP and has evolved a detailed procedure for FEP by spectrophotometry. Blood samples (5–10 ml) are centrifuged and the plasma removed. The cells are transferred to flat-bottomed tubes and mixed thoroughly with a 4:1 mixture of ethyl acetate:glacial acetic acid. The ethyl acetate employed here contains a small amount of EDTA to minimize interference by metals via chelation. Filtration of the extracts is followed by washing with dilute aqueous acetate. The porphyrin is transferred to 5 N HCl by multiple extraction, and the combined extracts are neutralized and reextracted with ethyl acetate. After washing the organic layer with water, sequential extraction using 0.24 N and 5 N HCl is employed, the former removing coproporphyrin and the latter taking up essentially FEP. The 5 N HCl extracts are then analyzed spectrophotometrically at 410 nm using the necessary correction factors (Wranne, 1960).

The spectrophotometric procedure of Heilmeyer (1966), which is an improved variation of that reported by Schwartz and Wikoff (1952), employs a different isolation sequence. Plasma is removed from heparinized blood followed by a double wash with isotonic saline. The cells are homogenized in a Teflon homogenizer with 10 volumes of ethyl acetate:glacial acetic acid (4:1). After filtration of the homogenates through a sintered glass funnel, the filtrates are washed with one-third volume of acetate solution. The balance of the procedure is essentially that of Wranne (1960), in which a distinction between coproporphyrin and FEP is desired.

A more recent spectrophotometric technique is that of Langer et al. (1972), which involves a change in the extraction medium for FEP isolation. Using 10 ml samples of heparinized venous blood, the plasma is removed and the original volume restored with distilled water. The lysed erythrocytes are extracted with an acetone:concentrated HCl solution (50:1). Treatment of the supernatants from this extraction with 1.4 *N* HCl precipitates hemin. The acid supernatant is then neutralized, extracted into ethyl ether, and after the usual washing sequence reextracted into 3 *N* HCl.

Using the correction factor noted above, the levels of FEP are calculated to be

$$\mu g \text{ FEP}/100 \text{ ml RBC} = [2E_{407} - (E_{380} + E_{430})] \times 1.226 \times 5 \times \frac{\text{Hct}}{100} \times 100$$

Fluorometric measurements of FEP. A number of reports have recently appeared describing microfluorometric procedures for determination of FEP. In all cases, the techniques are oriented to large-sample screening programs.

In the procedure of Granick et al. (1972), 2 μl whole blood freshly collected is taken via Drummond microdispenser and placed in 1 ml test tubes, which also serve as spectrofluorometric cuvettes. Ethyl acetate:glacial acetic acid (2:1) is then rapidly added followed by addition of 0.5 *N* HCl and vigorous agitation. The acidic bottom phase contains the porphyrins: 80% of the FEP, 86% of uroporphyrin, and 83% of the coproporphyrin. Using a recording spectrofluorimeter, the tubes are scanned from 560 to 680 nm, with excitation at 400 nm.

In acid solution two porphyrin fluorescence bands at 605 and 655 nm are observed, with the former more intense but the latter less vulnerable to the presence of contaminants. In the above procedure the ratio of these two band maxima is calculated for each sample; a value of ca. 2.1 indicates solely FEP, while a value that is less indicates the presence of copro- and/or uroporphyrin. In the latter case an added extraction step is carried out using the extracting medium and 0.05 *N* HCl on a second blood sample. In this step most of the uro- and coproporphyrin is removed but little of the FEP.

The method of Piomelli et al. (1973) uses 20 μl blood added to a 5% celite suspension in saline. The same organic extractant is employed as with the other procedures, but further work-up involves the use of 1.5 *N* HCl to generate the fluorescing acid layer. Excitation is carried out at 405 nm and fluorescence measured at 610 nm, using coproporphyrin as the standardizing agent in lieu of the more labile FEP.

Kammholz et al. (1972) have described a rapid procedure using 0.1 ml capillary blood and 5:1 ethyl acetate:glacial acetic acid as extracting medium. Supernatants are multiply (×3) extracted with 3 *N* HCl and made up to a final volume with more 3 *N* acid. Coproporphyrin is used to construct standard working curves. This method is not specific for FEP, but since the levels of the coproporphyrin are small relative to FEP, for screening purposes its influence can be discounted.

In a recent report by Lamola and co-workers (1975), it has been disclosed that FEP is actually zinc protoporphyrin (ZPP), and in the course of a detailed study of the spectroscopic properties of blood porphyrins, these investigators devised a rather rapid and sensitive fluorometric procedure for ZPP. Small samples of whole blood (20 μl) are diluted with phosphate buffer containing a detergent, dimethyldodecylamine oxide. The fluorescence spectrum is then recorded using an excitation setting of 424 nm and the fluorescence at 594 nm.

Measurements of Urinary δ-Aminolevulinic Acid in Lead Poisoning

Collection and sampling of urine. The stability of δ-ALA in urine has been studied in some detail (Haeger-Aronsen, 1960). δ-ALA is stable in acidified urine (pH 1–5) so that an acid preservative such as acetic or hydrochloric acid is satisfactory for use at the time of urine collection. Samples should be stored in the dark at 4°C until assayed. Freezing of the samples should be avoided. Urine samples handled in this manner maintain a constant δ-ALA content for several months.

Measurements of δ-ALA in urine. The classic procedure is that of Mauzerall and Granick (1956). This method involves condensation of δ-ALA with a β-diketone such as acetylacetone or ethyl acetoacetate to yield a substituted pyrrole derivative, which then is allowed to undergo reaction under acidic conditions with modified Ehrlich reagent (*p*-dimethylamino-benzaldehyde) to yield an Ehrlich color salt:

The condensation product in the case of acetylacetone was found to be 2-methyl-3-acetyl-4-(3-propionic acid) pyrrole. While the chromophoric Ehrlich salt formed can undergo reaction with a second unit of the pyrrolic product, the transformation is much slower than is the desired reaction, providing the acid concentration is optimal.

Aliquots of urine samples (pH 5–7) are chromatographed on columns of Dowex-2 resin; δ-ALA is then eluted with water along with urea. The eluates are transferred to Dowex-50, where urea is eluted with water. δ-ALA is then removed with acetate solution, the eluates diluted to a given volume after addition of acetylacetone. After containment in a boiling water bath to effect complete condensation of δ-ALA with the diketone, an aliquot of the reaction mixture is treated with modified Ehrlich reagent (*p*-dimethylaminobenzaldehyde in perchloric/acetic acid). The resulting Ehrlich color salt reaches maximum intensity at about 15 min after mixing, at which point it is inserted into a spectrophotometer and read at 553 nm. The limit of detection via this method is 3 nmol/ml or 3 μmol/liter urine, while the color intensity is stable for 15 min.

A number of variations of the basic Mauzerall and Granick procedure have appeared in the literature. Williams and Few (1967) have observed that the initial separation of porphobilinogen can be omitted so long as the possibility of false high readings (e.g., porphyria) is borne in mine. In a comparison of methods using urine samples from 39 lead workers, a coefficient of correlation of 0.99 was observed. In the method of Wada et al. (1969), urine samples are first subjected to an *n*-butanol extraction to remove interfering substances followed by extraction with a solution of ethylacetoacetate, which simultaneously reacts with δ-ALA to form the pyrrolic intermediate. Recoveries were 91% with a correlation coefficient of 0.98 with the Mauzerall and Granick procedure. Doss and Schmidt (1971) report that assays of δ-ALA using commercially available dual ion-exchange columns employed in tandem give results comparing favorably with the Mauzerall and Granick method.

Several reports cite the successful automating of procedures for δ-ALA. Grisler et al. (1969) using a simplified version of the standard technique have automated the method using the Technicon Autoanalyzer and the Clino-Mak Analyzer. The technicon unit employs a flow reaction scheme and is capable of 1 sample/min. In the method of Lauwerys et al. (1972), a Carlo Erba laboratory analyzer equipped with a Leeds recorder and using a reaction flow scheme is employed. The addition of an internal standard of δ-ALA to the urine samples avoids the necessity of preliminary chromatographic separations. A correlation coefficient of 0.98 is reported, relative to the basic technique using columns, for a large number of samples.

REFERENCES

Alexander, F. R. 1974. The uptake of lead by children in differing environments. *Environ. Health Perspect.* 7:155–160.

American Public Health Association. 1955. Committee on chemical procedures of the occupational health section on methods for determining lead in air and in biological materials. New York: Am. Public Health Assoc.

Anderson, W. N., Broughton, P. M. G., Dawson, J. B. and Fisher, G. W. 1974. An evaluation of some atomic absorption systems for the determination of lead in blood. *Clin. Chim. Acta* 50:129–136.

Barltrop, D., Strehlow, C. D., Thonton, I. and Webb, J. S. 1974. Significance of high soil lead concentrations for childhood lead burdens. *Environ. Health Perspect.* 7:74–82.

Baloh, R. W. 1974. Laboratory diagnosis of increased lead absorption. *Arch. Environ. Health* 28:198–208.

Barry, P. S. I. 1975. A comparison of lead in human tissues. *Br. J. Ind. Med.* 32:119–139.

Barry, P. S. I. and Mossman, D. B. 1970. Lead concentrations in human tissues. *Br. J. Ind. Med.* 27:339–351.

Berk, P. D., Tschudy, D. P., Shepley, L. A., Waggoner, J. G. and Berlin, N. I. 1970. Hematologic and biochemical studies in a case of lead poisoning. *Am. J. Med.* 48:137–144.

Bessis, M. C. and Jensen, W. H. 1965. Sideroblastic anemia, mitochondria and erythroblastic iron. *Br. J. Hemat.* 11:49–51.

Blanksma, L. A., Sachs, H. K. and Murrary, E. F. 1969. Incidence of high blood lead levels in Chicago children. *Pediatrics* 44:661–667.

Blumenthal, S. 1972. A comparison between two diagnostic tests for lead poisoning. *Am. J. Public Health* 62:1060–1064.

Bolanowska, W. 1968. Distribution and excretion of triethyllead in rats. *Br. J. Ind. Med.* 25:203–208.

Bonsignore, D., Calissano, P. and Cartasegna, C. 1965. Un semplice metode per la determinazione della δ-amino-leveilinicodeidratase nel sanque: Comportamento deil'enzima nell'intossicazione saturina. *Med. Lavoro* 56:199–205.

Calony, J. A., Knobloch, E. C. and Purdy, W. C. 1963. A comparative study of the dithizone and polarographic determinations for lead. *Am. J. Clin. Pathol.* 39:652–655.

Caprio, R. J., Margules, H. L. and Joselow, M. M. 1974. Lead absorption in children and its relationship to urban traffic densities. *Arch. Environ. Health* 28:195–197.

Carroll, K. E., Needleman, H., Tencay, O. C. and Shapiro, I. M. 1972. The distribution of lead in human deciduous teeth. *Experientia* 28:434–435.

Chisolm, J. J., Jr. 1962. Aminoaciduria as a manifestation of renal tubular injury in lead intoxication and a comparison with patterns of aminoaciduria seen in other diseases. *J. Pediatr.* 60:1–17.

Chisolm, J. J., Jr. 1968. The use of chelating agents in the treatment of acute and chronic lead intoxication in childhood. *J. Pediatr.* 73:1–38.

Chisolm, J. J., Jr. 1971. Screening techniques for undue lead exposure in children: Biological and practical considerations. *J. Pediatr.* 79:719–725.

Chisolm, J. J., Jr. and Harrison, H. E. 1956. The exposure of children to lead. *Pediatrics* 18:943–958.

Christian, G. D. and Feldman, F. J. 1970. *Atomic absorption spectroscopy. Applications in agriculture, biology and medicine.* New York: Wiley-Interscience.

Clark, A. J. and Hallett, G. W. 1971. Lead poisoning survey—Portland, Maine. *J. Maine Med. Assoc.* 62:5–7.

Clarkson, T. W. and Kench, J. E. 1956. Urinary excretion of amino acids by men absorbing heavy metals. *Biochem. J.* 4:361–372.

Coleman, D. L. 1966. Purification and properties of δ-aminolevulinate from tissues of two strains of mice. *J. Biol. Chem.* 241:5511–5517.

Cooke, R. E., Glynn, K. L., Ullmann, W. W., Lurie, N. and Lepow, M. 1974. Comparative study of micro-scale test for lead in blood, for use in mass screening programs. *Clin. Chem.* 20:582–585.

Coulston, F., Goldberg, L., Griffer, T. B. and Russell, J. C. 1972. The effects of continuous exposure to airborne lead. I. Exposure of rats and monkeys to particulate lead at a level of 21.5 mg/m^3. Final report to the EPA.

Cramer, K., Goyer, R. A., Jagenburg, O. R. and Wilson, M. H. 1974. Renal ultrastructure, renal function and parameter of lead toxicity in workers with different lengths of lead exposure. *Br. J. Ind. Med.* 31:113–127.

David, O., Clark, J. and Voeller, K. 1972. Lead and hyperactivity. *Lancet* ii:900–903.

De Barreiro, O. L. 1967. 5-Aminolevulinate hydrolyase from yeast: Isolation and purification. *Biochim. Biophys. Acta* 139:479–486.

de la Burde, B. and Choate, M. S. 1972. Does asymptomatic lead exposure in children have latent sequelae? *Pediatrics* 81:1088–1091.

Delves, H. T. 1970. A micro-sampling method for the rapid determination of lead in blood by atomic absorption spectrometry. *Analyst* 95:431–438.

DHEW. 1972. Childhood lead poisoning. A summary report of a survey for undue lead absorption and lead-based paint hazard in 27 cities, publication 73-10002. Cincinnati, Ohio: Bureau of Community Environmental Management.

DHEW. 1975. Increased lead absorption and lead poisoning in young children, publication 00-2629. Atlanta: Center For Disease Control.

Doss, M. and Schmidt, A. 1971. Quantitative determination of δ-aminolevulinic acid and porphobilinogen in urine with ready-made ion-exchange chromatographic columns. *Z. Klin. Chem. Klin. Biochem.* 9:99–102.

Ediger, R. D. and Coleman, R. L. 1972. Modified Delves cup atomic absorption procedure for the determination of lead in blood. *At. Absorpt. Newsl.* 11:33.

Editorial. 1968. Diagnosis of inorganic lead poisoning. *Br. Med. J.* 4:501.

Emmerson, B. T. 1973. Editorial. Chronic lead nephropathy. *Kidney Intern.* 4:1–5.

Fine, P. R., Thomas, C. W., Suhs, R. H., Cohnberg, R. E. and Flashner, B. A. 1972. Pediatric blood levels. A study of 14 Illinois cities of intermediate population. *J. Am. Med. Assoc.* 221:1475–1479.

Fullerton, P. M. and Harrison, M. J. 1969. Subclinical lead neuropathy in man. *Electroencephalog. Clin. Neurophysiol.* 7:718–719.

Gibson, K. D., Neuberger, A. and Scott, J. J. 1955. The purification and properties of δ-aminolaevulic acid dehydrase. *Biochem. J.* 61:618.

Gilsinn, J. F. 1972. Estimates of the nature and extent of lead paint poisoning in the United States. *NBS Tech. Note 746.*

Goyer, R. A. 1971. Lead and the kidney. *Curr. Top. Pathol.* 55:147–176.

Goyer, R. A. and Chisolm, J. J., Jr. 1972. Lead. In *Metallic contaminants and human health*, ed. D. H. K. Lee, chap. 3. New York: Academic Press.

Goyer, R. A. and Rhyne, B. C. 1973. Pathological effects of lead. *Int. Rev. Exp. Pathol.* 12:1–77.

Goyer, R. A., Knoll, A. and Kimball, J. P. 1968. The renal tubule in lead poisoning. II *In vitro* studies of mitochondrial structure and function. *Lab. Invest.* 19:71–77.

Goyer, R. A., Tsuchiya, K., Leonard, D. L. and Kahyo, H. 1972. Aminoaciduria in lead and cadmium industry workers. *Am. J. Clin. Pathol.* 57:635–642.

Granick, J. L., Sassa, S., Granick, S., Lavere, R. D. and Kappas, A. 1973. Studies in lead poisoning. II. Correlation between the ratio of activated to inactivated δ-aminolevulic acid dehydratase of whole blood and the whole blood level. *Biochem. Med.* 8:149–159.

Granick, S. and Levere, R. D. 1964. Hemesynthesis in erythroid cells. *Progr. Hematol.* 4:1–47.

Granick, S. and Mauzerall, D. 1958. Porphyrin biosynthesis in erythrocytes: II. Enzymes converting δ-aminolevulinic acid to coproporphyrinogen. *J. Biol. Chem.* 232:1119.

Granick, S., Sassa, S., Granick, J. L., Levere, R. D. and Kappas, A. 1972. Assays for porphyrins, δ-aminolevulinic acid dehydratase, porphyrinogen synthetase in microliter samples of whole blood: Applications to metabolic defects involving the heme pathway. *Proc. Natl. Acad. Sci. U.S.A.* 69:2381–2385.

Griggs, R. C. 1964. Lead poisoning: Hematologic aspects. *Progr. Hematol.* 4:117–137.

Grisler, R., Genchi, M. and Perini, M. 1969. Determination of urinary δ-aminolevulinic acid by continuous flux and sequential automatic analyzers. *Med. Lav.* 60:678–686.

Grotepass, W. 1937. Het prophyrine in normale bloedlichaampjes. *Ned. Tüdschr. Geneeskd.* 81:362.

Guinee, V. F. 1972. Lead poisoning. *Am. J. Med.* 52:283–288.

Haeger-Aronsen, B. 1960. Studies on urinary excretion of δ-aminolevulinic acid and other haem precursors in lead workers and lead-intoxicated rabbits. *Scand. J. Clin. Lab. Invest.* 12(Suppl. 47):1–128.

Haeger-Aronsen, B., Stathers, G. and Swah, G. 1968. Hereditary corproporphria: Study of a Swedish family. 69:221–227.

Haeger-Aronsen, B., Abdulla, M. and Fristedt, B. I. 1971. Effect of lead on δ-aminolevulinic acid dehydratase activity in red blood cells. *Arch. Environ. Health* 23:440–445.

Hammer, D. I., Finkler, J. F., Hendricks, R. H., Shy, C. M. and Horton, R. J. M. 1971. Hair trace metal levels and environmental exposure. *Am. J. Epidemiol.* 93:84–92.

Heilmeyer, L. 1966. *Disturbances in heme synthesis.* Springfield, Ill.: Charles C Thomas.

Hernberg, S., Nikkanen, J., Mellin, G. and Lilius, H. 1970. δ-Aminolevulinic acid dehydratase as a measure of lead exposure. *Arch. Environ. Health* 21:140–145.

Hernberg, S., Tola, S., Nikkanen, J. and Valkonen, S. 1972. Erythrocyted-aminolevulinic acid dehydratase in new lead exposure. *Arch. Environ. Health* 25:109–113.

Iida, C., Tanaka, T. and Yamasaki, K. 1966. Determination of lead in silicates by atomic absorption spectrophotometry. *Bunseki Kagaku* 15:1100–1104.

Ingalls, T. H., Tiboni, E. A. and Werrin, M. 1961. Lead poisoning in Philadelphia, 1955–1960. *Arch. Environ. Health* 3:575–579.

Jensen, W. N., Moreno, G. D., Bessis, M. 1965. An electron microscopic description of basophilic stippling in red cells. *Blood* 25:933–943.

Kahn, H. L. and Sebestyen, J. S. 1970. The determination of lead in blood and urine by atomic absorption spectrophotometry with the sampling boat system. *At. Absorpt. Newsl.* 9:33–34.

Kammholz, L. P., Thatcher, L. G., Blodgett, F. M. and Good, T. A. 1972. Rapid protoporphyrin quantitation for detection of lead poisoning. *Pediatrics* 50:625–631.

Klein, M., Namer, R., Harpur, E. and Corbin, R. 1970. Earthenware containers as a source of fetal lead poisoning. *N. Engl. J. Med.* 283:669–672.

Kolbye, A. C., Mahaffey, K. R., Finine, J. A., Corneluissen, P. C. and Jelinus, C. F. 1974. Food exposure to lead. *Environ. Health Perspect.* 7:65–74.

Kopito, L., Briley, A. M. and Shwachman, H. 1969. Chronic plumbism in children: Diagnosis by hair analysis. *J. Am. Med. Assoc.* 209:243–248.

Lamola, A. A., Joselow, M. and Yamane, T. 1975. Zinc protoporphyrin (ZPP): A simple, sensitive fluorometric screening test for lead poisoning. *Clin. Chem.* 21:93–97.

Landing, B. H. and Nakai, M. D. 1959. Histochemical properties of renal lead inclusions and their demonstrations in urinary sediment. *Am. J. Clin. Pathol.* 31:499–503.

Langer, E. E., Haining, R. G., Labbe, R. F., Jacobs, P., Crosby, E. F. and Finch, C. A. 1972. Erythrocyte protoporphyrin. *Blood* 40:112–128.

Lauwerys, R., Delbroeck, R. and Vens, M. D. 1972. Automated analysis of delta-aminolaevulinic acid in urine. *Clin. Chim. Acta* 40:443–447.

Lauwerys, R. R., Buchet, J. P. and Roels, H. A. 1973. Comparative study of effect of inorganic lead and cadmium on blood δ-aminolevulinate dehydratase in man. *Br. J. Ind. Med.* 30:359–364.

Lepow, M. L., Bruchman, L., Rubine, R. A., Markowitz, S., Gillette, M. and Kapish, J. 1974. Role of airborne lead in increased body burden of lead in Hartford children. *Environ. Health Perspect.* 7:99–102.

Lockeretz, W. 1975. Lead content of deciduous teeth of children in different environments. *Arch. Environ. Health* 30:583–587.

L'vov, B. V. 1970. *Atomic absorption spectrochemical analysis.* London: Adam Hilger.

Marcus, M., Hollander, M., Lucas, R. E. and Pfeiffer, N. C. 1975. Microscale lead determinations in screening: Evaluation of factors affecting results. *Clin. Chem.* 21:533–536.

Marsden H. B. and Wilson, V. K. 1955. Lead poisoning in children. *Br. Med. J.* 1:324–326.

Matson, W. R. 1968. Trace metals, equilibrium and genetics of trace metal complexes in natural media. M.A. Thesis, Massachusetts Institute of Technology.

Mauzerall, D. and Granick, S. 1956. The occurrence and determination of δ-aminolevulinic acid and porphobilinogen in urine. *J. Biol. Chem.* 219:435–446.

Miller, J. A., Battistine, V., Cumming, R. L. C., Comwell, R. and Goldberg, A. 1970. Lead and δ-aminolevulinic acid dehydratase levels in mentally retarded children and in lead poisoned suckling rats. *Lancet* 2:695–698.

Moore, J. F., Goyer, R. A. and Wilfon, M. H. 1973. Lead induced inclusion bodies: Solubility, amino acid content and acidic nuclear proteins. *Lab. Invest.* 29:488–494.

National Academy of Sciences. 1972. *Lead. Airborne lead in perspective.* Washington, D.C.

Needleman, H. L. and Shapiro, I. M. 1974. Dentine levels in asymptomatic Philadelphia school children. Subclinical exposure in high and low risk groups. *Environ. Health Perspect.* 7:27–31.

Ortzonsek, N. 1967. The activity of heme ferro-lyase in rat liver and bone marrow in experimental lead poisoning. *Int. Arch. Gewerbepathol.* 24:66–73.

Pentschew, A. 1965. Morphology and morphogenesis of lead encephalopathy. *Acta Neuropathol.* 5:133–160.

Perlstein, M. A. and Attala, R. 1966. Neurologic sequelae of plumbism in children. *Clin. Pediatr.* 5:292–298.

Piomelli, S. 1973. A micromethod for free erythrocyte porphyrins: The FEP test. *J. Lab. Clin. Med.* 81:932–940.

Piomelli, S., Davidow, B., Giunee, V. F., Young, P. and Gay, G. 1973. The

FEP (free erythrocyte porphyrins) test: A screening micromethod for lead poisoning. *Pediatrics* 51:254–259.

Pueschel, S. M., Kopito, L. and Schwachman, H. 1972. Children with an increased lead burden. A screening and follow-up study. *J. Am. Med. Assoc.* 222:462–466.

Rabinowitz, M. B., Wetherrill, G. W. and Kopple, J. D. 1973. Lead metabolism in the normal human: Stable isotope studies. *Science* 182:725–727.

Rabinowitz, M. B., Wetherrill, G. W. and Kopple, J. D. 1974. Studies of human lead metabolism by use of stable isotope tracers. *Environ. Health Perspect.* 7:145–154.

Richter, G. W., Kress, Y. and Cornwall, C. C. 1968. Another look at lead inclusion bodies. *Am. J. Pathol.* 53:189–217.

Rimington, C. and Sveinssen, S. L. 1950. The spectrophotometric determination of uroporphyrin. *Scand. J. Clin. Lab. Invest.* 2:209.

Sachs, H. K. 1975. Effect of a screening program on changing patterns of lead poisoning. *Environ. Health Perspect.* 7:41–46.

Sayre, J. W., Charney, R., Vostal, J. et al. 1974 House and land dust as a potential source of childhood lead exposure. *Am. J. Dis. Child.* 127:167–170.

Schlaepfer, W. W. 1969. Experimental lead neuropathy: A disease of the supporting cells in the peripheral nervous system. *J. Neuropathol. Exp. Neurol.* 28:401–418.

Schwartz, S. and Wikoff, H. M. 1952. Relation of erythrocyte coproporphyrin and protoporphyrin to erythropoiesis. *J. Biol. Chem.* 194:563–568.

Shapiro, I. M., Needleman, H. L. and Tuncay, O. C. 1972. The lead content of human deciduous and permanent teeth. *Environ. Res.* 5:467–470.

Shapiro, I. M., Mitchell, G., Davidson, I. and Katz, S. H. 1975. Lead content of teeth. *Arch. Environ. Health* 30:483–486.

Silbergeld, E. K. and Goldberg, A. M. 1974. Hyperactivity: A lead-induced behavior disorder. *Environ. Health Perspect.* 7:227–232.

Steinfeld, J. L. 1971. Medical aspects of childhood lead poisoning. *Pediatrics* 48:464–468.

Strehlow, D. C. 1972. The use of deciduous teeth as indicators of lead exposure. Doctoral dissertation, New York University.

Subcommittee on Accidental Poisoning of American Academy of Pediatrics. 1968. Statement on diagnosis and treatment of lead poisoning in childhood. *Pediatrics* 27:676–680.

Tepper, L. B. and Levin, L. S. 1972. A survey of air and population lead levels in selected American communities. Final report to the EPA. Publication of The Department of Environmental Health, College of Medicine, University of Cincinnati, Cincinnati, Ohio.

Ter Haar, G. and Aronow, R. 1974. New information on lead in dirt and dust related to the childhood lead burdens. *Environ. Health Perspect.* 7: 83–90.

Thomas, H. M. and Blackfan, K. D. 1914. Recurrent meningitis due to lead in a child of five years. *Am. J. Dis. Child.* 8:377–380.

Thompson, J. A. 1971. Balance between intake and output of lead in normal individuals. *Br. J. Ind. Med.* 28:189–194.

Tomio, J. M., Tuzman, V. and Grinstein, H. 1968. δ-aminolevulinate dehydratase from rat Harderian gland: Purification and properties. *Europ. J. Biochem.* 6:84–87.

Van den Bergh, A. A. H. and Hyman, A. J. 1928. Studien über Porphyrin. *Dtsch. Med. Wochenschr.* 54:1492.

Van den Bergh, A. A. H., Grotepass, W. and Revers, F. E. 1932. Beitrag über das Porphyrin in Blut und Galle. *Klin. Wochenschr.* 11:1534–1536.

Wada, O., Toyokawa, K., Urata, G. and Yano, Y. 1969. Quantitative analysis of urinary delta-aminolevulinic acid to evaluate lead absorption. *Br. J. Ind. Med.* 26:240–245.

Waldron, H. A. 1966. The anemia of lead poisoning: A Review. *Br. J. Ind. Med.* 23:83–100.

Weissberg, J. B., Lipschutz, F. and Oski, F. A. 1971. δ-Aminolevulinic acid dehydratase activity in circulating blood cells. *N. Engl. J. Med.* 284: 565–569.

Williams, H., Kaplan, E., Couchman, C. E. and Sayers, R. R. 1952. Lead poisoning in young children. *Public Health Rep.* 67:230–286.

Williams, M. K. and Few, J. D. 1967. A simplified procedure for the determination of urinary 5-aminolevulinic acid. *Br. J. Ind. Med.* 24:294.

Wilson, V. K., Thomson, M. L. and Dent, C. E. 1953. Aminoaciduria in lead poisoning. *Lancet* II:66–68.

With, T. K. 1955. Prophyrin concentration from ultraviolet extinction. *Scand. J. Clin. Lab. Inv.* 7:193–199.

Wranne, L. 1960. Free erythrocyte copro- and protoporphyrin. A methodological and clinical study. *Acta Pediat. Scand.* 49(Suppl. 124):1–78.

Chapter 3

TOXICOLOGY OF ENVIRONMENTAL ARSENIC

Bruce A. Fowler

Environmental Toxicology Branch
National Institute of Environmental Health Sciences
Research Triangle Park, North Carolina

INTRODUCTION

Arsenic, a common environmental toxicant, is found in soil, water, and air. Schroeder and Balassa (1966) have observed that it ranks 20th in elemental abundance in the earth's crust and 12th in the human body. To place the toxicology of this ubiquitous element in a proper context, the present review will attempt to examine arsenic from an overall environmental perspective. It is hoped that it will serve to supplement previous reviews by Vallee et al. (1960), Buchanan (1962), Schroeder and Balassa (1966), Frost (1967), Browning (1969), and Lisella et al. (1972). This summary will include sources of environmental arsenic, arsenic accumulation in biological ecosystems, and the pharmacokinetics, metabolism, and general toxicity of arsenicals with particular emphasis upon the known biochemical mechanisms of arsenic toxicity. The topic of arsenic carcinogenesis will not be discussed because recent findings have been well summarized elsewhere in this volume (see chapter by Sunderman).

SOURCES OF ENVIRONMENTAL ARSENIC

Natural Sources

Arsenic is found in rocks and soil from concentrations of less than 1 part per million (ppm) to hundreds of ppm (Wullstein and Snyder, 1971; Wells and Elliott, 1971). The main ores of arsenic are arsenopyrites (FeAsS), realagar ($As_2 S_2$), orpiment ($As_2 S_3$), and arsenolite ($As_2 O_3$). This element is usually not mined as such but is recovered as a by-product from the smelting of copper, lead, zinc, and other ores.

Leaching of arsenic from soils or rocks containing high levels into hot spring mineral waters has been reported in Japan (Kuroda, 1941; Tsuzuki et al., 1966), New Zealand (Sabadell and Axtmann, 1975), and the United States (Brannock et al., 1948). Considerable concentrations of arsenic have also been reported in drinking water supplies found in Chile (Borgano and Greiber, 1972), Silesia and Cordoba, Argentina (Hueper, 1966), and Taiwan (Tseng et al., 1968). Arsenic has also been found in drinking water supplies from various regions of the United States (Pringle, 1971; Berg and Burbank, 1972; Feinglass, 1973) and the United Kingdom (Thornton and Webb, 1970). Durum et al. (1971) evaluated surface waters of the United States for arsenic and found that 79% of the 727 samples had arsenic concentrations of less than 10 μg/liter, 21% had arsenic concentrations greater than 10 μg/liter while 2% had more than 50 μg/liter. The highest measured arsenic concentration in this series was 1,100 μg/liter. Sagner (1973) reviewed the toxicity of arsenic in drinking water and observed that most cases of human poisoning occurred when concentrations reached a range of 0.5–1.0 mg/liter.

Arsenic is also found in seawater at concentrations of 2–5 μg/liter (Johnson, 1972; Johnson and Pilson, 1972). The predominate oxidation state of inorganic arsenic in seawater is the pentavalent arsenate (Johnson and Pilson, 1972), but methylated forms have also been reported (Braman and Foreback, 1973).

The interaction of arsenicals with bacteria is thought to account for changes in the oxidation state and chemical form of arsenic. Mamelak and Quastel (1953) studied the effects of organoarsenicals on the enzymic activity of anerobic bacteria. These authors found that bacterial amino acid reductases were particularly sensitive to organoarsenoxides. Turner and co-workers (Turner, 1954; Turner and Legge, 1954) have succeeded in isolating five species of *Pseudomonas* capable of oxidizing arsenite from cattle dipping fluids. Turner and Legge (1954) have suggested that these species possessed an arsenite dehydrogenase system with cytochromes that performed the activity. Wood (1975) has postulated bacterial methylation of arsenic with the subsequent formation of mono- and dimethyl arsines. Cox and Alexander (1973) have reported that these compounds could also be produced by sewage treatment fungi.

Industrial Sources

Arsenic may be released into the environment through industrial processes such as the smelting of other metals (Anonymous, 1973; Milham and Strong, 1974; Crecelius et al., 1975), application of arsenical pesticides and herbicides (Miles, 1968; Wagner and Weswig, 1974), and the generation of power from coal (Duck and Himus, 1951; Headlee and Hunter, 1953; Lindberg et al., 1975) or geothermal sources (Axtmann, 1975; Sabadell and Axtmann, 1975).

Milham and Strong (1974) have studied arsenic concentrations in vacuum cleaner and attic dust at various distances from a copper smelter and reported values of 1,300 ppm less than 0.4 mi from the smelter stack to 70 ppm some 2.0–2.4 mi away. Hair and urine samples collected from persons in these areas showed a similar pattern. Haywood (1907) reported injury to vegetation and livestock living near a copper smelter in Montana. High levels of arsenic were measured in alfalfa and field grass consumed by cattle in the area.

Arsenic may be present in coal at variable concentrations (Headlee and Hunter, 1953; Von Lehmden et al., 1974). This element may be released into the environment by combustion of coal where it can accumulate in soils and plants. The factors influencing emission of elements such as arsenic from power stations are their concentrations in coal and the efficiency of flyash precipitator devices (Lindberg et al., 1975). Bencko (1966) compared arsenic concentrations in the hair of 10-yr-old boys living in a village near the E.N.O. power station with those of a similar population living in a noncontaminated area. The boys living near the power station had hair arsenic concentrations 3.5 times higher than the matched control population. Accumulation of arsenic in rabbits housed near the E.N.O. power station was subsequently described by Bencko et al. (1968).

The release of arsenic into effluent water from geothermal power plants in New Zealand has also been recently reported (Axtmann, 1975; Sabadell and Axtmann, 1975). Concentrations as high as 0.25 ppm were reported in rivers draining these plants during the summer months (Axtmann, 1975).

ACCUMULATION IN BIOTA

Plants

Arsenic is present in many plants and plant products. Schroeder and Balassa (1966) and Westöö and Rydälv (1972) have reported arsenic concentrations in numerous vegetables, grains, fruits, seafood, and meats. Arsenic is poorly absorbed from soil by some plants but is extensively accumulated in the bark and needles of Douglas fir grown in areas with high soil levels of arsenic (Warren et al., 1964, 1968), and the presence of arsenic in this species of tree has been suggested as a guide for the localization of gold, silver, and other ores usually associated with arsenic. Other species of trees are also known to accumulate arsenic in their bark to similar concentrations (Williams, 1972). Fruits and vegetables sprayed with arsenicals may also accumulate this element. Arsenic-containing dust may be released from trees treated with arsenical wood preservatives during the sawing process (Smith, 1970).

Schroeder and Balassa (1966) have compared arsenic concentrations in vegetables grown in a field that had not been sprayed with arsenicals with those of the same vegetable species purchased from stores. The purchased vegetables had higher levels. Wadsworth and McKenzie (1963) have found that some potatoes grown in fields defoliated with arsenite sprays had arsenic concentrations up to 1 ppm. Woolson (1972) has reported the effects of fertilizers on the uptake of arsenic by corn grown in soil containing arsenate. The strong influence of available soil phosphorous on arsenic accumulation was evident.

Utilization of arsenical pesticides during the early part of this century also increased its presence in cigarettes. The arsenic content in an American cigarette increased from 12.6 to 42 μg between 1930 and 1946 (Satterlee, 1956). Arsenic concentrations in tobacco products have been greatly decreased in more recent years due to utilization of organic insecticides (Frost, 1967).

Fish and Shellfish

Arsenic is concentrated from the marine environment by many species of fish and shellfish (Schroeder and Balassa, 1966; Lunde, 1970, 1972; Westöö and Rydälv, 1972; LeBlanc and Jackson, 1973; Peden et al., 1973; Windom et al., 1973; Fowler et al., 1975). This is particularly true of invertebrates such as crustaceans. Schroeder and Balassa (1966) have reported values from 10 to 19 ppm wet weight in crustacea. Peden et al. (1973) have measured arsenic levels in crabs and other marine species on the order of 40 ppm wet weight. We (Fowler et al., 1975) have recently measured arsenic concentrations in Dover sole and crabs collected in southern California waters

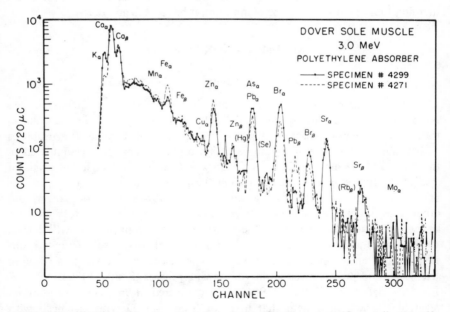

FIGURE 1 PIXEA spectra from two Dover sole *Microstomus pacificus* collected off Santa Barbara. Note prominent zinc, arsenic, lead, bromine, and strontium peaks (from Fowler et al., 1975).

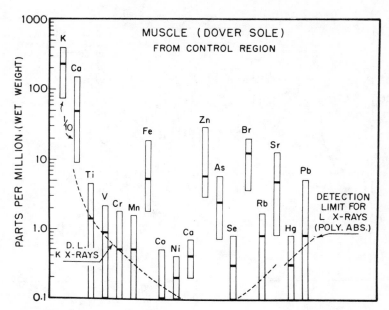

FIGURE 2 Summary of PIXEA data for muscle of Dover sole collected off Santa Barbara. Highest and lowest elemental values are indicated by ends of each bar. Means are indicated by horizontal black lines. The dotted line represents a 3 σ detection limit, which is a function of the peak of background ratio for X-rays of a given element (from Fowler et al., 1975).

by proton-induced X-ray emission analysis (PIXEA). Mean muscle concentrations of 2–4 ppm on a wet weight basis were found in Dover sole (*Microstomus pacificus*) and 14–43 ppm in the crab *Cancer anthonyi*. Typical PIXEA spectra from muscle samples of two Dover sole are presented in Fig. 1. Note that many other elements including lead, zinc, copper, and selenium are also present. Considerable interanimal variation with respect to tissue metal concentrations was observed in this study as shown in Fig. 2, which summarizes the overall range of values for the Dover sole. A PIXEA spectra for muscle of the crab *Cancer anthonyi* is given in Fig. 3. Much higher concentration ranges of arsenic were found in this species in comparison with Dover sole (Fig. 4). Synergistic or antagonistic interactions between these elements must be considered in evaluating trace metal toxicity from ingestion of fishery products. It is also worth noting that ingestion of fish and shellfish has been reported to markedly increase urinary excretion of arsenic in humans (Schrenk and Schreibeis, 1958; Westöö and Rydälv, 1972) due to the high levels of arsenic in these foods. The arsenic in marine organisms is apparently present as a highly stable organoarsenical (Lunde, 1975) whose properties need to be further defined in order to assess any potential human hazard.

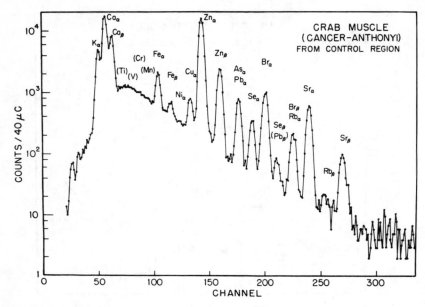

FIGURE 3 Proton-induced X-ray emission (PIXEA) spectra from the crab *Cancer anthonyi* collected in southern California coastal waters off Santa Barbara. Note prominent arsenic K_α peak and presence of other toxic trace metals.

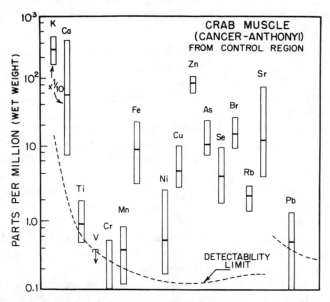

FIGURE 4 Summary of PIXEA data obtained for muscle samples of the crab (*Cancer anthonyi*) collected off Santa Barbara showing mean, high, and low values. Compare with concentrations of arsenic in Dover sole (from Fowler et al., 1975).

Freshwater species of fish in contrast to marine species have been found to possess very low muscle levels of arsenic (Lucas et al., 1970; Pakkala et al., 1972; Westöö and Rydälv, 1972). Further studies are needed to explain why marine species concentrate arsenic from their environment, while freshwater forms seemingly do not.

Poultry and Meat Products

Arsenicals, most notably arsanilic acid, have been used as feed additives for poultry and other livestock to promote growth and weight gain. Overby and Straube (1965) have found that nearly 100% of the arsanilic acid fed to chickens was excreted intact. Subsequently, Overby and Fredrickson (1965) have reported that more arsanilic acid was excreted as the intact compound than when comparable dosages of arsenate (As^{5+}) were given. They identified four metabolic products in the arsenate-fed animals.

The toxicity of arsenical feed additives is apparently low, although some incidents of livestock poisoning have occurred due to misuse. Baker and Parker (1969) have reported liver arsenic concentrations up to 12 ppm in growing broiler chickens given an overdose of aminarsenate. Dickinson (1972) has reported renal failure and kidney, liver, and muscle concentrations as high as 64, 27, and 10.4 ppm, respectively, in cattle treated with monosodium acid methanearsonate (10 mg/kg for 10 days). Case (1974) has reviewed the toxicity of various inorganic arsenicals to sheep and noted that sodium arsenite, arsenic acid, and copper arsenite were more toxic to this species than lead arsenate.

Selby and co-workers (1974) reported cases of beef cattle being poisoned by ingestion of field grass that had been previously sprayed with a sodium arsenite herbicide. These authors suggested that a potential human health hazard might exist from ingestion of beef cattle similarly contaminated with arsenic residues but without clinical manifestations.

HUMAN EXPOSURE TO ARSENIC

It should be clear that humans may be exposed to arsenic from a variety of sources. Lisella et al. (1972) have estimated the average daily human intake to be about 900 μg. "Normal" human whole blood and urinary levels are about 100 and 15 μg/liter, respectively. These concentrations may vary greatly, however, depending on environmental exposure.

Tarrant and Allard (1972) and Wagner and Weswig (1974) have reported increased urinary excretion of arsenic in forestry workers using organoarsenical silvicides. Rathus (1963) has described urinary arsenic concentrations of between 0.01 and 1 ppm in Australian banana growers. These findings were associated with ambient air concentrations of 0.1-0.5 ppm during the spraying process. Perry et al. (1948) have studied workers in a plant that manufactured sodium arsenite. They observed great variation in the mean particle size of dust generated during the process and measured atmospheric concentrations up to 1 mg As/m^3. Elevated hair and urinary arsenic concentrations were noted at various locations in the plant. An increased incidence of pigmented dermatoses and warts was also observed. Watrous and McCaughey (1945) have examined workers engaged in the manufacture of arsphenamine. Increased urinary elimination of arsenic was observed in plant workers and case reports of those individuals showing borderline arsenic poisoning were presented. No toxic effects of arsenic on the hematopoetic system could be detected.

PHARMACOKINETICS AND GENERAL TOXICITY OF ARSENICALS

In general it may be stated that organic arsenicals are more rapidly excreted than inorganic forms and that pentavalent arsenicals are cleared faster than trivalent. A further point is that the toxicity of a given arsenical is related to how fast it is cleared from the body and to what degree it may accumulate in tissues. The general pattern of toxicity is as follows:

$$AsH_3 > As^{3+} > As^{5+} > RAs-X$$

which is approximately the reverse order of excretion. Arsine gas (AsH_3) causes rapid hemolysis and subsequent renal damage that may persist for years following a single acute poisoning event (Muehrcke and Pirani, 1968; Fowler and Weissberg, 1974). Arsenite (As^{3+}) is next in general order of toxicity and degree of retention. This form is absorbed from the gastrointestinal tract and accumulates to high levels in tissues (Hunter et al., 1942; Lowry et al., 1942; Lanz et al., 1950). Holland et al. (1956) have reported uptakes of 4.8-8.8% arsenic-74 given as arsenic trioxide in cigarette smoke to terminal cancer patients, compared with 32-62% uptake for an aerosol of arsenic-74. About 45% of the inhaled arsenic was excreted in the urine and 2.5% in the feces after 10 days. Studies on rats (Morgareidge, 1963) also showed an extensive urinary arsenic excretion. Similar findings were reported by Hunter et al. (1942). They found that most of the arsenite was cleared from the blood of injected patients within 10 days but excretion continued for as long as 70 days after treatment. The major route of excretion was via the urine; little was excreted in the feces. Arsenic in the blood was concentrated in white blood cells.

Webster (1941) has examined nine male orchardists using lead arsenate and estimated that average intake of arsenic for these workers was 24 mg arsenic/day. In contrast to other reports, he noted that fecal excretion of arsenic was about 15% greater than urinary output. Mealy et al. (1959) have described arsenite studies in humans and found that most of the administered arsenic was present in the liver and kidney. They observed that arsenic was more concentrated in the red blood cells after 10 hr than in the plasma. Urinary elimination of arsenic was the major route of excretion, and they reported that most of this was in the pentavalent state due to possible atmospheric oxidation.

Animal Studies

Hunter et al. (1942) have studied the tissue distribution and excretion rates of arsenic administered as potassium arsenite to different species of animals. Arsenic in the blood of humans with leukemia was concentrated mainly in white blood cells. They observed that the rat, in contrast, had most of the arsenic in blood bound to red blood cells. The liver, spleen, kidney, and bone marrow had the highest arsenic concentrations in the animals studied. Lawton et al. (1945) have confirmed that arsenite administered by intraperitoneal injection to rats accumulated mainly in the liver and kidneys. DuPont et al. (1941) have studied the distribution of radioactive arsenic in rabbits injected with sodium arsenate. They reported rapid urinary excretion of arsenic and accumulation in the liver, kidney, and lungs during the first 3 hr after injection. Muscle, bone, and skin had relatively low arsenic concentrations, but because of their overall percentage of total body mass, they comprised the main storage sites. Chance (1945) has examined the tissue distribution and excretion of phenylarsenoxide, phenylarsonic acid, hydroxy-phenylarsenoxide, and hydroxyphenylarsonic acid in rabbits. These authors also reported extensive accumulation of arsenic by the kidneys and liver. Urinary elimination was the predominate route of arsenic excretion and the more toxic forms were observed to have a slower rate of clearance.

Ducoff et al. (1948) have administered arsenite to mice, rats, rabbits, and humans. Rats were observed to have a much slower arsenic excretion rate than humans or rabbits after 48 hr. They also found that rats had much higher arsenic blood concentrations than the other species for up to 96 hr following injection. It is unclear whether this phenomenon persists in a chronic context.

Schreiber and Brouwer (1964) have reported that most pentavalent arsenicals given by intravenous injection were rapidly and primarily excreted in the urine of fasted rats, while trivalent arsenicals were excreted at slower rates and mainly in the bile. Liver arsenic levels were higher in arsenite-treated animals, while renal levels were higher in arsenate-treated animals.

The fraction distribution of arsenic-74 in various mammalian tissues was conducted by Lowry et al. (1942). They noted that the bulk of this element was concentrated in the protein fraction followed by the acid soluble fraction and lipid fractions, respectively.

In more chronic studies, visceral concentrations of arsenic have been observed to rise quickly then fall after continued exposure. Katsura (1953) has studied the tissue distribution of arsenic in dogs given oral doses of arsenious acid for 4 months. He noted an initial rapid increase in kidney and liver levels for the first 1–2 months followed by a decline for the next 2 months. This change was attributed to decreased absorption from the gastrointestinal tract.

Bencko et al. (1968) have housed rabbits near the E.N.O. power plant for periods of up to 1 yr. They observed a continuous accumulation of arsenic in the viscera up to the final stage of treatment where a decrease in the arsenic content of muscle, liver, and lungs was noted. Bencko and Symon (1969) have reported arsenic accumulation studies in hairless mice given arsenite in drinking water up to 256 days. They observed a rapid rise in skin and liver concentrations of arsenic up to the 16th day of treatment, after which time concentrations declined. They speculated that this phenomenon was due to either decreased arsenic absorption, enhanced excretion, or both. Similar findings have been observed in rats fed both arsenate or arsenite for an extended time (B. A. Fowler, unpublished observations).

The effect of diet upon the pharmacokinetics of ingested arsenic has been studied by several investigators. Feeding studies, performed by Coulson et al. (1935), compared the accumulation of arsenic bound to shrimp tissues with similar levels of arsenic trioxide (As_2O_3) given in the diet. Animals receiving As_2O_3 in the diet accumulated over 100 times the liver arsenic concentrations of controls. About 18% of the administered arsenic was retained by the animals fed As_2O_3 during the first 3 months of feeding. Only 0.7% of the arsenic bound to shrimp tissues was retained during the same period. Urinary elimination was the major route of excretion. More recently, however, Bjornstad et al. (1974) have reported about 50% retention of arsenic fed to chicks as the organoarsenical present in a cod liver hydrolysate. It seems clear that the effects of cooking or processing on the pharmacokinetics of the arsenical present in fish should be studied further.

Morgareidge (1963) has studied the accumulation of arsenic by rats fed livers of turkeys given an organoarsenical in the diet in comparison with As_2O_3. He found that approximately 15% of the arsenic was retained by both groups with comparable liver and carcass levels. He observed somewhat higher excretion rates of arsenic in urine in comparison with the feces.

Sharpless and Metzger (1941) have supplemented rat diets with arsenic trioxide or arsenious acid and evaluated these animals for goiter. At intermediate doses, they observed decreases in growth and thyroid weight. It was concluded that a dietary arsenic level of 0.02% more than doubled the iodine requirement of these animals.

Tamura (1972) has reported studies on the absorption, distribution, and excretion of arsenite in rats given powdered milk or cereal diets. He noted that most of the arsenic remained in the feces. About 50% of the arsenic was excreted in the feces of cereal fed rats and 100% in the animals on the milk diet. Visceral concentrations of arsenic were higher in rats given the cereal diet than those raised on the powdered milk diet. The milk diet appeared to improve arsenic tolerance in comparison to cereal.

The binding of both organic and inorganic arsenic by the blood, kidney, and liver of rabbits was extensively studied by Hogan and Eagle (1944). They found that more toxic arsenicals were more firmly bound and slowly excreted than those of lesser toxicity. These authors observed *in vitro* that some arsenicals were more highly concentrated in the red blood cells, while others such as the arsphenamines, arsonic acids, acid-substituted phenyl arsenoxides, and inorganic arsenate were more concentrated in the plasma.

Studies of animals exposed to arsine indicate that this compound is extensively bound by red blood cells of all species studied and that large amounts of arsenic are eliminated initially with hemoglobinuria but that small amounts are excreted in the urine for long periods afterward (Muehrcke and Pirani, 1968; Levvy, 1947).

Human Poisonings

Human poisonings and deaths from arsenicals have occurred as the result of drinking beer contaminated with arsenic (Kelynack et al., 1900; Satterlee, 1960), use of arsenical pesticides (Conley, 1958; Hayes and Pirkle, 1966; Amarasingham and Lee, 1969; Chisolm, 1970; Deeths and Breeden, 1971), drinking arsenic-contaminated water (Terada et al., 1960; Yoshikawa et al., 1960), consumption of powdered milk (Eiji, 1955; Nagai et al., 1956; Okamura et al., 1956; Tokanehara et al., 1956) and arsine gas exposure (Fowler and Weissberg, 1974). To place arsenic toxicity into a proper perspective with respect to potential human risk and susceptible organ systems, it may be useful to review some of the major incidents and causes of human arsenic poisoning and the manifestations of toxicity.

The Manchester, England, arsenic poisoning epidemic. Around 1900, arsenic-contaminated sugar supplied to the breweries near Manchester, England, resulted in 6,000 poisonings and approximately 71 deaths (Kelynack, 1900; Satterlee, 1960). Most of the patients presented with a peripheral neuritis characterized by pain, muscular weakness, and parasthesias of the extremities, ataxia, anorexia, brown pigmentation, herpetic lesions, localized edema, and fatty degeneration of the heart.

Accidental deaths from arsenical pesticides. From 1946 to 1955 Conley (1958) has reported 410 deaths from arsenical pesticides in the United States with an average of 51 deaths per year. Hayes and Pirkle (1966) subsequently reported 54 arsenical pesticide deaths in 1956 and 29 in 1961. Arsenical weed killers were found to be the most commonly consumed poisons for suicides in Malaysia between 1963 and 1966 with a total of 308 cases or an average of about 62 per year (Amarasingham and Lee, 1969). Deeths and Breeden (1971) have reviewed 1,057 cases of poisoning in children between 1962 and 1968 and found that about 20 cases could be attributed to arsenic.

Morinaga milk poisoning incident. In 1955-1956 an outbreak of arsenic toxicosis occurred in Japan as the result of infants having consumed powdered milk contaminated with As_2O_3 at concentrations of 25-28 ppm. A total of 12,083 cases were reported with 128 deaths. The arsenic was introduced into the dried milk as a contaminant of a sodium phosphate stabilizer. Liver enlargement, anemia, and reduction of white blood cell numbers were observed in the infants with the highest blood levels of arsenic (Tokanehara et al., 1956). Melanosis, hyperpigmentation, white stria of the fingernails (Mee's lines), and abnormal electrocardiograms were also noted in some infants (Nagai et al., 1956). Liver swelling and abnormal liver function tests were also reported by these authors. Livers contained the highest visceral arsenic concentrations in those infants who died, and this was associated with fatty degeneration and necrosis of the liver parenchyma (Eiji, 1955). Clinical follow-up studies (Yamashita et al., 1972) have indicated residual functional impairment in children who survived this incident.

Arsenic contamination of well water in the Niigata Prefecture of Japan. In 1959 a second outbreak of 77 cases of arsenic toxicosis occurred in Japan as the result of well-water contamination by a plant manufacturing arsenic trisulfide (As_2S_3). Well water in the vicinity of this plant contained from 1 to 3 ppm arsenic. Melanosis, desquamation of skin, and keratosis were observed in the majority of these cases followed by anemia, liver swelling, electrocardiogram anomalies, proteinuria, and altered reflex actions (Terada et al., 1960). Liver biopsies on these patients disclosed some parenchymal swelling, but no cellular degeneration was observed (Yoshikawa et al., 1960).

Arsine poisoning. Arsine gas (AsH_3) is a potent hemolytic agent of particular occupational concern in the metal refining industry. Buchanan (1962) and Browning (1969) have extensively discussed the occupational circumstances under which arsine is commonly encountered. A review of the 207 cases and 57 deaths reported in the literature between 1928 and 1974 has also been recently published (Fowler and Weissberg, 1974). Most cases occurred in connection with the metal refining industry, while others occurred in chemical plants, holds of ships, and tank cars.

Persons poisoned by arsine usually present with nausea, headache, signs of shock, anemia, icterus, decreased hemoglobin levels, hemoglobinuria, and coppery skin pigmentation within an hour after exposure. Oliguria or anuria commonly develop at about 24 hr depending on the level of arsine exposure. Renal failure is the usual cause of death in most cases. These manifestations of arsine toxicity are uniform throughout the clinical literature and were observed in nearly all of the reported studies. A summary of the 224 cases that have been reported from 1928 to date is given in Table 1. An overall mortality figure for this series is about 23%.

Factors that influence arsine toxicity to humans include concentration of the gas (Dernehl et al., 1944; Morse and Setterlind, 1950; Spolyar and Harger, 1950), proximity to the source of evolution (Morse and Setterlind, 1950; Spolyar and Harger, 1950; Derot et al., 1963), duration of exposure (Morse and Setterlind, 1950; Spolyar and Harger, 1950; Derot et al., 1963), and individual differences in susceptibility (Nau, 1948). Differential effects of arsine on other body organ systems have also been reported. High concentrations of arsenic have been noted (Binet et al., 1941; Assouly and Griffon, 1949; Macauley and Stanley, 1956) in the livers, lungs, and kidneys of persons

TABLE 1 Summary of Reported Human Arsine Poisoning
Incidents between 1929 and 1974

Year	Author	Number of cases	Number of deaths	Treatment
1929	Grigg	1	1	–
1929	Manceau	3	1	–
1929	Meyer and Heubner	7	1	–
1931	Loning	11	4	Atropine, O_2 inhalation, transfusion
1931	Kogan	2	0	–
1932	Bomford and Hunter	2	0	–
1932	Nuck and Jaffe	14	7	–
1939	Smith and Rardin	2	1	–
1940	Bulmer et al.	14	0	–
1941	Binet et al.	8	0	Transfusion
1943	Wilson and Mangum	3	1	Alkali, transfusion
1944	Dernehl et al.	2	0	Transfusion
1944	Nau et al.	3	0	Transfusion
1948	Wills	1	1	BAL, transfusion
1949	Assouly and Griffon	2	1	O_2 inhalation, BAL
1949	Gaultier et al.	1	1	Peritoneal dialysis
1949	Garlick	1	0	BAL
1949	De Scoville	1	0	–
1950	Pinto et al.	13	4	BAL
1950	Steel and Feltham	1	0	Transfusion
1950	Triosi	1	0	–
1950	Spolyar and Harger	13	4	–
1950	Morse and Setterlind	2	2	BAL
1951	Mohacek and Stajduhar	1	1	–
1953	Johnson	1	0	–
1955	Gramer	1	1	–
1956	Macaulay and Stanley	6	2	BAL
1957	McKinstry and Hickes	8	4	O_2 inhalation, BAL
1958	Doig	3	1	Methyl blue, cortisone, ascorbic acid
1958	Greig et al.	2	0	–
1958	Ullrich	5	0	Peritoneal dialysis
1961	Neuwirtova et al.	3	2	Dialysis
1961	Lasch	5	1	–
1963	Derot	5	1	BAL, transfusion, peritoneal dialysis
1969	DePalma	3	0	BAL, transfusion, dialysis
1964	Dalgaard and Gregersen	5	5	–
1964	Kipling and Fothergill	5	0	Transfusion
1965	Jenkins et al.	1	0	BAL, transfusion
1966	Konzen and Dodson	1	0	Transfusion
1967	Ikegami et al.	1	1	Peritoneal dialysis
1967	Elkins and Fahy	2	0	BAL
1968	Nielsen	14	0	Peritoneal dialysis
1968	Franco	3	1	–

TABLE 1 (*continued*) Summary of Reported Human Arsine
Poisoning Incidents between 1929 and 1974

Year	Author	Number of cases	Number of deaths	Treatment
1968	Muehrcke and Pirani	1	0	Transfusion, hemodialysis
1969	Teitelbaum and Kier	5	1	Transfusion, hemodialysis
1969	Coles et al.	2	0	Transfusion dialysis
1970	Hocken and Bradshaw	1	0	Transfusion
1970	Uldall et al.	3	0	Dialysis
1970	Levinsky et al.	3	0	BAL, transfusion, dialysis
1970	Guajardo et al.	9	0	Dialysis
1974	Anonymous	17	0	Transfusion, dialysis
	Total	224	51	

fatally poisoned by arsine. The kidneys and lungs have been observed (Kipling and Fothergill, 1964; Jenkins et al., 1965; Hocken and Bradshaw, 1970) to be more sensitive to arsine toxicity than the liver.

Treatment of arsenic poisoning. British antilewisite (BAL) is commonly used to combat arsenic poisoning in humans (Heyman et al., 1956; Chisolm, 1970; Harvey, 1970). For arsine, however, this compound has not proven useful and supportive therapy, coupled with hemodialysis to prevent renal failure, has been more effective (Fowler and Weissberg, 1974).

CLINICAL MANIFESTATIONS OF ARSENIC POISONING IN HUMANS

In more chronic cases of arsenic poisoning, parathesias, progressive weakness, motor palsies, and a painful neuritis usually develop (Cannon, 1936; Dinman, 1960; Satterlee, 1960). In addition to these findings, Barry and Herndon (1962) have reported electrocardiographic changes characterized by T-wave inversion and persistent prolongation of the $Q-T$ interval in patients who had ingested acute doses of arsenicals. Glazener et al. (1968) have confirmed both of the above electrocardiographic changes in patients ingesting arsenic. These alterations could not be correlated with changes in serum electrolytes, and the authors suggested that arsenic was exerting a direct toxic effect on the myocardium.

Frejaville et al. (1972) have reported four cases of arsenical intoxication. They noted widespread intravascular coagulation and swollen liver and kidney mitochondria by electron microscopy. Franklin et al. (1950) reported marked cirrhosis and ascites in patients consuming therapeutic doses of Fowler's solution. These changes were thought to be related to the hepatotoxicity of inorganic arsenite. Similar cirrhotic changes have also been noted in the livers of winemakers using arsenical herbicides (Butzengeiger, 1940, 1949; Luchtrath, 1972).

Kyle and Pease (1965) have studied hematologic changes in six cases of arsenic intoxication and noted anemia, leukopenia, prominent basophilic stippling, and altered bone marrow erythropoesis.

The most prominent effects of arsine gas poisoning are usually observed in the blood and kidneys. Massive hemolysis followed by leukocytosis and increased hemopoesis have been frequently reported (Pinto et al., 1950; Gramer, 1955; Macauley and Stanley, 1956; Neuwirtova et al., 1961) in connection with arsine poisoning. At present it is unclear whether this effect is due to the direct action of arsine or whether it is secondary to hemolysis (Jenkins et al., 1965).

A coppery skin pigmentation frequently observed in persons poisoned by arsine is thought to represent the presence of methemoglobin (Pinto et al., 1950; Macauley and Stanley, 1956) rather than jaundice since normal serum bilirubin levels have been reported (Jenkins et al., 1965) in patients with this condition.

Electrocardiographic changes characterized by increased T-wave height were reported in arsine-poisoned individuals by Josephson et al. (1951). They felt these changes were due to the direct action of arsine on the heart rather than hyperkalemia secondary to hemolysis. Macauley and Stanley (1956) have subsequently described six cases of arsine poisoning in which these electrocardiographic changes were absent.

Kidneys of persons poisoned by arsine characteristically contain hemoglobin casts in the tubule lumens (Neuwirtova et al., 1961; Nielsen, 1968; Hocken and Bradshaw, 1970; Uldall et al., 1970). Cloudy swelling and necrosis of the proximal tubule cells with attendant albuminuria, azotemia, altered serum electrolyte concentrations, and decreased T_m PAH values have also been reported (Neuwirtova et al., 1961; Nielsen 1968). These effects have been found to continue many months after the occurrence of arsine poisoning by numerous investigators (Neuwirtova et al., 1961; Muehrcke and Pirani, 1967; Nielsen, 1968; Hocken and Bradshaw, 1970; Uldall et al., 1970).

It is clear from the above that environmental or occupational arsenic exposure may cause toxicity to humans under a variety of circumstances. The nature and severity of toxicity, however, is highly dependent on the chemical form and oxidation state of the arsenical involved. The following sections attempt to evaluate the toxicity of environmental arsenic under experimental conditions and the biochemical mechanisms by which toxicity is produced.

ANIMAL TOXICITY STUDIES

The toxicity of arsenicals to experimental animals depends on species, sex, age, dose, and duration of exposure. This section will summarize the known effects of arsenicals on laboratory animals and livestock as a function of their chemical form and oxidation state.

Inorganic Arsenic

Arsenate (As^{5+}). Arsenate is probably the most common environmental form of inorganic arsenic (Frost, 1967; Braman and Foreback, 1973). It is considered to have a lower toxicity than arsenite (As^{3+}) and many of the organoarsenicals (Johnstone, 1963; Byron et al., 1967; Harvey, 1970; Dickinson, 1972). The 96-hr LD50 for arsenate in rats is around 1,000 mg/kg by oral administration (Schroeder and Balassa, 1966). Byron et al. (1967) have found that both rats and dogs survived a 2-yr feeding study with sodium arsenate or sodium arsenite in the diet at concentrations up to 400 ppm. Enlargement of the common bile duct was observed in rats but not dogs given both arsenate and arsenite.

Ferm and Carpenter (1968) have reported a high incidence of exencephaly in golden hamster embryos of mothers given a 20 mg/kg intravenous dose of sodium arsenate on day 8 of gestation. Ferm et al. (1971) have subsequently demonstrated that these anomalies and a high incidence of fetal resorptions were dependent on the time of injection during critical stages of embryogenesis. Exencephaly and many other malformations have been reported in offspring of mice given intraperitoneal injections of arsenate (25 mg/kg) during gestation (Hood and Bishop, 1972). These effects were found to be alleviated by pretreatment with BAL (Hood and Pike, 1972). More recently, Beaudoin (1974) has reported arsenate teratogenesis in rats at dose levels of 30 mg/kg.

Arsenite (As^{3+}). Inorganic arsenite is considerably more toxic to animals than is arsenate. Its LD50 following injection is around 10 mg/kg in most species. Levvy (1947), however, reported a LD50 for mice of 5 mg/kg. Harrisson et al. (1958) have evaluated the acute oral LD50 of crude and purified arsenic trioxide in rats and several strains of mice. They found the 96-hr LD50 for rats to be 23.6 mg As/kg body weight for crude arsenic trioxide and 15.1 mg As/kg for the purified compound. The acute 96-hr LD50 in all strains of mice studied was 42.9 mg As/kg body weight for crude and 39.4 mg As/kg body weight for purified arsenic trioxide. Done and Peart (1971) have found the oral LD50 of sodium arsenite in rats to be 24 mg/kg while that of arsenic trioxide was 293 mg/kg. These differences in the toxicity of these two inorganic trivalent arsenicals appeared to be related to their relative purity, solubility, and absorption.

In other animal studies, the liver and kidney have been found to show marked histological and functional alteration following trivalent arsenic exposure. Finner and Calvery (1939) have observed renal and hepatic degeneration in rats and dogs fed diets containing various inorganic arsenicals up to 12 wk. Westernhagen (1970) has reported pathological and histochemical changes in the inner ear of guinea pigs exposed to arsenic trioxide at 1 mg/kg dose levels for 28 days. Shibuya (1971) has injected rabbits with arsenious acid and evaluated short- and long-term effects on blood chemistries. He noted that functional alteration of the liver seemed to occur somewhat earlier than that of the kidney. In these animals, the administered arsenic was excreted mainly in the urine and was accumulated in the bone marrow, skin, liver, and kidney.

The effects of trivalent arsenic on the growth and reproduction of rats have also been evaluated by several authors. Morris et al. (1938) have studied growth and reproduction in rats fed arsenic trioxide. They observed maternal transfer of arsenic to fetuses and that the arsenic content of the newborn animals was higher than at 15 days of age. Kiyono et al. (1974) have administered arsenic trioxide to infant male and female rats. Males were found to be somewhat more sensitive than females. Spontaneous mobility was increased in arsenic-treated animals.

Arsine gas (AsH$_3$). The toxicity of arsine gas has received relatively little attention in comparison to other arsenicals, even though it is the most toxic form of inorganic arsenic known. The LD50 dose for laboratory animals following inhalation exposure has been estimated to be 0.67 mg As/kg (Levvy, 1947). The LD50 following intraperitoneal injection was found to be 2.5 mg As/kg (Levvy, 1946). In the latter study, Levvy also observed that half of the injected dose was eliminated within 24 hr. In short, the lower toxicity of injected arsine appears to be related to decreased interaction with red blood cells.

Organoarsenicals

The toxicity of an organoarsenical varies greatly with the valence state of the incorporated arsenic and those chemical properties of the molecule that influence its absorption, distribution, and excretion. The pharmacokinetics of the feed additive arsanilic acid have been extensively studied (Overby and Fredrickson, 1965; Underwood, 1971). These workers have reported that this arsenical is rapidly excreted with little tissue accumulation. It has also been shown to be excreted primarily as the intact compound. More recent reports (Keenan and Oe 1973; Ledet et al., 1973) have shown it to be toxic to pigs following high-level exposure. Ledet et al. (1973) have observed quadriplegia and neurological degeneration in pigs fed a diet containing 1,000 ppm arsanilic acid for 18 days. Total arsenic levels were highest in the kidneys and livers of these animals. These authors also noted little neurological recovery in the animals following cessation of treatment.

Al-Timimi and Sullivan (1972) have evaluated the toxicity of a number of organoarsenicals in young turkeys. They reported a LD50 dietary dose level of about 0.1% for arsanilic acid in young turkeys after 28 days of exposure. In subsequent reports, these authors evaluated carbarsone (Sullivan and Al-Timimi, 1972a) with a dietary LD50/28 days of 0.32%, nitarsone (Sullivan and Al-Timimi, 1972b) with a LD50/28 days of 0.05%, and roxarsone (Sullivan and Al-Timimi, 1972c) with a LD50/28 days of 0.03%. In other studies Baker and Parker (1969) have reported liver and kidney damage in broiler chickens that received an overdose of sodium aminarsonate in the drinking water. The average liver concentration arsenic was about 9 ppm.

Organoarsenical herbicides also seem to cause renal and hepatic damage. Dickinson (1972) has studied the toxicity of monosodium acid methanearsonate (MSMA) in cows and found that a total dose of 100 mg As/kg given orally at 10 mg As/kg-day killed 4 of 5 cattle in 10 days. Arsenic concentrations were found to be as high as 64 ppm in kidneys and 27 ppm in livers. Extensive renal damage and hemorrhagic gastritis were observed in these animals although other viscera appeared normal.

A pharmacological basis for the differential toxicity of organoarsenicals has been provided by Hogan and Eagle (1944). The more toxic organoarsenicals were generally more extensively bound by red blood cells than the less toxic forms. Concentrations of arsenic in the liver and kidney after intravenous administration of organoarsenoxides or organoarsenic acids were roughly proportional to their relative toxicity. The rate of arsenical excretion was also inversely proportional to toxicity. Organoarsenicals of low toxicity were more rapidly excreted than highly toxic forms.

McKee and Wolfe (1963) have reviewed the LD50 values for various arsenicals in both mammals and aquatic species. They noted wide species variation to arsenical toxicity and have given extensive data in table form concerning the toxicity of inorganic arsenic to various aquatic species.

CELLULAR MECHANISMS OF ARSENICAL TOXICITY

Because arsenic toxicity is highly dependent upon its chemical form and oxidation state, arsenite (As^{3+}), arsenate (As^{5+}), arsine gas (AsH_3), and organoarsenicals vary in their degree of toxicity. Trivalent arsenite is considered more toxic than pentavalent arsenate, although arsenate is the more common environmental form. Arsine gas is a potent hemolytic agent of occupational concern in some industries. An understanding of the mechanisms by which these arsenicals induce toxicity is complicated by their *in vivo* metabolism. Arsenate has been reported to be reduced to arsenite in the kidney of laboratory animals and urinary excretion of methylated arsenic has also been described in humans. The hemolytic action of arsine gas is thought to be mediated by an oxidized intermediate.

One manifestation of toxicity shared by all the above arsenicals is inhibition of cellular respiration. In the following section, the cellular and biochemical effects of arsenic will be examined with particular emphasis on their effects on cellular respiration and energy-producing systems.

Arsenate

Arsenate has long been known as an uncoupler of mitochondrial oxidative phosphorylation. The mechanism by which this occurs is thought to be related to competitive arsenate substitution for inorganic phosphate with subsequent formation of an unstable arsenate ester that spontaneously decomposes. This process, referred to as arsenolysis, is diagrammatically depicted in Fig. 5. This inhibitory effect of arsenate on mitochondria function has been observed both *in vitro* (Crane and Lipman, 1953; Azzone and Ernster, 1961; Estabrook, 1961; Packer, 1961; Wadkins, 1961; Ter Welle and Slater, 1967) and *in vivo* (Brown et al., 1976; Fowler, 1975).

Crane and Lipman (1953) have found that arsenate stimulated succinate-mediated respiration of rat liver mitochondria in the absence of phosphate. This phenomenon was competitively inhibited by addition of phosphate to the reaction mixture. The ADP:O ratio was decreased and incorporation of ^{32}P in ATP inhibited when arsenate was added to the medium. Inhibitory effects of arsenate on respiration at higher concentrations were attributed to the reduction of arsenate to arsenite with subsequent enzyme inhibition. Stimulation of mitochondrial ATPase by arsenate has been reported by several investigators (Azzone and Ernster, 1961; Wadkins, 1961). Wadkins (1961) has found that this process required Mg^{2+} and was inhibited by phosphate. Azzone and Ernster (1961) have reported that arsenate stimulation of ATPase activity was inhibited by succinate. These authors also found that arsenate decreased the $P_i \leftrightharpoons ATP$ exchange reaction.

Arsenate induces mitochondrial swelling both *in vitro* (Packer, 1961; DeMaster and Mitchell, 1970; Mitchell et al., 1971) and *in vivo* (Fowler, 1974, 1975; Brown et al., 1976). Packer (1961) has noted that arsenate causes mitochondrial swelling *in vitro*, which is not readily reversed by addition of ADP. DeMaster and Mitchell (1970) have studied the arsenate $\leftrightharpoons H_2O$ exchange in rat liver mitochondria and found that this exchange is not inhibited by dinitrophenol or oligomycin. Mitchell et al. (1971) have subsequently suggested that arsenate inhibition of mitochondrial respiration occurs by two modes: competition with phosphate during oxidative phosphorylation and the phosphate $\leftrightharpoons ATP$ exchange; and arsenate inhibition of energy-linked reduction of NAD by succinate, which is not ameliorated by phosphate and for which ADP was required. This second inhibitory effect was not thought to depend on the mechanism of arsenolysis. Ter Welle and Slater (1967) have found that

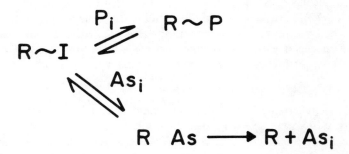

FIGURE 5 Proposed scheme for arsenolytic mechanism of arsenate inhibition of cellular respiration. $R \sim I =$ a high energy intermediate, $P_i =$ inorganic phosphate, $As_i =$ inorganic arsenate.

arsenate stimulated succinate oxidation in a phosphate-free medium and that this effect was inhibited by low concentrations of phosphate in the absence of ADP. In the absence of phosphate, ADP stimulated arsenate-mediated respiration. On the basis of these data, they suggested that at some step in oxidative phosphorylation, a high-energy intermediate exists with P_i or $ADP + P_i$. Ernster et al. (1967) have subsequently discussed this hypothesis in greater detail.

In vivo studies (Brown et al., 1976) have demonstrated similar effects of chronic arsenate administration upon mitochondrial efficiency in the kidney. Decreased state 3 respiration and respiratory control ratios were observed in renal mitochondria and associated with mitochondrial swelling (Fig. 6). Similar effects have been observed on state 3 respiration and respiratory control ratios in liver mitochondria (Table 2) and associated with *in situ* swelling of liver mitochondria (Fig. 7).

Studies by Kagawa and Kagawa (1969) on arsenate-76 accumulation by rat liver mitochondria have indicated that this is an active process. Chan et al. (1969) have reported the formation and isolation of an arsenic-binding component within liver mitochondria. The above findings suggest that arsenate is avidly accumulated by mitochondria and may have an intramitochondrial metabolism related to that of phosphate.

Chromosomal effects. As noted earlier in this review, arsenate is a known teratogen and has the ability to substitute for phosphate in some cellular processes. It has been suggested by Petres and co-workers (Petres and Hundeiker, 1968; Petres et al., 1970) that arsenate causes chromosomal abnormalities by simply substituting for phosphate in the DNA chain. Arsenate is also known to inhibit normal DNA repair processes. Jung (1969, 1971) has reported autoradiographic studies that demonstrated decreased enzymic DNA repair following uv irradiation and incubation of skin grafts in an arsenate solution. He concluded that arsenate inhibited "dark repair" of DNA in these cells. Sibatani (1959) has found that arsenate decreased incorporation of ^{32}P into lymphatic cell DNA to a greater extent than in RNA, suggesting possible inhibition of DNA repair or synthesis. Paton and Allison (1972) have reported similar findings and felt that both of these mechanisms might be operating. Both arsenate and arsenite have been observed to decrease the extractability of DNA from tumor cells (Grunicke et al., 1973), suggesting that DNA–protein cross-linking may also be occurring.

Arsenite

Trivalent arsenicals including inorganic arsenite are regarded as being primarily sulfhydryl (SH) reagents. These arsenicals have been shown to cause enzymic inhibition of the tricarboxylic acid (TCA) cycle. In addition, they inhibit a number of other enzyme systems in various tissues (Webb, 1966).

The pyruvate oxidase system has been shown to be especially sensitive to trivalent arsenicals because of their apparent interaction with the disulfhydryl

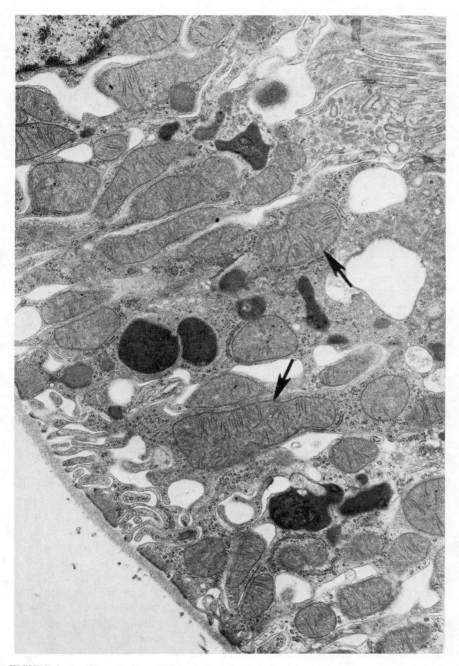

FIGURE 6 Swollen renal proximal tubule cell mitochondria (arrows) from a rat given 85 ppm arsenate in its drinking water for 6 wk. ×11,172.

TABLE 2   Effect of Ingested Arsenate (As^{5+}) on Rat Liver
Mitochondrial Parameters Following 6 wk of Exposure

Arsenic concentration (ppm) in drinking water	State 4^a	State 3^a	RCR	ADP:O
	Succinate substrate			
Control	52.02	161.87	4.66	1.45
20	60.72	155.19	3.28	1.36
40	63.87	166.06	3.06	1.36
85	57.05	156.89	3.16	1.30
	Pyruvate/malate substrate			
Control	17.36	75.22	5.20	2.60
20	23.63	60.77	4.50	2.20
40	25.57	70.33	3.66	2.20
85	21.39	46.35	2.66	2.25

$^a\mu g$ = atoms O_2 consumed/min-g protein. Respiration measurements obtained polarographically using a Clark-type oxygen electrode. Mitochondrial reaction mixture contained 50 mM Tris–Cl (pH 7.5), 5 mM MgSO$_4$, 100 mM KCl, and 50 mM K$_2$HPO$_4$.

lipoic acid moiety of this system. This interaction is schematically presented in Fig. 8.

The net result of this reaction is the inhibition of lipoic acid activity with subsequent decreases in pyruvate oxidation leading to accumulation of pyruvate in blood (Peters, 1955). Inhibition of pyruvate oxidation leads to decreased mitochondrial respiration.

Fluharty and Sanadi (1960) have reported that arsenite stimulated mitochondrial succinate oxidation but decreased the ADP:O ratio. In contrast, mitochondrial oxidation of β-hydroxybutyrate was strongly inhibited by the same arsenite concentration, while the ADP:O ratio declined only slightly. They also found that addition of equivalent amounts of BAL and arsenite decreased succinate oxidation. If a 10-fold excess of BAL was added, then the activity of the system was restored. These authors noted that arsenite stimulated ATPase activity and decreased the ATP-^{32}P exchange reaction. Fluharty and Sanadi (1962) have further evaluated the effects of arsenite–BAL on liver oxidative phosphorylation and compared these effects with mitochondria swelling. They concluded that swelling was secondary to the uncoupling of mitochondrial oxidative phosphorylation. Fletcher et al. (1962) have confirmed earlier studies by Fluharty concerning BAL–arsenite uncoupling of mitochondrial oxidative phosphorylation. They found that rat liver mitochondrial particles were more sensitive to this complex than heart mitochondrial particles. Hassinen and Hallman (1967) have reported that BAL–arsenite inhibited NADH$_2$ oxidase, reduction of cytochrome *b*, and DNP-activated ATPase.

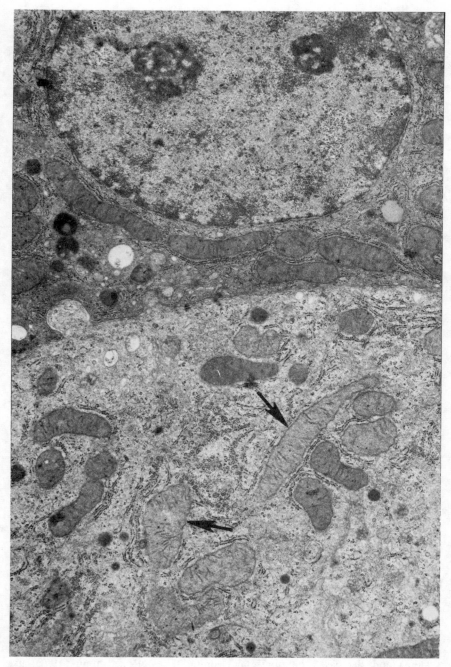

FIGURE 7 Swollen mitochondria (arrows) within a liver parenchymal cell from a rat
exposed to 85 ppm arsenate in its drinking water for 6 wk. ×10,500.

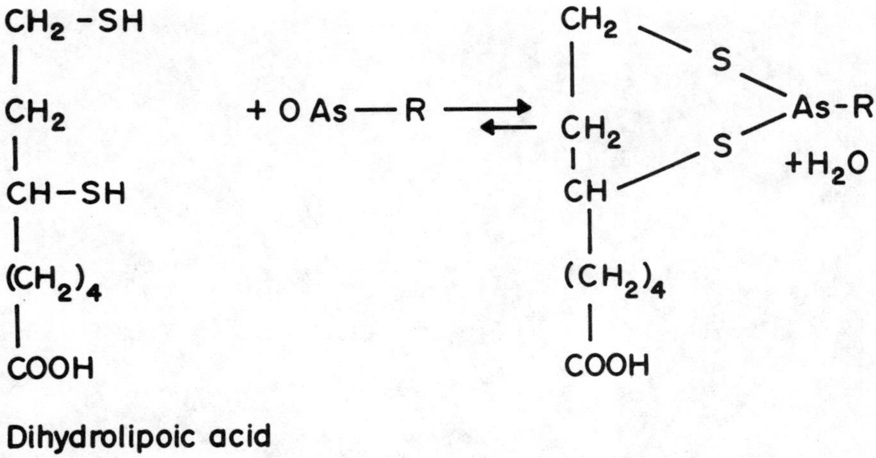

$$CH_2-SH$$
$$|$$
$$CH_2$$
$$|$$
$$CH-SH$$
$$|$$
$$(CH_2)_4$$
$$|$$
$$COOH$$

$$+\, O\, As - R \rightleftarrows$$

$$CH_2 \diagdown_S$$
$$| \qquad\qquad As-R$$
$$CH_2 \diagup_S$$
$$| \qquad\qquad +H_2O$$
$$CH$$
$$|$$
$$(CH_2)_4$$
$$|$$
$$COOH$$

Dihydrolipoic acid

FIGURE 8 Proposed schematic interaction between dihydrolipoic acid and a trivalent arsenical showing possible ring formation.

Bencko and Simane (1968) have studied respiration of liver homogenates from hairless mice given arsenic trioxide in drinking water. Decreased tissue oxygen consumption was observed, but it was noted that the homogenate reaction mixtures from treated mice were less sensitive than controls when As_2O_3 was added directly *in vitro*. An intracellular adaptation mechanism to arsenic was suggested. In further studies Bencko et al. (1970) have reported that arsenite added to the drinking water at a concentration of 250 mg/liter significantly reduced respiration of liver parenchyma of mice after 4-64 days of exposure, while a dose of 50 mg/liter did not. Konings (1972) found that arsenite inhibited the respiration of rat liver mitochondria but not that of thymus nuclei. He concluded that this effect resulted from a decreased ability of arsenite to penetrate the nuclear membrane in comparison to the mito-chondrial membrane.

Other enzyme systems are also known to be affected by arsenite. Cottam and Srere (1969) have reported a 50% inhibition of citrate cleavage enzymes by $3 \times 10^{-4} M$ arsenite. Bartley and Dean (1969) have found decreased formation of phosphoenolpyruvate (PEP) by rat liver mitochondria by both As^{5+} and As^{3+} at 12.5 mM As^{5+} and 2 mM As^{3+} concentrations, respectively. Coughlan et al. (1968) have observed reversible binding with subsequent spectral alterations of xanthine oxidase by As^{3+}. They felt that this effect was the result of As^{3+} interaction with the molybdenum com-ponent at the active center of the enzyme. In a preliminary report, Wagstaff (1972) has reported that 1,000 ppm arsenic trioxide added to the diet of rats for 15 days significantly increased the activities of microsomal enzyme systems and decreased hexobarbital sleeping times. Rozenshtein (1970) has described inhibition of blood cholinesterase by arsenite and accumulation of pyruvate in the blood of rats exposed to aerosols of arsenic trioxide.

Arsine Gas

In discussing the biochemical mechanisms of arsine toxicity, one must examine both the mechanisms of arsine hemolysis that may lead to anoxia in visceral tissues and its direct effect on cellular respiration in these tissues.

Numerous experimental studies on arsine were conducted in Europe between 1930 and 1950. The majority of these investigations noted that there is wide species variation in susceptibility to arsine toxicity and that arsine is more toxic than inorganic arsenic (arsenite). Little work concerning the mechanisms of arsine action has been conducted since.

Thauer (1934) investigated species differences to arsine toxicity among mice, dogs, cats, and rabbits. He found mice to be most sensitive to arsine and dogs most resistant. Kiese (1934) examined the effects of chronic arsine exposure on mice, rats, guinea pigs, cats, and dogs for 8 hr/day for 4 wk. He reported that inhalation of as little as 4 μg arsine/liter air caused a decrease in the osmotic resistance of erythrocytes, slight reduction of hemoglobin levels and erythrocyte numbers, coupled with an increase in reticulocytes. A strong increase in numbers of reticulocytes and leukocytes at higher exposure levels was noted immediately following initiation of arsine administration. This increase in white blood cells corresponded in time with decreased red blood cell osmotic stability, red blood cell counts, and hemoglobin levels. He noted a subsequent adaptation to the effects of low-level arsine exposure and suggested that the animals reached a steady state condition. Heubner and Wolf (1936) also discussed the adaptation (Gewöhnung) of organisms to chronic low levels of arsine exposure with time.

Binet et al. (1941) have observed individual and species variation in susceptibility to arsine hemolysis. They found that the red cell volumes of their animals were decreased by approximately 50% 1 hr after treatment and maximum depression occurred at 5-7 hr. Leukocytosis did not become marked until 3 hr following exposure.

Levvy (1946) has investigated the effects of arsine–hydrogen gas mixtures injected into the peritoneal cavity of mice, rabbits, cats, and sheep. The median lethal dose was observed to be much larger than when arsine was administered by inhalation. He later reported (Levvy, 1947) LD50 values for mice and rabbits exposed to arsine at various levels in the air and noted that arsine was more toxic than arsenite at equal dose levels. The liver, lungs, and kidneys of exposed animals were found to have the highest concentrations of arsine. One of his basic conclusions was that the absorption of arsine varied with concentration and duration of exposure and that about 60% of the gas inhaled was absorbed. The toxicity of arsine to exposed animals was greatly reduced with injection of dithiols shortly after treatment. He also suggested that the body must possess a means of detoxifying arsine and that some arsine reaches other body organs prior to absorption by red blood cells.

Gebert (1937) and Jung (1939) have investigated the reaction between arsine and blood and various solutions. They reported values for arsine solubility in water, acids, bases, egg white albumin solutions, and serum proteins. These authors also observed a protective effect by hydrogen sulfide against arsine hemolysis. Graham et al. (1946) have examined the binding of arsine to blood cells. They found that most of the arsenic fixed to blood cells was in a nondialyzable form and that it was apparently in the trivalent oxidation state.

Kensler et al. (1946) have studied the mechanism of arsine hemolysis and tested the efficacy of various compounds against arsine toxicity. They observed that dithiol compounds offered protection against arsine hemolysis during *in vitro* experiments on human erythrocytes when added to the solution 5 min after arsine exposure. The effectiveness of dithiols was tested on rabbits, monkeys, and dogs exposed to arsine and found to protect against arsine hemolysis and renal failure when administered shortly after exposure and prior to the onset of hemolysis.

The mechanism by which arsine produces hemolysis is obscure, but several suggestions concerning this phenomenon have been given. Gramer (1955) has stated that arsenic acid was formed from the *in vivo* oxidation of arsine and that this compound was responsible for hemolysis. This idea is somewhat supported by the observation that oxygen (Pernis and Magistretti, 1960) seems to be necessary for arsine to exert a hemolytic effect.

Arsine interaction with sulfhydryl groups could also cause hemolysis by inhibiting proteins responsible for maintenance of the osmotic balance within red cells (Levinsky et al., 1970). This concept is also supported by the work of Pernis and Magistretti (1960) who investigated the mechanism of arsine hemolysis in washed human erythrocytes. They noted a decrease in the glutathione content of these cells following arsine exposure, which was correlated with hemolysis. It was also found that addition of reduced glutathione to the medium at a concentration nearly 10 times greater than the amount of arsine added protected the erythrocytes from lysis. Younger erythrocytes, which contained more reduced glutathione than older cells, were also found to be more resistant to arsine hemolysis. Whatever the mechanism, the rapid and fulminating nature of arsine-induced hemolysis seems to set it apart from hemolytic anemias produced by other agents. Hughes and Levvy (1947) have also examined the direct effects of arsine on kidney and liver tissue slice respiration. They observed that both arsine and arsenite inhibited oxygen uptake by the kidney and liver, although arsine was more toxic to liver slices than arsenite.

Organoarsenicals

There are many organoarsenicals ranging in toxicity from the war gas lewisite to feed additives such as arsanilic acid. In general the direct toxicity of these agents is similar to that of inorganic compounds with cellular respiratory systems being highly sensitive. Differences in the toxicity of these agents are determined by their respective chemical properties, absorption, distribution, and excretion characteristics and whether more toxic metabolites are released *in vivo*.

Early work by Voegtlin and co-workers demonstrated the interaction of arsenicals with SH groups of proteins (Rosenthal and Voegtlin, 1930) and their inhibition of cellular respiration (Voegtlin et al., 1931).

Voegtlin et al. (1931) studied the effects of various arsenicals on respiration of several tissues and found that pentavalent organoarsenicals had little effect on tissue respiration, while trivalent forms were strongly inhibitory. Addition of glutathione in a 10:1 ratio with arsenoxide prevented reduction in tissue respiration. Barron and Singer (1945) observed that trivalent organoarsines were more potent inhibitors of cellular respiratory enzymes such as pigeon breast muscle succinoxidase than arsenite. They further demonstrated that pyruvate oxidase, carboxylase, α-ketoglutarate oxidase, malate oxidase, and ATPase were strongly inhibited by these agents, whereas lactic dehydrogenase, isocitric dehydrogenase, acid phosphatase, carbonic anhydrase, catalase, cytochrome oxidase, and flavoprotein were not inhibited. It was subsequently reported (Barron et al., 1947) that lewisite was also a more potent inhibitor of cellular respiratory enzymes and serum cholinesterase than arsenite.

Thompson (1946) studied the effects of lewisite and arsenite on the respiration of skin and found that lewisite was the more potent inhibitor. He also noted that pyruvate oxidation was more sensitive to arsenical inhibition (86–100% inhibition) than that of succinate, which was only slightly affected (0–10% inhibition) at the same concentration of arsenic. In a subsequent report Thompson (1948) further confirmed the sensitivity of the pyruvate oxidase system to trivalent arsenicals and noted increased blood pyruvate levels in pigeons and mice injected with arsenite. To explain the greater sensitivity of the pyruvate oxidase system to arsenicals in comparison with the succinic oxidase system, Thompson first observed that both are sulfhydryl-containing enzyme systems and that the protective effect of BAL is due to formation of a stable dithiol ring with arsenic. He suggested that the pyruvate oxidase system possessed a dithiol component (lipoic acid), while the succinate oxidase did not. Gordon and Quastel (1948) examined the effects of various organoarsenicals on different enzymes systems in comparison to arsenite. These toxicants were more effective in inhibiting urease, cholinesterase, and pyruvate oxidase than arsenite. Cytochrome oxidase, catalase, and invertase were not greatly affected. Pentavalent organoarsenicals were also shown to have little effect on the enzymes studied.

Stocken and Thompson (1949) have reviewed the toxicity of arsenicals and their formation of stable dithiol rings with BAL. Extensive reviews of the biochemical effects of arsenicals on SH-containing enzyme systems have also been published by Peters (1955) and more recently by Webb (1966).

The above studies indicate that trivalent organoarsenicals produce toxicity like inorganic arsenite by reacting with SH groups of enzymes, in particular those involved in cellular respiration. The degree of this toxicity probably depends on the chemical properties of the arsenicals involved (i.e., lipophilicity). Whether the toxicity of these trivalent arsenicals is due to the *in vivo* release of inorganic arsenite has received little attention, and work in this area is needed. Pentavalent organoarsenicals seem to possess a very low toxicity and a mechanism of action separate from that of the trivalent forms. The possible reduction of pentavalent organoarsenicals to more toxic trivalent forms has also been discussed by Gordon and Quastel (1948).

IN VIVO INTERCONVERSION OF ARSENATE AND ARSENITE

It is obvious from the preceding discussion that the toxicity and mode of cellular arsenic action are highly dependent on oxidation state. The *in vivo* transformation of arsenicals has been studied by several groups of investigators (Lanz et al., 1950; Winkler, 1962; Ginsburg and Lotspeich, 1964; Peoples, 1964; Ginsburg, 1965). The results of these studies are somewhat conflicting for arsenate and arsenite.

Lanz et al. (1950) have reported that 10–15% of radioactive arsenate injected into rats was reduced *in vivo* to arsenite and excreted in the urine. Arsenate and arsenite were separated in these studies using a differential precipitation method. Ginsburg and Lotspeich (1964) and Ginsburg (1965) have observed that 10–15% of the arsenic in the urine of dogs given arsenate by injection was also in the form of arsenite—a finding similar to that of Lanz et al. (1950).

Winkler (1962) has reported an extensive extraction procedure for separating the two oxidation states of arsenic in tissue and found that most of the arsenic in livers of rats fed arsenite was in the pentavalent state suggesting *in vivo* oxidation. Peoples (1964) using the technique developed by Winkler has studied arsenic acid metabolism in cows. He noted that rapid urinary elimination of arsenic occurs and that tissue arsenic residues are all in the pentavalent state thus supporting the findings of Winkler.

Several factors could account for these somewhat variable findings that seem to indicate that arsenate may be reduced or unchanged *in vivo* while arsenite is oxidized.

Winkler studied animals fed arsenite; Lanz et al. and Ginsburg injected arsenate. This means that oxidation of arsenite to arsenate could have occurred in the gastrointestinal tract of the arsenite-fed animals or during the digestion and analytical processes. Another possibility is that the metabolism of arsenic in the liver could favor its oxidation to arsenate while that of the kidney favors its reduction to arsenite during the process of urine formation as suggested by Ginsburg and Lotspeich (1964) and Ginsburg (1965).

More recently Braman and Foreback (1973) and Crecelius (1975) have reported excretion of arsenate, arsenite, methyl arsonic acid, and dimethyl arsinic acid in urine of humans. Crecilius (1975) observed that most of an ingested dose of arsenite (As^{3+}) was excreted in the urine of a human subject as methyl arsonic acid and dimethyl arsinic acid, while only 10% was eliminated as arsenite. The biological half-lives were estimated to be 10 hr for arsenite and 30 hr for the methylated forms of arsenic. These findings suggest that methylation may be a more important pathway than oxidation in the metabolism of arsenite.

A final understanding of *in vivo* arsenic metabolism will only occur following improved analytical methodology and more complete studies on *in vivo* arsenic metabolism and absorption in the intestine.

ARSENICAL INTERACTION WITH SELENIUM

Arsenic and selenium have been reported to have antagonistic inter-actions with respect to their individual toxicities. Arsenic has been reported to decrease the retention of selenium (Moxon and Dubois, 1939; Dubois et al., 1940) and the overall toxicity of selenium (Palmer and Bonhorst, 1957; Levander and Argett, 1969). This decrease in toxicity is associated with increased excretion of selenium from the liver into the bile (Levander and Bauman, 1966) and decreased exhalation of volatile selenium compounds (Levander and Argett, 1969). Hemolytic anemia induced in rats by exposure to selenium has also been decreased by concomitant administration of arsenite (Halverson et al., 1970).

Conversely, Holmberg and Ferm (1969) reported that the concomitant injection of a nonteratogenic dose of sodium selenite with sodium arsenate protected against the teratogenicity of this arsenical in golden hamsters.

METHODS OF ARSENIC ANALYSIS

There are many methods of arsenic analysis, some of which measure total arsenic and others that determine oxidation state and chemical form. This section will review only the more common available techniques and cite their respective characteristics.

The Gutzeit method involves the reduction of arsenic to arsine with concentration determined by volumetric titration. The molybdenum blue procedure utilizes spectrophotometric measurement of an arsenic–molybdenum blue complex at 850 nm. A more recent colorimetric method is the silver diethyl dithiocarbamate method developed by Powers et al. (1959). This method involves the reduction of tissue arsenic to arsine, which is subsequently reacted with a solution of silver diethyldithiocarbamate–pyrine to form a stable red complex that is read at 540 nm.

Dubois and Monkman (1961) have compared the precision of the Gutzeit method, silver diethyldithiocarbamate method, and an iodine microtitration method with electrical indication of the endpoint. They concluded that overall the silver diethyldithiocarbamate method was superior to the others.

More recent advances in atomic absorption spectrophotometry have greatly aided the precision and reliability of this technique. Development of an arsine generator by Manning (1971) allows collection of arsine gas from acid-treated samples and subsequent measurement of the total vaporized arsenic by atomic absorption.

Physical techniques such as neutron activation (Heydorn, 1970; Orvini et al., 1974), X-ray fluorescence, or proton-induced X-ray emission analysis (Walter et al., 1974) may be readily used to measure total arsenic.

Procedures for separating the different oxidation states and chemical forms of arsenic have also been developed by Braman and Foreback (1973), Daugtrey et al. (1975), and Fitchett et al. (1975). These methods may be effectively applied to water, urine, and other fluid matrices not requiring digestion procedures. Further studies are needed to evaluate their applicability to analysis of tissue matrices. Luh et al. (1973) have reviewed a number of methods for arsenic analysis.

FUTURE RESEARCH NEEDS

From the preceding discussion, it may be concluded that arsenic is a common element in the environment and that much needs to be done toward understanding its metabolism and potential toxicity. These research needs may be divided into several general areas.

Analysis

The analytical techniques currently available are capable of yielding reproducible results for arsenic, but the ability to distinguish different chemical forms and oxidation states of arsenic in biological tissues is essential to assessing potential toxicity and the setting of meaningful regulations for anthropogenic sources.

Fate of Arsenicals in the Environment

Basic studies are needed to evaluate the oxidation, reduction, and methylation of arsenic in marine, aquatic, and terrestrial ecosystems. Data generated from such research would be useful in assessing the potential toxicity of environmental arsenic to humans.

Pharmacokinetics of Arsenic

The pharmacokinetics of environmental arsenic in humans, plants, and animals are also poorly understood. Research in this area should evaluate the absorption, distribution, metabolism, and excretion of arsenical metabolites in biota as well as interactions between arsenicals and other xenobiotics.

Cellular Mechanisms of Toxicity

Arsenicals are general protoplasmic poisons whose level of toxicity may vary greatly with oxidation state and chemical form. Sound toxicological and biochemical research into the mechanisms of arsenic toxicity is essential to understanding the deleterious effects of this ubiquitous environmental toxicant.

SUMMARY

Arsenic is a common environmental agent whose toxic properties have been known for centuries. The toxicology of arsenic may be divided into three general areas: direct inhibition of cellular respiration, mutagenic effects, and hemolysis. Arsenical toxicity with respect to these areas is highly dependent on chemical form, oxidation state, and route of exposure. Trivalent arsenicals are usually more toxic than pentavalent forms.

Arsenic may be released into the environment from natural or anthropogenic sources. Arsenic may be inhaled or ingested in drinking water or food. Food products such as fish and shellfish are known to concentrate arsenic from their environment. Arsenic in tissues of these organisms appears to be organically bound and may be rapidly excreted following ingestion depending on whether the tissues have been cooked or processed. Arsenicals have also been used as feed additives for poultry and livestock in order to enhance growth, and hence a small amount of arsenic finds its way into the human diet in this manner. Plants and other crops sprayed or grown in soil treated with arsenicals may also incorporate this element, thus contributing to human exposure.

Binding of arsenic by the blood appears to show species variation. In leukemic blood of humans, arsenic is concentrated in white blood cells and plasma. Rats transport most of their blood arsenic in erythrocytes. An absorbed arsenic dose is distributed to the kidney, liver, spleen, skin, hair, and nails in that order. Some arsenic may remain in these tissues long after it has disappeared from the blood, urine, and feces. In humans, a single administered dose of arsenite is rapidly excreted in the first few days following treatment, and a lower elimination rate occurs in the succeeding 2 wk.

Acute ingested arsenic poisoning in animals and humans is characterized by gastrointestinal damage, convulsions, and hemorrhage. Fatty degeneration of the liver and kidneys is frequently found at autopsy. Peripheral neuropathy may also be observed. Acute inhalation of arsine gas is followed by extensive hemolysis, hemoglobinuria, and death from renal failure. In more chronic forms of arsenic intoxication, anorexia, anemia, and disturbances in renal, hepatic, and neurological function may occur.

The cellular toxicity of arsenic is usually attributed to inhibition of cellular respiration. Trivalent arsenicals inhibit the pyruvate oxidase system leading to the accumulation of pyruvate in the blood and urine. Pentavalent arsenic may substitute for phosphate during oxidative phosphorylation leading to formation of arsenate esters that spontaneously decompose resulting in decreased oxidative phosphorylation. These effects have been correlated with swelling of mitochondria in both liver and kidney.

In conclusion, arsenic must be regarded as an important environmental toxicant due to its general toxic properties and extensive presence in the environment. Much remains to be learned about the basic toxicology and biochemistry of this interesting and ancient toxicant.

REFERENCES

Al-Timimi, A. A. and Sullivan, T. W. 1972. Safety and toxicity of dietary organic arsenicals relative to performance of young turkeys. I. Arsanilic acid and sodium arsanilate. *Poult. Sci.* 51:111–116.

Amarsingham, R. D. and Lee, H. 1969. A review of poisoning cases examined by the department of chemistry, Maylaysia, from 1963–1967. *Med. J. Malaya* 23:220–227.

Anonymous. 1973. Increased heavy metals around Sudbury's smelters. *Water Pollu. Control* 111:48–49.

Anonymous. 1974. Acute arsine poisoning. *Lancet* 2:1433.

Assouly, M. and Griffon, M. 1949. L'intoxication par l'hydrogene arsenie au cours du detartrage par les acides. *Arch. Mal. Prof.* 10:337–346.

Axtmann, R. C. 1975. Environmental impact of a geothermal power plant. *Science* 186:795–803.

Azzone, S. F. and Ernster, L. 1961. Compartmentation of mitochondrial phosphorylation as disclosed by studies with arsenate. *J. Biol. Chem.* 236:1510–1517.

Baker, F. C. and Parker, A. J. 1969. Accidental arsenic poisoning in the broiler chicken. *Vet. Rec.* 85:632.

Barron, E. G. and Singer, T. P. 1945. Studies on biological oxidations XIX. Sulfhydryl enzymes in carbohydrate metabolism. *J. Biol. Chem.* 157:221–253.

Barron, E. G., Miller, Z. B., Bartlett, G. R. Meyer, J. and Singer, T. P. 1947. Reactivation by dithiols of enzymes inhibited by lewisite. *Biochem. J.* 41:69–74.

Barry, K. G. and Herndon, E. G., Jr. 1962. Electrocardiographic changes associated with acute arsenic poisoning. *Med. Ann. D.C.* 31:65–66.

Bartley, W. and Dean, B. 1969. The effect of inhibitors on the formation of phosphoenolpyruvate by rat liver mitochondria. *Biochem. J.* 115:903–912.

Beaudoin, A. R. 1974. Teratogenicity of sodium arsenate in rats. *Teratology* 10:153–158.

Bencko, V. 1966. Arsenic in hair of a non-professionally exposed population. *Cesk. Hyg.* 11:539–543.

Bencko, V. and Simane, F. 1968. The effect of chronical intake of arsenic on the liver tissue respiration in mice. *Experientia* 24:706.

Bencko, V. and Symon, K. 1969. Suitability of hairless mice for experimental work and their sensitivity to arsenic. *J. Hyg. Epidemiol. Microbiol. Immunol.* 13:1–6.

Bencko, V., Cmarko, V. and Polan, S. 1968. The cumulation dynamics of arsenic in tissues of rabbits exposed in the area of the ENO plant. *Cesk. Hyg.* 13:18–22.

Bencko, V., Mokry, Z. and Nemeckova, H. 1970. The metabolic oxygen consumption by mouse liver homogenate during drinking water arsenic exposure. *Cesk. Hyg.* 15:305–308.

Berg, J. W. and Burbank, F. 1972. Correlations between carcinogenic trace metals in water supplies and cancer mortality. *Ann. N.Y. Acad. Sci.* 199:249–264.

Binet, L., Bour, H. and Larcorne, H. 1941. Fe foie dans l'intoxication par l'hydrogene arsenie. *Arch. Mal. Prof.* 3:307–318.

Bjornstad, J., Opstvedt, J. and Lunde, G. 1974. Unidentified growth factors in fish meal: Experiments with organic arsenic compounds in broiler diets. *Br. Poult. Sci.* 15:481–487.

Bomford, R. R. and Hunter, S. 1932. Arseniuretted hydrogen poisoning due to the action of water on metallic arsenides. *Lancet* 2:1446–1449.

Borgono, J. M. and Greiber, R. 1972. Epidemiological study of arsenicism in the city of Antofagasta. In *Trace substances in environmental health*, ed. D. Hemphill, vol. 5, pp. 13–24. Columbia: University of Missouri.

Braman, R. S. and Foreback, C. C. 1973. Methylated forms of arsenic in the environment. *Science* 182:1247–1249.

Brannock, W. W., Fix, P. F., Gianella, V. P. and White, D. E. 1948. Preliminary geochemical results at Steamboat Springs, Nevada. *EOS Trans. Am. Geophys. Union* 29:211–226.

Brown, M. M., Rhyne, B. C., Goyer, R. A. and Fowler, B. A. 1976. Intracellular effects of chronic arsenic administration on renal proximal tubule cells. *J. Toxicol. Environ. Health* 1:505–514.

Browning, E. 1969. *Toxicity of industrial metals*, pp. 39–60. London: Butterworths.

Buchanan, W. D. 1962. *Toxicity of arsenic compounds*. Elsevier Monograph, ed. E. Browning. Amsterdam: Elsevier.

Bulmer, F. M. R., Rothwell, H. E., Polack, S. S. et al. 1940. Chronic arsine poisoning among workers employed in the cyanide extraction of gold: A report of fourteen cases. *J. Ind. Hyg. Toxicol.* 22:111–123.

Butzengeiger, K. H. 1940. On peripheral circulatory disorders during arsenic intoxication. *Klin. Wochenschr.* 19:523–527.

Butzengeiger, K. H. 1949. Chronic arsenic poisoning I. *Dtsch. Arch. Klin. Med.* 194:1–16.

Byron, W. R., Bierbower, G. W., Brouwer, J. B. and Hansen, W. H. 1967. Pathologic changes in rats and dogs from a two-year feeding of sodium arsenite or sodium arsenate. *Toxicol. Appl. Pharm.* 10:132–147.

Cannon, A. B. 1936. Chronic arsenical poisoning: Symptoms and sources. *N.Y. State J. Med.* 36:219–241.

Case, A. A. 1974. Toxicity of various chemical agents to sheep. *Am. Vet. Med. Assoc. J.* 164:277–283.

Chan, T. L., Thomas, R. R. and Wadkins, C. L. 1969. The formation and isolation of an arsenylated component of rat liver mitochondria. *J. Biol. Chem.* 244:2883–2890.

Chance, A. C. 1945. The fate of arsenic in the body following treatment of rabbits with certain organic arsenicals. *Q. J. Exp. Phys. Cogn. Med. Sci.* 33:137–147.

Chisolm, J. J., Jr. 1970. Poisoning due to heavy metals. *Pediatr. Clin. N. Am.* 17:591–615.

Coles, G. A., Daley, D., Davies, H. J. and Mallick, N. P. 1969. Acute intravascular hemolysis and renal failure due to arsine poisoning. *Postgrad. Med. J.* 45:170–172.

Conley, B. E. 1958. Morbidity and mortality from economic poisons in the United States. *Arch. Ind. Health* 18:126–133.

Cottam, G. L. and Srere, R. A. 1969. The sulfhydryl groups of citrate cleavage enzyme. *Arch. Biochem. Biophys.* 130:304–311.

Coughlan, M. P., Rajagopalan, K. V. and Handler, P. 1968. The role of molybdenum in xanthine oxidase and related enzymes. *J. Biol. Chem.* 244:2658–2668.

Coulson, E. J., Remington, R. E. and Lynch, K. M. 1935. Metabolism in the rat of naturally occurring arsenic. *J. Nutr.* 10:255–270.

Cox, D. P. and Alexander, M. 1973. Production of trimethylarsine gas from various arsenic compounds by three sewage fungi. *Bull. Environ. Contam. Toxicol.* 9:84–87.

Crane, R. K. and Lipman, R. 1953. The effect of arsenate on aerobic phosphorylation. *J. Biol. Chem.* 201:235–243.

Crecelius, E. A. 1975. Chemical changes in arsenic following ingestion by man. *The biological implications of metals in the environment.* Proc. Fifteenth Ann. Hanford Life Sci. Symp. Richland, Wash., Sept. 29–Oct. 1.

Crecelius, E. A., Bothner, M. H. and Carpenter, R. 1975. Geochemistries of arsenic, antimony, and related elements in sediments of Puget Sound. *Environ. Sci. Technol.* 9:325–333.

Dalgaard, J. B. and Gregersen, M. 1964. Forgiftnungsfaren under søtransport af ferrosilicuim. *Nordisk Med.* 71:271–273.

Daugtrey, E. H., Jr., Fitchett, A. W. and Mushak, P. 1975. Quantitative measurements of inorganic and methyl arsenicals via gas–liquid chromotography. *Anal. Chem. Acta* 79:199–206.

Deeths, T. M. and Breeden, J. T. 1971. Poisoning in children—A statistical study of 1,057 cases. *J. Pediatr.* 78:299–305.

DeMaster, E. G. and Mitchell, R. A. 1970. The insensitivity of mitochondrial-catalyzed arsenate-water oxygen exchange reaction to dinitrophenol and to oligomycin. *Biochem. Biophys. Res. Commun.* 39:199–203.

DePalma, A. E. 1969. Arsine intoxication in a chemical plant: A report of three cases. *J. Occup. Med.* 11:582–587.

Dernehl, C. U., Stead, F. M. and Nau, C. A. 1944. Arsine, stibine and hydrogen sulfide: Accidental generation in a metal refinery. *Ind. Med.* 13:361–362.

Derot, M., Legrain, M., Jacobs, C., Prunier, P. and Hazebroucq, E. 1963. Intoxication par l'hydrogene arsenie: Etude des formes renales a propos d'une intoxication collective (5 cas). *J. Urol. Nephrol.* 169:407–433.

De Scoville, A. 1949. Etude histo-pathologique et histo-physiologique d'un cas d'intoxication par l'hydrogene arsenie: Considerations sur le mecanisme de son anurie. *Rev. Belg. Pathol. Med. Exp.* 19:195–213.

Dickinson, J. O. 1972. Toxicity of the arsenical herbicide monosodium acid methanearsonate in cattle. *Am. J. Vet. Res.* 33:1889–1892.

Dinman, B. D. 1960. Arsenic: Chronic human intoxication. *J. Occup. Med.* 2:137–141.

Doig, A. T. 1958. Arseniuretted hydrogen poisoning in tank cleaners. *Lancet* 2:88–92.

Done, A. K. and Peart, A. J. 1971. Acute toxicities of arsenical herbicides. *Clin. Toxicol.* 4:343–355.

Dubois, L. and Monkman, J. G. 1961. Determination of arsenic in air and biological materials. *Am. Ind. Hyg. Assoc. J.* 22:292–295.

Dubois, K. P., Moxon, A. L. and Olson, O. E. 1940. Further studies on the effectiveness of arsenic in preventing selenium poisoning. *J. Nutr.* 19:477–482.

Duck, N. W. and Himus, G. W. 1951. Arsenic in coal and its mode of occurrence. *Fuel* 30:267–272.

Ducoff, H. S., Neal, W. B., Straube, R. L., Jacobson, L. O. and Brues, A. M. 1948. Biological studies with arsenic excretion and tissue localization. *Proc. Soc. Exp. Biol. Med.* 69:548–554.

DuPont, O., Ariel, I. and Warren, S. L. 1941. The distribution of radioactive arsenic in the normal and tumor-bearing (brown Pearce) rabbits. *Am. J. Syph. Gon. Ven. Dis.* 26:96–118.

Durum, W. H., Hem, J. D. and Heidel, S. G. 1971. Reconnaissance of selected minor elements in surface waters of the U.S. *Geol. Surv. Circ.* 643:1–49.

Eiji, H. 1955. Infant arsenic poisoning by powdered milk. *Nihon iji shimpō* No. 1649:3–12. (EPA translation No. TR105-74.)

Elkins, H. B. and Fahy, J. P. 1967. Arsine poisoning from aluminum tank cleaning. *Ind. Med. Surg.* 36:747–749.

Estabrook, R. W. 1961. Effect of oligomycin on the arsenate and DNP stimulation of mitochondrial oxidations. *Biochem. Biophys. Res. Commun.* 4:89–91.

Feinglass, E. J. 1973. Arsenic intoxication from well water in the United States. *New Engl. J. Med.* 288:828–830.

Ferm, V. H. and Carpenter, S. 1968. Malformations induced by sodium arsenate. *J. Reprod. Fertil.* 17:199–201.

Ferm, V. H., Saxon, A. and Smith, B. M. 1971. The teratogenic profile of sodium arsenate in the golden hamster. *Arch. Environ. Health* 22: 557–560.

Finner, L. L. and Calvery, H. O. 1939. Pathologic changes in rats and in dogs fed diets containing lead and arsenic compounds. *Arch. Pathol.* 27: 433–446.

Fitchett, A. W., Daughtrey, E. H., Jr., and Mushak, P. 1975. Quantitative measurements of inorganic and organic arsenic by flameless atomic absorption spectrophotometry. *Anal. Chem. Acta* 79:93–99.

Fletcher, M. J., Fluharty, A. L. and Sanadi, D. R. 1962. On the mechanism of oxidative phosphorylation. V. Effects of arsenite and cadmium ions in mitochondrial fragments. *Biochim. Biophys. Acta* 60:425–427.

Fluharty, A. L. and Sanadi, D. R. 1960. Evidence for a vicinal dithiol in oxidative phosphorylation. *Proc. Natl. Acad. Sci. U.S.A.* 46:608–616.

Fluharty, A. L. and Sanadi, D. R. 1962. On the mechanism of oxidation phosphorylation. IV. Mitochondrial swelling caused by arsenite in combination with 2,3-dimercaptopropanol and by cadmium ion. *Biochemistry* 1:276–281.

Fowler, B. A. 1974. The morphologic effects of mercury, cadmium, lead and arsenic on the kidney, pp. 65–76. In *Trace metals in water supplies: Occurrence, significance and control.* Proceedings: Sixteenth Water Quality conf. Champaign–Urbana: Department of Engineering, University of Illinois.

Fowler, B. A. 1975. The ultrastructural and biochemical effects of arsenate on the kidney. *Proc. XVIII Int. Congr. Occup. Health* Brighton, U.K. Abstr.

Fowler, B. A. and Weissberg, J. B. 1974. Arsine Poisoning. *New Engl. J. Med.* 291:1171–1174.

Fowler, B. A., Fay, R. C., Walter, R. L., Willis, R. D. and Gutknecht, W. F. 1975. Levels of toxic metals in marine organisms collected from southern California coastal waters. *Environ. Health Perspect.* 12:71–76.

Franco, P. 1968. Un episodio di avvelenamento collectivo da gas tossici a bordo die una nave mercantile. *Ann. Sanit. Pubbl.* 29:409–420.

Franklin, M., Bean, W. B. and Hardin, R. C. 1950. Fowler's solution as an etiologic agent in cirrhosis. *Am. J. Med. Sci.* 219:589–596.

Frejaville, J. P., Bescol, J., Leclerc, J. P., Guillam, L. Crabie, P., Consco, F., Gervais, P. and Faultier, M. 1972. Intoxication aigue par les derives arsenicaux; (a propos de 4 observations personnelles); troubles de l'hematose, etude ultramicroscope du foie et du rein. *Ann. Med. Intern.* 123:713–722.

Frost, D. B. 1967. Arsenicals in biology—Retrospect and prospect. *Fed. Proc.* 26:194–208.

Garlick, H. W. 1949. A case of industrial poisoning with arsine treated with British anti-lewisite. *Med. J. Austral.* 1:69–70.

Gaultier, M. Tanret, P., and Reymond, J. C. 1949. Commentaires dur deux cas d'intoxication grave par l'hydrogene arsenie. *Bull. Soc. Med. Hop. Paris* 65:951–959.

Gebert, F. 1937. Uber die Reaktion zwischen Arsenwasserstoff und Hämoglobin. *Biochem. Z.* 293:157–186.

Ginsburg, J. M. 1965. Renal mechanisms for excretion and transformation of arsenic in the dog. *Am. J. Physiol.* 208:832–840.

Ginsburg, J. M. and Lotspeich, W. W. 1964. Interrelations of arsenate and phosphate transport in the dog kidney. *Am. J. Physiol.* 205:707–714.

Glazener, F. S., Ellis, J. G. and Johnson, P. K. 1968. Electrocardiographic findings with arsenic poisoning. *Calif. Med.* 109:158–162.

Gordon, J. J. and Quastel, J. H. 1948. Effects of organic arsenicals on enzyme systems. *Biochem. J.* 42:337–350.

Graham, A. F., Crawford, T. B. B. and Marrian, G. F. 1946. The action of arsine on blood: Observations on the nature of the fixed arsenic. *Biochem. J.* 40:256–260.

Gramer, L. 1955. Uber eine tödliche perakute Arsenwasserstoffvergiftung. *Arch. Gewerbepathol. Gewerbehyg.* 13:601–610.

Greig, H. B. W., Bradlow, B. A., Harrison, C. and Dalton, J. P. 1958. Arsine poisoning in industry: A report of two cases. *S. Afr. Med. J.* 32:101–104.

Grigg, F. J. T. 1929. Distribution of arsenic in the body after a fatal case of poisoning by hydrogen arsenide. *Analyst* 54:659–660.

Grunicke, H., Bock, K., Becker, H., Gang, V., Schnierda, J. and Puschendorf, B. 1973. Effect of alkylating antitumor agents on the binding of DNA to protein. *Cancer Res.* 33:1048–1053.

Guarjardo, A. S., Garcia, A. M., Heredia, E. M. and Sicilia, L. S. 1970. Emolisi e blocco renale acute nell'intossicanzione collective da arsina: Studio di cinque casi. *Minerva Med.* 61:5799–5808.

Halverson, A. W., Tsay, D.-T., Triebwasser, K. C. and Whitehead, E. I. 1970. Development of hemolytic anemia in rats fed selenite. *Toxicol. Appl. Pharm.* 17:151–159.

Harrisson, J. W. E., Packman, E. W. and Abbott, D. D. 1958. Acute oral toxicity and chemical and physical properties of arsenic trioxides. *Arch. Ind. Health* 17:118–123.

Harvey, S. C. 1970. Heavy metals. In *The pharmacological basis of therapeutics*, 4th ed., ed. L. S. Goodman and A. Gilman, pp. 958–965. New York: Macmillan.

Hassinen, I. and Hallman, M. 1967. Comparison of the effects of disulfiram and dimercaptopropanol arsenite on mitochondrial structure and function. *Biochem. Pharmacol.* 16:2155–2161.

Hayes, W. J. and Pirkle, C. I. 1966. Mortality from pesticides in 1961. *Arch. Environ. Health* 12:43–55.

Haywood, T. K. 1907. Injury to vegetation and animal life by smelter fumes. *J. Am. Chem. Soc.* 19:998–1008.

Headlee, A. J. W. and Hunter, R. G. 1953. Elements in coal ash and their industrial significance. *Ind. Eng. Chem.* 45:548–551.

Heubner, W. and Wolf, K. 1936. Uber die Wirkungsweise des Arsenwasserstoffs. *Naunyn Schmiedebergs. Arch. Exp. Pathol. Pharmakol.* 181:149–150.

Heydorn, K. 1970. Environmental variations of arsenic levels in human blood determined by neutron activation analysis. *Clin. Chim. Acta* 28:349–357.

Heyman, A., Pfeiffer, J. B., Willett, R. W. and Taylor, H. M. 1956. Peripheral neuropathy caused by arsenical intoxication: A study of 41 cases with observations on the effects of BAL (2,3-dimercapto-propanol). *New Engl. J. Med.* 254:401–409.

Hocken, A. G. and Bradshaw, G. 1970. Arsine poisoning. *Br. J. Ind. Med.* 27:56–60.

Hogan, R. B. and Eagle, A. 1944. The pharmacologic basis for the widely varying toxicity of arsenicals. *J. Pharmacol. Exp. Ther.* 80:93–113.

Holland, R. H., McCall, M. S. and Lanz, H. C. 1959. A study of inhaled arsenic-74 in man. *Cancer. Res.* 19:1154–1156.

Holmberg, R. E. and Ferm, V. H. 1969. Interrelationships of selenium, cadmium, and arsenic in mammalian teratogenesis. *Arch. Environ. Health* 18:873–877.

Hood, R. and Bishop, S. L. 1972. Teratogenic effects of sodium arsenate in mice. *Arch. Environ. Health* 24:62–65.

Hood, R. D. and Pike, C. T. 1972. BAL alleviation of arsenate-induced teratogenesis in mice. *Teratology* 6:235–238.

Hueper, W. C. 1966. *Occupational and environmental cancers of the respiratory system*, pp. 30–38. New York: Springer-Verlag.

Hughes, W. and Levvy, G. A. 1947. The toxicity of arsine solutions for tissue slices. *Biochem. J.* 41:8–11.

Hunter, F. T., Kip, A. F. and Irvine, J. W. 1942. Radioactive tracer studies on arsenic injected as potassium arsenite I. Excretion and localization in tissues. *J. Pharmacol. Exp. Ther.* 76:207–220.

Ikegami, K., Nakamura, T. and Kanbara, T. 1967. A case of acute renal failure due to arseniuretted hydrogen poisoning undergone a long-term peritoneal dialysis. *Jap. J. Urol.* 58:369–380.

Jenkins, G. C., Ind, J. E., Kazantzis, G. and Owen, R. 1965. Massive hemolysis with minimal impairment of renal function. *Br. Med. J.* 2:78–80.

Johnson, D. L. 1972. Bacterial reduction of arsenate in sea water. *Nature* 240:44–45.

Johnson, D. L. and Pilson, M. E. O. 1972. Arsenate in the western north Atlantic and adjacent regions. *J. Mar. Res.* 30:140–149.

Johnson, G. A. 1953. An arsine problem: Engineering notes. *Am. Ind. Health Assoc. Q.* 14:188–190.

Johnstone, R. M. 1963. Sulfhydryl agents: Arsenicals. In *Metabolic inhibitors*, ed. R. M. Hochster and J. H. Quastel, vol. 2, pp. 99–112. New York: Academic Press.

Josephson, C. J., Pinto, S. S. and Petronella, S. J. 1951. Arsine: Electrocardiographic changes produced in acute human poisoning. *Arch. Ind. Hyg. Occup. Med.* 4:43–52.

Jung, E. 1969. Arsenic as an inhibitor of the enzymes concerned in cellular recovery (dark repair). *Ger. Med. Mon.* 14:614–616.

Jung, E. 1971. Molecular biological investigation of chronic arsenic poisoning. *Z. Haut Geschlechtskr.* 46:35–36.

Jung, F. 1939. Löslichkeit und Reaktionsweise des Arsenwasserstoffs in Blut. *Biochem. Z.* 302:294–309.

Kagawa, Y. and Kagawa, A. 1969. Accumulation of arsenate-76 by mitochondria. *J. Biochem. (Tokyo)* 65:105–112.

Katsura, K. 1953. Medicolegal studies on arsenical poisoning. *Shikoku Igaku Zasshi* 12:706–720.

Keenan, D. M. and Oe, R. D. 1973. Acute arsanilic acid intoxication in pigs. *Aust. Vet. J.* 49:229–231.

Kelynack, T. N., Kirkly, W., Delepine, S. and Tattersall, C. H. 1900. Arsenical poisoning from beer drinking. *Lancet* 2:1600–1602.

Kensler, C. J. Abels, J. C. and Rhoads, C. P. 1946. Arsine poisoning, mode of action and treatment. *J. Pharmacol. Exp. Med.* 88:99–118.

Kiese, M. 1934. Chronische Arsenwasserstoffvergiftung. *Naunyn-Schmiedebergs Arch. Exp. Pathol. Pharmakol.* 176:531–549.

Kipling, M. D. and Fothergill, R. 1964. Arsine poisoning in a slag-washing plant. *Br. J. Ind. Med.* 21:74–77.

Kiyono, S., Hasui, K., Takasu, K. and Seo, M. 1974. Toxic effect of arsenic trioxide in infant rats. *J. Physiol. Soc. Jap.* 36:253–254.

Kogan, B. 1931. Zur Klinik der akuten Arsenwasserstoffvergiftung. *Naunyn-Schmiedebergs Arch. Exp. Pathol. Pharmakol.* 161:310–324.

Konings, A. W. 1972. Inhibition of nuclear and mitochondrial respiration by arsenite. *Experientia* 28:883–884.

Konzen, J. L. and Dodson, J. N. 1966. Arsine: A cause of hemolytic anemia. *J. Occup. Med.* 8:540–541.

Kuroda, K. 1941. Analyse des Mineralwassers von Kinkei in der Provinz Totigi. *Chem. Soc. Japan Bull.* 16:234–237.

Kyle, P. A. and Pease, G. L. 1965. Hematologic aspects of arsenic intoxication. *New Engl. J. Med.* 273:18–23.

Labes, V. R. 1937. Die oxydative und reduktive Entstehung von kolloiden Elementen der Arsen und Tellurgruppe als Ursache zahlreicher Giftwirkungen AsH_3, H_2S, TeO_2 usw. auf Zellstrukturen und Fermente des Tierkorpers. *Kolloid-Z.* 79:1–10.

Lanz, H., Jr., Wallace, P. C. and Hamilton, J. G. 1950. The metabolism of arsenic in laboratory animals with As^{74} as a tracer. *Univ. Calif. Publ. Pharmacol.* 2:263–282.

Lasch, R. 1961. Erfolgreiche Behandlung einer akuten schweren Arsenvergiftung. *Med. Klin.* 57:62–63.

Lawton, A. H., Ness, A. T., Brady, F. J. and Cowie, D. B. 1945. Distribution of radioactive arsenic following intraperitoneal injection of sodium arsenite into cotton rats injected with *Littomosoides carinii. Science* 102:120–122.

LeBlanc, P. J. and Jackson, A. L. 1973. Arsenic in marine fish and invertebrates. *Mar. Pollut. Bull.* 4:89–90.

Ledet, A. E., Duncan, J. R., Buck, W. B. and Rampey, F. K. 1973. Clinical, toxicological, and pathological aspects of arsanilic acid poisoning in swine. *Clin. Toxicol.* 6:439–457.

Levander, O. A. and Argett, L. C. 1969. Effect of arsenic, mercury, thallium, and lead on selenium metabolism in rats. *Toxicol. Appl. Pharm.* 14:308–314.

Levander, O. A. and Bauman, C. A. 1966. Selenium metabolism. VI. Effect of arsenic on the excretion of selenium in the bile. *Toxicol. Appl. Pharm.* 9:106–115.

Levinsky, W. J., Smalley, R. V., Hillyer, P. N. and Shindler, R. L. 1970. Arsine hemolysis. *Arch. Environ. Health* 20:436–440.

Levvy, G. A. 1946. The toxicity of arsine administered by intraperitoneal injection. *Br. J. Pharmacol.* 1:287–290.

Levvy, G. A. 1947. A study of arsine poisoning. *Q. J. Exp. Physiol.* 34:47–67.

Lindberg, S. E., Andrew, A. W., Raridon, R. J. and Fulkerson, W. 1975. Mass balance of trace elements in walker branch watershed—The relation to coal fired steam plants. *Environ. Health Perspect.* 12:9–18.

Lisella, F. S., Long, K. R. and Scott, H. G. 1972. Health aspects of arsenicals in the environment. *J. Environ. Health* 34:511–518.

Loning, F. 1931. Klinischer Bericht über den Verlauf von elf akuten AsH_3-Vergiftungen. *Zentralbl. Invere Med.* 52:833–837.

Lowry, O. H., Hunter, F. T., Kip, A. F. and Irvine, J. W., Jr. 1942. Radioactive tracer studies on arsenic injected as potassium arsenite. II. Chemical distribution in tissues. *J. Pharmacol. Exp. Ther.* 76:221–225.

Lucas, H. F., Edgington, D. N. and Colby, P. J. 1970. Concentrations of trace elements in great lakes fishes. *J. Fish. Res. Board Can.* 27:677–684.

Luchtrath, H. 1972. Liver cirrhosis in chronic arsenic intoxication of vintagers. *Dtsch. Med. Wochenschr.* 97:21–22.

Luh, M.-D., Baker, R. A. and Henley, D. E. 1973. Arsenic analysis and toxicity. A review. *Sci. Total Environ.* 2:1–12.

Lunde, G. 1970. Analysis of arsenic and selnium in marine raw materials. *J. Sci. Food Agric.* 21:242–247.

Lunde, G. 1972. The analysis of arsenic in the lipid phase from marine and limnetic algae. *Acta Chem. Scand.* 26:2642–2644.

Lunde, G. 1975. A comparison of arseno-organic compounds from different marine organisms. *J. Sci. Food Agric.* 26:1257.

Macaulay, D. G. and Stanley, D. A. 1956. Arsine poisoning. *Br. J. Ind. Med.* 13:217–221.

Mamelak, R. and Quastel, J. H. 1953. Amino acid interactions in strict anaerobes. *Biochim. Biophys. Acta* 12:103–120.

Manceau, P. P. 1929. Un cas curieux et inedit d'empoisonment par l'hydrogene arsenie. *J. Pharmacol. Chim.* 10:133. Abstr.

Manning, D. C. 1971. A high sensitivity arsenic-selenium sampling system for atomic absorption spectroscopy. *At. Absorpt. Newsl.* 10:123–124.

Mealy, J., Jr., Bronwell, G. L. and Sweet, W. H. 1959. Radioarsenic in plasma, urine, normal tissues and intracranial neoplasms; distribution and turnover after intravenous injection in man. *Arch. Neurol. Psychiatr.* 81:310–320.

Meyer, E. and Heubner, W. 1929. Beobachtungen über Arsenwasserstoffvergiftung. *Biochem. Z.* 206:212–222.

McKee, J. E. and Wolf, H. W. 1963. *Water quality criteria.* Publ. 3-A, pp. 140–14. Sacramento: State Water Pollution Control Board of California.

McKinstry, W. J. and Hickes, J. M. 1957. Arsine poisoning. *Arch. Ind. Health* 16:32–34.

Miles, J. R. W. 1968. Arsenic residues in agricultural soils of southwestern Ontario. *J. Agr. Food Chem.* 16:620–622.

Milham, S., Jr. and Strong, T. 1974. Human arsenic exposure in relation to a copper smelter. *Environ. Res.* 7:176–182.

Mitchell, R. A., Chang, B. F., Huang, C. H. and DeMaster, E. G. 1971. Inhibition of mitochondrial energy-linked functions by arsenate. Evidence for a nonhydrolytic mode of inhibitor action. *Biochemistry* 10:2049–2053.

Mohacek, I. and Stajduhar, Z. 1951. A case of fatal acute arsine poisoning. *Arch. Ind. Hyg. Occup. Med.* 4:610.

Morgareidge, K. 1963. Metabolism of two forms of dietary arsenic by the rat. *J. Agric. Food Chem.* 1:377–378.

Morris, H. P., Laug, E. P., Morris, H. J. and Grant, R. L. 1938. The growth and reproduction of rats fed diets containing lead acetate and arsenic trioxide. *J. Pharmacol. Exp. Ther.* 64:420–445.

Morse, K. M. and Setterlind, A. S. 1950. Arsine poisoning in the smelting and refining industry. *Arch. Ind. Hyg. Occup. Med.* 2:148–169.

Moxon, A. L. and Dubois, K. P. 1939. The influence of arsenic and other elements on the toxicity of seleniferous grains. *J. Nutr.* 18:447–457.

Muehrcke, R. C. and Pirani, C. L. 1968. Arsine induced anuria: A correlative clinicopathological study with electron microscopic observations. *Ann. Int. Med.* 68:853–866.

Nagai, H., Okuda, R., Nagami, H., Yagi, A., Mori, C. and Wada, H. 1956. Subacute–chronic "arsenic" poisoning in infants—Subsequent clinical observations. *Shonika kiyo* 2:124–132. (EPA translation No. TR113-74)

Nau, C. A. 1948. The accidental generation of arsine gas in an industry. *South. Med. J.* 41:341–344.

Nau, C. A., Anderson, W. and Cone, R. E. 1944. Arsine, stibine and hydrogen sulfide. *Ind. Med.* 13:308–310.

Neuwirtova, R., Chytil, M., Valek, A., Daum, S. and Valach, V. 1961. Acute renal failure following an occupational intoxication with arsine (AsH_3) treated by the artificial kidney. *Acta Med. Scand.* 170:535–546.

Nielsen, B., ed. 1968. Arsine poisoning in a metal refining plant: Fourteen simultaneous cases. *Acta Med. Scand. Suppl.* 496:1–31.

Nuck, D. and Jaffe, R. 1932. Sonderfälle von Vergiftungsmöglichkeiten durch Arsenwasserstoff. *Arch. Gewerbepathol. Gewerbehyg.* 3:496–508.

Okamura, K., Ota, T., Horiuchi, K., Hiroshima, H., Takai, T., Sakurane, Y. and Baba, T. 1956. Symposium on arsenic poisoning by powdered milk. *Shinryo* 9:155–162. (EPA translation No. TR117-74)

Orvini, E., Gills, T. E. and Lafleur, P. D. 1974. Method for determination of selenium, arsenic, zinc, cadmium and mercury in environmental matrices by neutron activation analysis. *Anal. Chem.* 46:1294–1297.

Overby, L. R. and Fredrickson, R. L. 1965. Metabolism of arsanilic acid. II. Localization and type of arsenic excreted and retained by chickens. *Toxicol. Appl. Pharm.* 7:855–867.

Overby, L. R. and Straube, L. 1965. Metabolism of arsanilic acid. I. Metabolic stability of doubly labeled arsanilic acid in chickens. *Toxicol. Appl. Pharm.* 7:850–854.

Packer, L. 1961. Metabolic and structural states of mitochondria. II. Regulation by phosphate. *J. Biol. Chem.* 236:214–220.

Pakkala, I. S., White, M. W., Lisk, D. J., Burdick, G. E. and Harris, E. J. 1972. Arsenic content of fish from New York State waters. *N.Y. Fish Game J.* 19:12–31.

Palmer, I. S. and Bonhorst, C. W. 1957. Modification of selenite metabolism by arsenite. *J. Agric. Food Chem.* 12:928–930.

Paton, G. and Allison, A. C. 1972. Chromosome damage in human cell culture induced by metal salts. *Mutat. Res.* 16:332–336.

Peden, J. D., Crothers, J. A., Waterfall, C. E. and Beasley, J. 1973. Heavy metals in Somerset marine organisms. *Marine Pollut. Bull.* 4:7–9.

Peoples, S. A. 1964. Arsenic toxicity in cattle. *Ann. N.Y. Acad. Sci.* 111:644–649.

Pernis, B. and Magistretti, M. 1960. A study of the mechanism of acute hemolytic anemia from arsine. *Med. Lav.* 51:37–41.

Perry, K., Bowler, R. G., Buchell, H. M., Durett, H. A. and Schilling, 1948. Studies in the incidence of cancer from inorganic arsenic compound. II. Clinical and environmental investigations. *Br. J. Ind. Med.* 5:6–15.

Peters, R. A. 1955. Biochemistry of some toxic agents. I. *Bull. Johns Hopkins Hosp.* 97:1–20.

Petres, J. and Hundeiker, M. 1968. Chromosome pulverization induced *in vitro* in cell cultures by sodium diarsonate. *Arch. Klin. Exp. Dermatol.* 231:366–370.

Petres, J., Schmidt-Ullrich, K. and Wolf, U. 1970. Chromosomenaberrationen an Menschlichen lymphozyten bei chronischen arsenschäden. *Dtsch. Med. Wochenschr.* 95:79–80.

Pinto, S. S., Petronella, S. J., Johns, D. R. and Arnold, M. F. 1950. Arsine poisoning: A study of thirteen cases. *Arch. Ind. Hyg. Occup. Med.* 1:437–451.

Powers, G. W., Jr., Martin, R. L., Piehl, F. J. and Griffin, J. M. 1959. Arsenic in napthas. *Anal. Chem.* 31:1589–1593.

Pringle, B. H. 1971. Health aspects of environmental pollutants as they pertain to potable water: Solution to the problem. *Proc. Int. Symp. Ident. Measure. Environ. Pollut.* Ottawa. pp. 163–166.

Rathus, E. M. 1963. Study of arsenical spray hazard to banana growers. *Queensland Agric. J.* June:366–367.

Rosenthal, S. M. and Voegtlin, C. 1930. Biological and chemical studies of the relationship between arsenic and crystalline glutathione. *J. Pharmacol. Exp. Ther.* 39:347–367.

Rozenshtein, I. S. 1970. Sanitary toxicological assessment of low concentrations of arsenic trioxide in the atmosphere. *Hyg. Sanit.* 35:16–21.

Sabadell, J. E. and Axtmann, R. C. 1975. Heavy metal contamination from geothermal sources. *Environ. Health Perspect.* 12:1–8.

Sagner, S. 1973. The toxicology of arsenic in drinking water. *Schrift. Vereins Wasser Boden Lufthyg.* 40:189–208.

Satterlee, H. S. 1956. The problem of arsenic in American cigarette tobacco. *New Engl. J. Med.* 254:1149–1154.

Satterlee, H. S. 1960. The arsenic poisoning epidemic of 1900: Its relation to lung cancer in 1960—An exercise in retrospective epidemiology. *New Engl. J. Med.* 263:676–684.

Schreiber, M. and Brouwer, E. A. 1964. Metabolism and toxicity of arsenicals. *Fed. Proc.* 23:199. Abstr.

120   *B. A. Fowler*

Schrenk, H. A. and Schreibeis, L. 1958. Urinary arsenic levels as an index of industrial exposure. *Am. Ind. Hyg. Assoc. J.* 19:225–228.

Schroeder, A. A. and Balassa, J. J. 1966. Abnormal trace metals in man: Arsenic. *J. Chron. Dis.* 19:85–106.

Selby, L. A., Case, A. A., Dorn, C. R. and Wagstaff, D. J. 1974. Public health hazards associated with arsenic poisoning in cattle. *Am. Vet. Med. Assoc. J.* 165:1010–1014.

Sharpless, G. R. and Metzger, M. 1941. Arsenic and goiter. *J. Nutr.* 21: 341–346.

Shibuya, Y. 1971. Studies on experimental arsenious acid poisoning. *Tokyo Jikeikai Ika Daigaku Fasshi* 86:563–575.

Sibatani, A. 1959. *In vitro* incorporation of ^{32}P into nucleic acids of lymphatic cells. *Exp. Cell Res.* 17:131–143.

Smith, K. D. and Rardin, T. E. 1939. Arsine poisoning: Report of two cases. *Ohio State Med. J.* 35:157–159.

Smith, R. S. 1970. Responsibilities and risks involved in the use of wood protecting chemicals. *Occup. Health Rev.* 21:1–6.

Spolyar, L. W. and Harger, R. N. 1950. Arsine poisoning: Epidemiologic studies of an outbreak following exposure to gases from metallic dross. *Arch. Ind. Hyg. Occup. Med.* 1:419–436.

Steel, M. and Feltham, D. V. G. 1950. Arsine poisoning in industry: Report of a case. *Lancet* 1:108–110.

Stocken, L. A. and Thompson, R. H. S. 1949. Reactions of British anti-lewisite wih arsenic and other metals in living systems. *Physiol. Rev.* 29:168–194.

Sullivan, T. W. and Al-Timimi, A. A. 1972a. Safety and toxicity of dietary organic arsenicals relative to performance of young turkeys. 2. Carbarsone. *Poult. Sci.* 51:1498–1501.

Sullivan, T. W. and Al-Timimi, A. A. 1972b. Safety and toxicity of dietary organic arsenicals relative to performance of young turkeys. 3. Nitarsone. *Poult. Sci.* 51:1582–1586.

Sullivan, T. W. and Al-Timimi, A. A. 1972c. Safety and toxicity of dietary organic arsenicals relative to performance of young turkeys. 4. Roxarsone. *Poult. Sci.* 51:1641–1644.

Tamura, S. 1972. Arsenic metabolism. *Folia Pharmacol. Jap.* 68:586–601.

Tarrant, R. F. and Allard, J. 1972. Arsenic levels in urine of forest workers applying silvicides. *Arch. Environ. Health* 24:277–280.

Teitelbaum, D. T. and Kier, L. C. 1969. Arsine poisoning: Report of five cases in the petroleum industry and a discussion of the indications for exchange transfusion and hemodialysis. *Arch. Environ. Health* 19:133–143.

Terada, H., Katsuta, K., Sasakawa, T., Saito, H., Shrota, H., Fukuchi, K., Sekiya, E., Yokoyama, Y., Hirokawa, S., Watanabe, G., Hasegawa, K., Oshina, T. and Sekiguchi, E. 1960. Clinical observations of chronic toxicosis by arsenic. *Nihon rinsho* 18:2394–2403. (EPA translation No. TR106-74)

Ter Welle, H. and Slater, E. C. 1967. Uncoupling of respiratory-chain phosphorylization by arsenate. *Biochim. Biophys. Acta* 143:1–17.

Thauer, R. 1934. Zur Analyse der Arsenwasserstoffvergiftung. *Naunyn-Schmiedebergs. Arch. Exp. Pathol. Pharmakol.* 176:531–549.

Thompson, R. H. S. 1946. The effect of arsenical vesicants on the respiration of skin. *Biochem. J.* 40:525–535.

Thompson, R. H. S. 1948. The reactions of arsenicals in living tissues. *Biochem. Soc. Symp.* 2:28–38.

Thornton, I. and Webb, J. S. 1970. Regional geo-chemical reconnaissance and trace element mapping for agriculture. *Indian Soc. Soil Sci.* 18:357–365.

Tokanehara, S., Akao, S. and Tagaya, S. 1956. Blood findings in infantile arsenic toxicosis caused by powdered milk. *Shonika rinsho* 9: 1078–1084. (EPA translation No. TR110-74)

Triosi, F. M. 1950. Intossicazone da idrogeno arsenicale durante il ravoro di pulitura interna di una caldaia cornovaglia. *Med. Lav.* 41:166–171.

Tseng, W. P., Chu, A. M., How, S. W., Fong, J. M., Lin, C. S. and Yeh, S. 1968. Prevalence of skin cancer in an epidemic area of chronic arsenicism in Taiwan. *J. Natl. Cancer Inst.* 40:453–463.

Tsuzuki, T., Nakaya, S. and Katsuta, N. 1966. Chemical compositions of mineral springs in Hokkaido. Arsenic and fluorine in hot spring water. *Hokkaido-Ritus Eisei Kenkyrisholio* 16:103–110.

Turner, A. W. 1954. Bacterial oxidation of arsenite. *Aust. J. Biol. Sci.* 7:452–478.

Turner, A. W. and Legge, J. W. 1954. Bacterial oxidation of arsenite. *Aust. J. Biol. Sci.* 7:479–495.

Uldall, P. R., Khan, H. A., Ennis, J. E., McCallum, R. I. and Grimson, T. A. 1970. Renal damage from industrial arsine poisoning. *Br. J. Ind. Med.* 27:372–377.

Ullrich, G. 1958. Verlauf und Behandlung einer akuten toxischen Niereninsuffizienz nach Arsenwasserstoffvergiftung. *Wiener Klin. Wochenschr.* 70:538–544.

Underwood, E. J. 1971. *Trace elements in human and animal nutrition*, 3d ed., pp. 427–431. New York: Academic Press.

Vallee, B. L., Ulmer, D. D. and Wacker, W. E. C. 1960. Arsenic toxicology and biochemistry. *Arch. Environ. Health* 21:132–151.

Voegtlin, C., Rosenthal, S. M. and Johnson, J. M. 1931. The influence of arsenicals and crystalline glutathione on the oxygen consumption of tissues. *Publ. Health Rep.* 46:339–354.

Von Lehmden, D. J., Jungers, R. H. and Lee, R. E. 1974. Determination of trace elements in coal, fly ash, fuel oil and gasoline. *Anal. Chem.* 46:239–245.

Wadkins, C. L. 1961. Stimulation of adenosine triphosphatase activity of mitochondria and submitochondrial particles by arsenate. *J. Biol. Chem.* 235:3300–3303.

Wadsworth, G. R. and McKenzie, J. C. 1963. The potato with special reference to its use in the United Kingdom. *Nutr. Abstr. Rev.* 33: 327–344.

Wagner, S. L. and Weswig, P. 1974. Arsenic in blood and urine of forest workers. *Arch. Environ. Health* 28:77–79.

Wagstaff, D. J. 1972. Arsenic trioxide: Stimulation of liver enzyme detoxication activity. *Toxicol. Appl. Pharmacol.* 22:310. Abstr.

Walter, R. L., Willis, R. D., Gutknecht, W. F. and Joyce, J. M. 1974. Analysis of biological, clinical, and environmental samples using proton-induced X-ray emission. *Anal. Chem.* 46:843–855.

Warren, H. V., Delavault, R. E. and Barakso, J. 1964. The role of arsenic as a pathfinder in biogeochemical prospecting. *Econ. Geol.* 59:1381–1385.

Warren, H. V., Delavault, R. E. and Barakso, J. 1968. The arsenic content of Douglas fir as a guide to some gold, silver and base metal deposits. *Can. Min. Metall.* 61:860–866.

Watrous, R. M. and McCaughey, M. D. 1945. Occupational exposure to arsenic. *Ind. Med. Surg.* 14:639–646.

Webb, J. L. 1966. *Enzyme and metabolic inhibitors*, vol. 3, pp. 595–793. New York: Academic Press.

Webster, S. H. 1941. Lead and arsenic ingestion and excretion in man. *Publ. Health Rep.* 56:1359–1368.

Wells, J. D. and Elliott, J. F. 1971. Geochemical reconnaissance of the Cortez-Buckhorn area, Southern Cortez mountains, Nevada. *U.S. Geol. Surv. Bull.* 1312P:1–18.

Westernhagen, B. V. 1970. Histochemical demonstrable metabolic changes in the inner ear of the guinea pig after chronic arsenic poisoning. *Arch. Klin. Exp. Onien-Nasen Hehl-Kopfheilkunde* 197:7–13.

Westöö, G. and Rydälv, M. 1972. Arsenic levels in foods. *Var föda* 24:21–40.

Williams, A. I. 1972. The use of atomic-absorption spectrophotometry for the determination of copper, chromium and arsenic in preserved wood. *Analyst* 97:104–110.

Wills, R. A. 1948. Arsine gas poisoning. *Ind. Med. Surg.* 17:208.

Wilson, R. and Mangum, G. H. 1943. Acute hemolytic anemia in fertilizer workers: A new industrial hazard. *South. Med. J.* 36:212–218.

Windom, H., Stickney, R. and Smith, D. 1973. Arsenic, cadmium, copper, mercury and zinc in some species of north Atlantic finfish. *J. Fish. Res. Board Can.* 30:275–279.

Winkler, W. O. 1962. Identification and estimation of the arsenic residue in livers of rats ingesting arsenicals. *J. Assoc. Off. Anal. Chem.* 45:80–91.

Wood, J. M. 1975. The methylation of arsenic compounds. *Science* 187:765.

Woolson, E. A. 1972. Effects of fertilizer materials and combinations on the phytotoxicity, availability, and content of arsenic in corn (maize). *J. Sci. Food Agric.* 23:1477–1481.

Wullstein, L. H. and Snyder, K. 1971. Arsenic pollutants in the ecosystem. *Proceedings of the second international clean air congress.* Washington, D.C. pp. 295–301. National Society for Clean Air.

Yamashita, N., Doi, M., Nishio, M., Hojo, H. and Tanaka, M. 1972. Current state of Kyoto children poisoned by arsenic tainted morinaga dry milk. *Nihon Eiseigaku Zasshi* 27:364–399. (EPA translation No. TR108-74)

Yoshikawa, T., Utsumi, J., Okada, T., Moriuchi, Masana, M., Ozawa, K. and Kaneko, Y. 1960. Concerning the mass outbreak of chronic arsenic toxicosis in Niigata prefecture. *Chiryō* 42:1739–1749. (EPA translation No. TR109-74)

Chapter 4

TOXICOLOGY OF COPPER

C. H. Hill
Department of Poultry Science
North Carolina State University
Raleigh, North Carolina

Copper, because of the high reactivity of its ions, might be expected to be extremely toxic. As this metal is encountered in nature, however, its toxicity is relatively low except in two special circumstances: humans with Wilson's disease are especially sensitive to copper, and sheep can be afflicted with copper toxicosis under practical husbandry practices. Other species are very resistant to the deleterious effects of this metal. Some aspects of copper toxicity have recently been reviewed (Underwood, 1971; Bremner, 1974).

Under normal conditions ingested copper is absorbed from the stomach and upper part of the small intestine. The element passes into the blood stream and is relatively loosely bound, probably to albumin, and passes to the liver. The liver concentrates the copper first in the cytosol, then in the mitochondria, and finally in the nuclei, depending upon the dose and the length of time of ingestion (Gregoriadia and Sourkes, 1967; Evans et al., 1970; Lal and Sourkes, 1971a). Copper reappears in the plasma in the form of ceruloplasmin, a copper-containing protein, which has ferridoxase activity (Frieden, 1971). The function of this protein in copper metabolism is not clear although there is some evidence to suggest that it can exchange copper during the course of oxidative activity (Owen, 1965). The major excretory pathway is via the bile, although some copper evidently is excreted directly into the intestine. A small proportion of copper is excreted via the urine, ranging from 1 to 4% of the absorbed copper.

Wilson's disease represents a special case of copper toxicosis. This condition is an inborn error of human metabolism in which the liver, kidney, brain, and cornea accumulate large amounts of copper. Studies with ^{64}Cu have shown that normally 80% of an oral dose of the isotope can be recovered in the feces and less than 1% in the urine. In patients with Wilson's disease, on the other hand, excretion via the urine is increased, but this is more than offset by a decrease in excretion in the feces. This indicates that in such patients a positive copper balance would be achieved

123

at lower levels of dietary copper than would be the case in normal subjects.

The plasma copper levels of patients with Wilson's disease are lower than normal because of the reduced amount of ceruloplasmin found in most of these cases. The rate of copper uptake by the liver from the plasma is slower in patients than with normal controls (Osborne et al., 1963), but over a period of time more total copper is deposited in the affected organs. The increased liver copper concentration leads to cirrhosis, while the increased concentration of copper in the brain leads to death of neurons and resulting neurological symptoms. In the kidneys renal tubular damage results from the increased copper concentration. The concentration of copper in the cornea leads to a characteristic brown or green ring called the Kayser-Fleischer ring. The toxic manifestations of copper in patients with this condition are present with a normal dietary intake of copper of 2–5 mg daily, which leads to liver levels of ca. 200 ppm copper and brain levels of ca. 20 ppm copper on a wet weight basis.

Except for Wilson's disease, copper toxicosis is seldom encountered in humans. Deliberate or accidental ingestion of large amounts of copper has been reported. For instance, Chuttani et al. (1965) state that acute copper sulfate poisoning is common in Delhi since it is frequently used to commit suicide. Occasionally children are accidently poisoned with this salt. In cases reported the doses were estimated to range from 1 to 100 g. Under these conditions the symptoms included vomiting, diarrhea, jaundice, impaired liver function, hemoglobinuria, hematuria, oliguria, hypotension, coma, and death. All the symptoms were not observed in all patients, depending on the dose and the promptness of treatment.

Domestic animals are much more likely to sustain copper toxicosis than are humans because of agricultural practices, which include copper-containing drenches and salt licks and the deliberate feeding of high levels of copper to swine. In addition, the extreme sensitivity of sheep to copper can lead to copper toxicosis in this species under conditions in which other animals would not be affected.

In sheep chronic copper poisoning is characterized by the gradual accumulation of copper in the liver over a period of several weeks or months with no outward symptoms of toxicity (Todd and Thompson, 1963). This period is followed by a precipitous onset of hemolysis, hemoglobinemia, and hemoglobinuria accompanied by the development of jaundice. These occur coincident with the release of copper from the liver.

Ishmael et al. (1971, 1972) have studied the sequence of events leading up to the hemolytic crisis in sheep by dosing the animals with 1 g copper sulfate 5 days per week until the crisis occurred, 28–60 days later. Raised blood copper levels were found only immediately prior to or during the hemolysis. An increase in the concentration of sorbitol dehydrogenase, arginase, and glutamate oxaloacetate transaminase was detected 1 wk after the start of dosing. A marked increase in enzyme activity was detected immediately prior to the hemolytic crisis. Histological studies of biopsies found necrosis of isolated parenchymal cells and occasional swollen PAS-positive, diastase-resistant, copper-containing Kupffer cells, which were rich in acid phosphatase, some 6 wk prior to the hemolytic crisis. Thus, while the animals exhibit no outward signs of toxicity, it is evident that the liver begins to be adversely affected almost immediately. In acute toxicity studies in which as little as 25 mg copper was given intravenously to sheep, necrotic lesions were found in the liver 48 hr after dosing (Ishmael and Gopinath, 1972). In the animals that survived, these lesions were absent 6 days after dosing. Todd and Thompson (1964) have reported that intravenous injection of copper acetate to supply 80 or 160 mg copper caused the death of almost all the animals but that the symptoms resembled those of gastroenteritis rather than chronic copper poisoning.

In calves chronic copper poisoning was not observed until they were fed 500 ppm copper, some tenfold more than is required for sheep to be affected (Todd and Thompson, 1965). The clinical symptoms in the calves were much the same as seen in sheep: jaundice, methemoglobinemia, and hemoglobinuria. Plasma glutamic oxaloacetate transaminase and lactic dehydrogenase activities were increased 1 wk before the hemolytic crisis. Blood copper concentrations remained normal until the hemolytic crisis at which time five- to tenfold increases were recorded.

Swine have also been studied in relation to the effects of high copper levels. This investigation is particularly important in this species since it is a common practice in Great Britain to feed up to 200 ppm copper as a growth stimulant. Suttle and Mills (1966a) have reported that the addition of from 425 to 750 ppm copper to the diet of pigs produced a toxicosis characterized by decreased growth, microcytic hypochromic anemia, jaundice, increased serum aspartate transaminase, and increased serum copper. The latter two manifestations were transitory indicating that the pigs adapted somewhat to the high copper intake, but the anemia and growth depression persisted. The addition of 500 ppm zinc or 750 ppm iron to the diets containing 750 ppm copper eliminated the jaundice and elevated aspartate transaminase, and the addition of the iron supplement eliminated the anemia. These findings suggest that in swine part of the etiology of copper toxicosis may be related to an induced deficiency of zinc and iron.

Some studies have been conducted on the administration of high levels of copper to rats. Elliot and Bowland (1972) have found no deleterious effects by feeding 250 ppm copper to this species. Lal and Sourkes (1971b) administered copper intraperitoneally to rats over an 18-wk period. Even though hepatic copper concentration increased to 710 μg/g wet weight at the highest dose, 3.75 mg/kg body weight, after 18 wk the body weights were reduced but not markedly so. Rats are evidently not sensitive to copper toxicosis.

Smith et al. (1975) reported that yearling ponies fed up to 791 ppm copper for 6 months showed no toxic effects even though the liver concentration was 3,445 to 4,294 ppm copper on a dry weight basis. Blood copper concentration, serum glutamic oxaloacetate transaminase, and lactic dehydrogenase were not elevated indicating that no liver damage had occurred.

The findings of Suttle and Mills (1966a) indicating that the effects of toxic levels of copper could be alleviated by additional zinc and iron point up another aspect of copper toxicology that deserves further comment. Dietary constituents do not always act alone but are affected by the rest of the diet. This is true of copper, especially in sheep.

In Australia chronic copper poisoning of sheep has been observed especially among the British breeds grazing on pastures in which clover, *Trifolium subterraneum*, predominated. This plant contains only 10–15 ppm copper but is extremely low in molybdenum, 0.1–0.2 ppm. The low molybdenum level leads to a high copper status of the animals with consequent copper toxicosis. Providing a source of molybdenum leads to a reduction of copper levels of the liver and prevents the toxicosis under these conditions (Dick and Bull, 1945). The interaction between molybdenum and copper is apparently complicated and may lie partly in the formation of a copper-molybdenum complex that is unavailable to the animals (Dowdy and Matrone, 1968). There is also the possibility that molybdenum acts together with rumen bacterial sulfite reductase leading to the formation of sulfide, which could in turn form the cupric sulfide that may be available to the animal (Huisingh and Matrone, 1975). This suggests that sulfate levels may also affect the levels of copper necessary for toxicity.

Suttle and Mills (1966b) have reported that the kind of dietary protein also affects the degree of copper toxicosis in swine. Even though the diets were similar in total protein, zinc, and iron content and the copper content was ca. 600 ppm, the inclusion of white fish meal as part of the protein supplement led to a greater increase in serum copper, serum aspartic transaminase, and jaundice than when the protein supplements were soybean meal or dried skim milk. These differences were observed in spite of the fact that the liver copper levels of all the animals were the same.

It would seem, then, that copper toxicosis is not a major problem. There are a number of reasons for this. In humans the ingestion of excessive copper salts leads immediately to vomiting and the expulsion of the copper. The highly reactive ions of copper result in complex formations in nature so that the free ion is seldomly encountered. In addition, the animal has a great capacity to store copper in the liver without resultant damage so that short-term ingestion of considerable copper can be well tolerated.

REFERENCES

Bremner, I. 1974. *Q. Rev. Biophys.* 7:75–124.
Chuttani, H. K., Gutpa, P. S., Gulati, S. and Gupta, D. N. 1965. *Am. J. Med.* 39:849–854.

Dick, A. T. and Bull, L. B. 1945. *Austral. Vet. J.* 21:70–72.

Dowdy, R. P. and Matrone, G. 1968. *J. Nutr.* 95:191–197.

Elliot, J. I. and Bowland, J. P. 1972. *Can. J. Anim. Sci.* 52:97–101.

Evans, G. W., Myron, D. R. and Cornatzer, W. E. 1970. *Am. J. Physiol.* 218:298–300.

Frieden, E. 1971. In *Bioinorganic chemistry, Advances in chemistry series*, ed. R. F. Gould, pp. 292–321. Washington, D.C.: American Chemical Society.

Gregoriadia, G. and Sourkes, T. L. 1967. *Can. J. Biochem.* 45:1841–1851.

Huisingh, J. and Matrone, G. 1975. In *Proceedings of Symposium on molybdenum in the environment*, ed. W. Chattel and K. Peterson. New York: Marcel-Dekker. In press.

Ishmael, J. and Gopinath, G. 1972. *J. Comp. Pathol.* 82:47–57.

Ishmael, J., Gopinath, G. and Howell, J. McC. 1971. *Res. Vet. Sci.* 12: 358–366.

Ishmael, J., Gopinath, G. and Howell, J. McC. 1972. *Res. Vet. Sci.* 13:22–29.

Lal, S. and Sourkes, T. L. 1971a. *Toxicol. Appl. Pharmacol.* 18:562–572.

Lal, S. and Sourkes, T. L. 1971b. *Toxicol. Appl. Pharmacol.* 20:269–283.

Osborne, S. B., Roberts, C. N. and Walshe, J. M. 1963. *Clin. Sci.* 24:13–22.

Owen, D. A., Jr. 1965. *Am. J. Physiol.* 209:900–904.

Smith, J. D., Jordan, R. M. and Nelson, M. L. 1975. *J. Anim. Sci.* 41:1645–1649.

Suttle, N. F. and Mills, C. F. 1966a. *Br. J. Nutr.* 20:135–148.

Suttle, N. F. and Mills, C. F. 1966b. *Br. J. Nutr.* 20:149–161.

Todd, J. R. and Thompson, R. H. 1963. *Br. Vet. J.* 119:189–198.

Todd, J. R. and Thompson, R. H. 1964. *J. Comp. Pathol. Therap.* 74: 542–551.

Todd, J. R. and Thompson, R. H. 1965. *Br. Vet. J.* 121:90–97.

Underwood, E. J. 1971. *Trace elements in human and animal nutrition*, 3rd ed., pp. 101–106. New York: Academic Press.

Chapter 5

NICKEL TOXICITY

Forrest H. Nielsen
United States Department of Agriculture
Agricultural Research Service
Human Nutrition Laboratory
Grand Forks, North Dakota

INTRODUCTION

The first reported investigation of the toxic effects of nickel was published in 1826 by Gmelin, who found that the administration of large doses of nickel sulfate to rabbits and dogs by stomach tube resulted in severe gastritis and fatal convulsions and that a sublethal dosage of nickel sulfate by gavage produced cachexia and conjunctivitis. In the 150 yr following this report, numerous descriptions of the toxic effects of nickel have appeared. In recent years there have been several reviews on the subject. In this chapter the most pertinent material discussed in the excellent and comprehensive review by Sunderman et al. (1975) will be summarized and updated. Since a chapter on metal carcinogenesis appears elsewhere in this volume, nickel carcinogenesis will not be discussed here.

Nickel enters the body by four different routes—oral intake, inhalation, parenteral administration, and percutaneous absorption. The signs of nickel toxicity, the mechanisms behind the toxicological action of nickel, and the therapeutic measures to counteract nickel toxicity for these four routes of entry will be discussed.

NICKEL TOXICITY THROUGH ORAL INTAKE

The toxicity of nickel, or nickel salts, through oral intake is low, ranking with such elements as zinc, chromium, and manganese. Nickel salts exert their action mainly by gastrointestinal irritation and not by inherent toxicity (Schroeder et al., 1961). The basis for the relative nontoxicity of nickel appears to be a mechanism in mammals that limits intestinal absorption of nickel. In addition, nickel has little tendency to accumulate in tissues during lifetime exposure.

Large doses of oral nickel salts are necessary to overcome the homeostatic control of nickel. Nickel carbonate, nickel soaps, and nickel catalyst supplemented in the diet of young rats at 250, 500, and 1,000 μg/g for 8 wk, or nickel catalyst supplemented in the diet at 250 μg/g for 16 months, had no effect on the general condition or growth rate (Phatak and Patwardhan, 1950, 1952). These high levels of dietary nickel did result in higher tissue nickel contents. The highest concentration (140-360 μg/g) was found in bone. Other tissues contained 10-50 μg nickel/g.

Nickel acetate and nickel chloride may be somewhat more toxic to rats than the aforementioned materials. Whanger (1973) found that feeding 500 μg nickel acetate/g diet to weanling rats for 6 wk significantly depressed growth. High dietary levels of nickel acetate (500-1,000 μg/g) also significantly depressed hemoglobin concentrations, packed cell volumes, plasma alkaline phosphatase activity, and heart cytochrome oxidase activity. Other findings were increased nickel, iron, and zinc content in both plasma and red blood cells, increased nickel content in heart, liver, testes, and kidney (levels of 2-40 μg nickel/g tissue were found), and a two- to threefold increase in iron content in some of these tissues. Nickel accumulated in the soluble fraction of the liver and in the soluble fraction, nuclei, and debris of the kidney. An accumulation of iron occurred in all of the cellular fractions of these two tissues.

Clary (1975) found that feeding 225 μg nickel chloride/ml drinking water for 4 months was toxic to rats. The signs of toxicity were depressed growth, lower serum triglyceride and cholesterol concentrations, and less total urine, urinary calcium, and urinary zinc. High levels of nickel sulfate (25 mg/kg daily) orally administered to male rats for 120 days resulted in marked dystrophic histological changes in the testes, and the activities of succinic dehydrogenase, DPN-diaphorase and steroid-3β-dehydrogenase were depressed in this organ (Vulcheva et al., 1970).

Schroeder et al. (1974) and Schroeder and Nason (1974) found a much lower level of dietary soluble nickel (5 μg/ml drinking water) fed to rats of both sexes for life to be virtually innocuous, not affecting survival, longevity, incidence of tumors, or specific lesions. Feeding this level of nickel was associated with increased concentrations of chromium in heart and spleen and of manganese in kidney and decreased concentrations of copper in lung and spleen, zinc in lung, and manganese in spleen. Nickel did not accumulate in any tissue examined. Schroeder and Mitchener (1971) reported that a similar exposure of soluble nickel in drinking water to breeding rats may be moderately toxic during reproduction. When carried through three generations, 5 μg nickel/ml drinking water apparently caused increased perinatal death and the occurrence of runting. The size of the litters decreased with each generation, and few males were born in the third generation.

Nickel is apparently less toxic to mice than to rats. Weber and Reid (1969) found that 1,100 μg nickel/g diet in the form of nickel acetate depressed the growth of female mice, but 1,600 µg nickel/g diet was needed to reduce growth in males. Dietary nickel at the 1,100 µg/g level decreased the activities of liver cytochrome oxidase and isocitric dehydroganse. At the 1,600 µg/g level the following enzyme activities were depressed: liver NADH-cytochrome *c* reductase, heart cytochrome oxidase, and malic dehydrogenase and kidney malic dehydrogenase. These investigators also observed that these high levels of dietary nickel had no influence on body weights of adult mice; if these mice were allowed to breed, however, 1,600 µg/g dietary nickel significantly reduced the number of pups weaned per litter. Dietary nickel at 1,100 µg/g had no effect on reproduction, which is in contrast to the observations on rats reported by Schroeder and Mitchener (1971). Mice tolerate low levels of nickel acetate in their drinking water (5 µg nickel/ml) over their lifetime without apparent difficulty. In terms of growth, survival, and tumor incidence, nickel administered in this manner was essentially inert (Schroeder et al., 1963, 1964).

Nickel chloride administered orally to young male rabbits at 500 µg/day for 5 months resulted in decreased liver glycogen, increased muscle glycogen, and a prolonged hyperglycemia after a glucose load (Gordynya, 1969). Oral administration of 0.2 or 0.5 mg nickel sulfate/day for 150 days resulted in a dose-dependent increase in nickel in all organs and tissues of rabbits, with the highest increase observed in liver and muscle (Moiseeva, 1973).

When chicks were fed diets containing nickel, as either the sulfate or the acetate, a significant decrease in growth was observed at a level of 700 µg/g or above (Weber and Reid, 1968). Up to 300 µg/g the body weights were normal. Nitrogen retention decreased progressively above 500 µg/g. The higher dosages of nickel that caused growth depression also reduced food consumption. Thus when food consumption was equalized by pair feeding, 1,100 µg/g dietary nickel did not affect growth, although nitrogen retention was decreased.

O'Dell et al. (1970b, 1971b) studied nickel toxicity in young bovines employing four levels of supplemental dietary nickel: 0, 62.5, 250, and 1,000 µg/g as nickel carbonate. The signs of toxicity were similar to those for chicks. Feed intake and growth were slightly retarded by 250 µg nickel/g diet. At the dietary level of 1,000 µg/g feed intake and growth were greatly reduced. The calves, however, were not emaciated, but did appear to be younger than the calves fed the lower levels of supplemented nickel. The kidneys of the experimental animals were nephritic, and the degree of severity increased with nickel toxicity. The high dietary level of nickel lowered nitrogen retention and caused anorexia. The 1,000 µg/g dietary nickel also resulted in an increased nickel content in many tissues, with the exception of the liver and heart. The increase in tissue nickel concentration occurred even though food intake was reduced; thus the total nickel intake was not much greater than when the diet was supplemented with 250 µg nickel/g. Apparently, 1,000 µg nickel/g diet will overwhelm homeostatic control mechanisms.

In other experiments O'Dell et al. (1970a) found that nickel chloride was more toxic to calves than was nickel carbonate. They also found that either 100 μg nickel/g as the chloride or 500 μg nickel/g as the carbonate reduced the palatability of the diet for calves, but did not do so at half these amounts. Feeding lactating dairy cows 250 μg nickel/g concentrate ration had no significant effect on milk production, milk composition, animal health, or feed consumption (O'Dell et al., 1971a). Milk from the cows contained less than 0.1 μg nickel/ml.

Diets containing nickel at 250, 500, and 1,000 μg/g as nickel catalyst, nickel soap, and nickel carbonate were innocuous when fed to monkeys for 24 wk as judged by growth, behavior, hemoglobin concentration, red cell count, and white cell count (Phatak and Patwardhan, 1950). Nickel content of tissues or organ histopathology was not examined.

It appears that many of the signs of nickel toxicity are the result of reduced food intake, partially caused by reduced palatability. However, homeostatic mechanisms controlling nickel absorption and excretion can be disrupted, thus increasing the tissue levels of nickel. Since nickel is a divalent cation, it reacts with active groups on proteins. Such reactions might explain the findings of reduced tissue enzyme activities in animals fed high levels of nickel.

The chances of nickel toxicity occurring under usual circumstances by the oral route are remote given the large amount of nickel required to produce toxic effects by ingestion. However, if it does occur, an appropriate therapeutic measure is the reduction or elimination of the unnatural source of nickel.

NICKEL TOXICITY VIA PARENTERAL ADMINISTRATION

In contrast to nickel salts administered orally, nickel salts administered intravenously or subcutaneously are highly toxic. Most available data on parenteral LD_{50} for various nickel compounds have been tabulated (Sunderman et al., 1975). The LD_{50} ranges from 6 mg nickel oxide/kg given intravenously to dogs to 600 mg nickel disodium EDTA/kg administered intraperitoneally to mice. Gross signs of nickel toxicity as a result of parenterally administered nickel salts include gastroenteritis, tremor, convulsions, paralysis, coma, anaphylactoid edema, and death.

Sublethal, but toxic doses, of parenterally administered nickel salts had effects on several biochemical parameters. Subacute injection of nickel L-glutamate increased total lipids, phosphatides, free fatty acids, and cholesterol in the serum or plasma of rabbits (Fiedler and Hoffmann, (1970). Subacute injection of nickel chloride (5 mg nickel/kg) into rabbits inhibited ADP-induced platelet aggregation in plasma samples taken 4 and 24 hr postinjection (Fiedler and Herrmann, 1971). This effect was correlated with a parallel increase in plasma lipids. Joó (1968, 1969) found that 0.01, 0.015, or

0.25 g/kg nickel chloride given intravenously changed the molecular organization of the basement membrane in rat brain capillaries. Structural looseness and finger-like hypertrophy occurred. This was followed by various degrees of formation of collagen-like fibers in the substance of the basement membrane. At the time these changes occurred, he found that nickel chloride specifically inhibited ATPase activity in the capillary walls. Joó suggested that the inhibition of ATPase was the primary cause of the changes in the blood brain barrier.

Intravenous injection of nickel chloride into cats resulted in severe disturbances of heart excitability, conductivity, and automatism, and decreased Ca^{2+} and K^+ in the right atrium and Ca^{2+} in the left ventricle (Malyshev, 1971).

Parenterally administered nickel also affects glucose metabolism. Freeman and Langslow (1973) found that intraperitoneal nickel chloride (10 mg/kg) in chicks resulted in a transient but significant increase in plasma glucose after 15 min followed by a progressive fall to the hypoglycemic range 60 and 120 min postinjection. In chicks starved 16 hr prior to injection, a hyperglycemic response occurred within 15 min, which persisted throughout the 2-hr experimental period. Clary (1975) reported that a single intraperitoneal or intratracheal injection of nickel chloride to rats caused a rapid transient increase in serum glucose, a decrease in serum insulin, and glucosuria. When exogenous insulin was given simultaneously with the nickel challenge, the increase in serum glucose was prevented. Glucose turnover studies revealed that the effect of nickel appeared to be an inhibition of insulin release by the pancreas. High concentrations of nickel were found in the pituitary gland. Therefore it was suggested that the inhibition of insulin release was related to a secretion of certain pituitary hormones—growth hormone and adrenocorticotropin—in response to nickel. Support for this hypothesis was provided by *in vitro* studies of LaBella et al. (1973a). They found that nickel increased the rate of release of growth hormone and adrenocorticotropin and decreased the rate of release of prolactin by the pituitary. LaBella et al. (1973b) also reported that intravenous administration of nickel chloride (300–600 μg/kg) to chlorpromazine-treated male rats resulted in a 40% decrease in serum prolactin 30 min postinjection. This finding suggests a direct, specific inhibitory action of nickel on prolactin-secreting cells of the anterior pituitary.

Dormer et al. (1973) reported that nickel inhibited the release of amylase from rat parotids, insulin from mouse pancreatic islets, and growth hormone from bovine pituitary slices when release was evoked by a variety of stimuli both physiological and pharmacological. They suggested that nickel might have blocked exocytosis by interfering with secretory granule migration or membrane fusion and microvillus formation. An antagonism between nickel and calcium ions in stimulus–secretion coupling might also have been in part responsible.

Jasmin (1974) found that injections of nickel sulfide, chloride, or sulfate

induced anaphylactoid edema in rats. Intraperitoneally or intravenously injected nickel salts (5 mg/100 g) evoked erythema and exoserosis within 1–10
min. Mast cells in the subcutaneous tissue of edematous ear lobes were often
degranulated. Since it appears that nickel does not directly induce histamine
release from, or degranulation of, mast cells (Taubman and Malnick, 1975), it
has been suggested that nickel complexes with a preexisting protein to give a
substance capable of mast cell degranulation.

Possible mechanisms of action of parenteral nickel were suggested
previously, They include enzyme inhibition (ATPase) that might cause neurological abnormalities including tremor, convulsions, and coma; inhibition or
stimulation of hormone release or action; internal rearrangement of Ca^{2+} ions
in muscle that might cause paralysis and abnormal heart rhythm; and indirect
stimulation of mast cell degranulation that might result in anaphylactoid
edema.

Nickel toxicity induced through parenteral intake occurs only under
experimental conditions. Thus there has been essentially no investigation into
the therapeutic measures to be used in this type of toxicity. Injected nickel is
rapidly cleared from plasma or serum (Onkelinx et al., 1973). Thus if the
animal survives the initial toxic effect of the nickel salt, it will recover within
a few days.

Two means of parenteral nickel administration have not been discussed
in this section. The injection of nickel carbonyl gives results similar to those
caused by its inhalation (Hackett and Sunderman, 1967). This will be
discussed in the inhalation toxicity section. Nickel introduced in the body via
metallic devices and prostheses may result in dermatological responses. This
toxic response is similar to that observed with the percutaneous administration
of nickel to sensitive individuals and will be discussed in the section on
percutaneous absorption.

NICKEL TOXICITY CAUSED BY PERCUTANEOUS ABSORPTION

For more than 60 yr it has been known that contact with nickel, or
with solutions of nickel salts, can result in dermatitis. Herxheimer described
nickel dermatitis in industrial workers in 1912. The prevalence of nickel
dermatitis in modern society has been reviewed (Sunderman et al., 1975).
Nickel allergy, an important problem in everyday life, has the highest
incidence among women. The first described cases of nickel dermatitis were
observed in nickel miners, smelters, and refiners and was termed "nickel itch."
The clinical manifestations including an itching or burning papular erythema
in the web of the fingers, which spread to the fingers, the wrists, and the
forearms. Nickel dermatitis is usually papular or papulovesicular and has
characteristics similar to those of atopic dermatitis, rather than of eczematous
contact dermatitis.

Baer et al. (1973) reported that the incidence of allergic contact sensitivity to nickel, as determined by reaction to nickel sulfate on selected patient populations, has been quite constant over a period of 33 yr (1937–1970)—approximately 12%. In other surveys (Sunderman et al., 1975) the percentage of nickel reactivity ranged from 4 to 13%. These data indicate that a large number of persons are sensitive to cutaneously applied nickel.

Occupational sources of exposure to nickel include nickel mining, extraction and refining, plating, casting, grinding and polishing, nickel powder metallurgy, nickel alloys and nickel–cadmium batteries, chemical industry, electronics and computers, food processing, and nickel waste disposal and recycling (Sunderman et al., 1975). Persons in other occupations in which they may contact nickel include nickel catalyst makers, ceramic makers and workers, duplicating-machine workers, dyers, ink makers, jewelers, spark plug makers, and rubber workers.

Due to technological improvements and advances, nickel dermatitis is seen infrequently in major industries. However, it still is a problem in electroplating shops (Kadlec, 1969).

There are many sources of contact with nickel for persons who are sensitive to this metal. These include jewelry, coins, clothing fasteners, tools, cooking utensils, stainless steel kitchens and kitchenware, detergents, prostheses, and other medical appliances. That such sources of exposure to nickel can cause dermatitis was shown by Fisher and Shapiro (1956). Among the sources of nickel they reported as causing dermatitis were earrings, garter clasps, metal chairs, thimbles, needles, scissors, nickel coins, fountain pens, car door handles, necklace clasps, zippers, watch bands, bracelets, bobby pins, spectacle frames, safety pins, shoe eyelets, metal arch supports, eyelash curlers, and handbags. They found that they could patch test for nickel sensitivity by using nickel coins. Rostenberg and Perkins (1951) reported an individual who reacted to wearing a jacket having metal clasps. Stoddart (1960) reported two case studies of patients who reacted to nickel cannulae used for routine intravenous infusions. Tinckler (1972) reported a case of skin sensitivity to surgical skin clips. Reports of "internal" allergic contact dermatitis caused by nickel-containing prostheses have also been published (McKenzie et al., 1967; Barranco and Soloman, 1972).

The mechanism of sensitization for nickel is probably similar to that of other simple chemicals. That is, nickel ions contact the surface of the skin, penetrate the epidermis, and combine with a body protein. The body then reacts to this conjugated protein. It has been suggested that in all likelihood, the specificity of the reaction is determined primarily by the haptenic portion of the molecule, nickel; however, the carrier protein necessary to make the complex antigen need not be inert and may be the immunologic determinant (Sunderman et al., 1975). Several studies have shown that nickel ions can be leached from household and everyday items, medical equipment, and prostheses (Ferguson et al., 1962; Mears, 1966; Samitz and Klein, 1973; Katz and Samitz, 1975; Samitz and Katz, 1975) by physiological saline, blood, sweat, and other body fluids. Wells (1956) reported that nickel ions can penetrate the skin via the sweat ducts and hair-follicle ostia and that they have a special affinity for keratin. Based on histochemical evidence, Wells suggested that nickel is bound by the carboxyl groups of keratin. Kolpakov (1963), using cadaver skin as an experimental model, found the malpighian layer of the epidermis, the dermis, and the hypodermis were readily permeable to nickel sulfate. The greatest accumulation of nickel was found in the malpighian layer, the sweat glands, and the walls of the blood vessels. The stratum corneum was a barrier to the penetration of nickel sulfate. Spruit et al. (1965), also using cadaver skin, reported that nickel ions penetrate and are bound by the dermis. He suggested nickel bound by the dermis can serve as a reservoir for subsequent release of nickel ions. Since nickel is an ion of the first transition series, it is not unexpected that it would bind to a variety of proteins. Evidence for binding of nickel to albumin, lysine vasopressin, conalbumin, alpha-chymotrypsin, ribonuclease, myoglobin, pseudoglobin, and keratin has been reported (Cotton, 1964; Bryce et al., 1966; Tsangaris et al., 1968; Callan and Sunderman, 1973). Recent evidence has shown that nickel might participate in a latter step of the series of reactions necessary for the interaction of guinea pig complement with immune aggregate (Amiraian et al., 1974).

Once the diagnosis of nickel contact dermatitis has been established, removal of the patient from the source of nickel contact should be followed by the alleviation of the malady. Prevention of further dermatitis requires prevention of contact with nickel-containing items. Cloth or plastic substitutes should be used. Some protection may be given by introducing a physical barrier between the skin and the metal. Fingernail polish, lacquer (Fisher, 1967), or polyurethane coating (Moseley and Allen, 1971) has been used for such a barrier. Fisher (1964) reported that a film formed by spraying the skin with topical aerosol dexamethasone was effective in preventing nickel dermatitis. Kurtin and Orentreich (1954) reported an active chelating agent (such as disodium EDTA) applied in ointment form will inhibit the appearance of a positive nickel patch test applied over the ointment site.

NICKEL TOXICITY CAUSED BY INHALATION

Because of its industrial importance, nickel carbonyl toxicity has been extensively studied. Nickel carbonyl is a colorless, volatile liquid (boiling point 43°C) that is particularly dangerous if inhaled. A variety of pathological lesions can result from the inhalation of nickel carbonyl, depending on amount inhaled and the duration of exposure. Chronic exposure to nickel carbonyl (and respirable particles of nickel, nickel subsulfide, and nickel oxide) can cause cancer. This was reviewed (Sunderman et al., 1975) recently and is also discussed elsewhere in this volume. Signs and symptoms that occur in animals following acute exposure to nickel carbonyl include dyspnea, tachypnea, cyanosis, fever, apathy, anorexia, vomiting, diarrhea, and, occasionally, hind limb paralysis. Generalized convulsions are frequently a terminal event. The pathological lesions following experimental acute exposure in animals are summarized in Table 1. In lung (the primary target organ for nickel carbonyl) findings include edema, focal hemorrhage, capillary congestion, alveolar cell degeneration, inflammation, and interstitial fibrosis. The lesions in other organs are less severe than those in the lung. They include edema, congestion, focal hemorrhage, hydropic degeneration, inflammation, and vacuolization in such organs as adrenals, brain, kidney, liver, pancreas, or spleen. If death occurs, it usually happens between the third and fifth day after exposure.

The clinical manifestations of nickel carbonyl poisoning in 25 men and the relative occurrence of the symptoms and signs are shown in Table 2. Other symptoms and signs observed in human nickel carbonyl poisoning (Sunderman, 1970) include cough, hypernea, cyanosis, leukocytosis, and increased temperature and pulse rate. Terminally, patients frequently become delirious. Tseretili and Mandzhavidze (1969) reported hyperglycemia, glucosuria, and heptomegaly in patients with severe nickel carbonyl poisoning. The pathological findings in men who died from nickel carbonyl inhalation were similar to those of experimental animals (Sunderman et al., 1975). In humans death usually occurs 3–13 days after exposure and is attributed primarily to respiratory failure. Cerebral edema and punctate cerebral hemorrhages might also be contributing causes of death. Patients who recover from nickel carbonyl poisoning often have a long period of convalescence (see Table 2) during which time they fatigue easily.

Very small amounts of nickel carbonyl were required to produce the toxic manifestations in animals (Table 1). Armit (1908) found that inhaled nickel carbonyl is approximately 100 times more toxic than carbon monoxide. A lethal atmospheric concentration of nickel carbonyl for a man is 30 μg/ml for a 30-min exposure (Kincaid et al., 1956).

Inhaled or injected nickel carbonyl does not immediately decompose and can pass across the alveolar membrane in either direction without metabolic alteration (Sunderman et al., 1968; Sunderman and Selin, 1968; Kasprzak and Sunderman, 1969). Hackett and Sunderman (1967, 1968) and Sunderman and Selin (1968) suggested that pathological lesions in the lungs result from damage produced during transit of the nickel carbonyl across the alveolar epithelium. Sunderman, Jr. (1971) also proposed that acute toxicity of nickel carbonyl might be partly due to an inhibition of ATP utilization.

On the basis of clinical experience, sodium diethyldithiocarbamate (Dithiocarb) is currently the drug of choice for the treatment of nickel carbonyl poisoning (Sunderman and Sunderman, 1958; Sunderman, 1964, 1971). After the initiation of oral Dithiocarb therapy, the urinary excretion of nickel is promptly increased, and the clinical manifestations of nickel carbonyl poisoning are relieved in a few hours. With continued Dithiocarb therapy, the

TABLE 1 Pathologic Lesions after Acute Exposure of Experimental Animals to Nickel Carbonyl[a]

Authors	Route of administration	Animal	Dose	Observation period after exposure	Observations
Armit, 1908	Inhalation	Rabbit	1.4 mg/liter for 50 min	1–5 days	*Lungs:* intraalveolar hemorrhage, edema, and exudate and alveolar cell degeneration; *adrenals:* hemorrhages; *brain:* perivascular leukocytosis and neuronal degeneration
Barnes and Denz, 1951	Inhalation	Rat	0.9 mg/liter for 30 min	2 hr to 1 yr	*Lungs:* at 2–12 hr capillary congestion and interstitial edema; at 1–3 days massive intraalveolar edema; at 4–10 days pulmonary consolidation and interstitial fibrosis
Kincaid et al, 1953	Inhalation	Rat	0.24 mg/liter for 30 min	0.2 hr to 6 days	*Lungs:* at 1 hr pulmonary congestion and edema; at 12 hr to 6 days interstitial pneumonitis with focal atelectasis and necrosis, and peribronchial congestion; *liver, spleen, kidneys, and pancreas:* parenchymal cellular degeneration with focal necrosis
Sunderman et al, 1961	Inhalation	Rat Dog	1.0 mg/liter for 30 min	1–6 days 1–7 days	*Lungs:* at 1–2 days intraalveolar edema and swelling of alveolar lining cells; at 3–5 days inflammation, atelectasis, and interstitial fibroblastic proliferation; *kidneys and adrenals:* hyperemia and hemorrhage
Hackett and Sunderman, 1967	Intravenous	Rat	65 mg/kg	0.1 hr to 21 days	*Lungs:* at 1–4 hr perivascular edema; at 2–5 days severe pneumonitis with intraalveolar edema, hemorrhage, subpleural consolidation, hypertrophy and hyperplasia of alveolar lining cells, and focal adenomatous proliferation; at 8 days interstitial fibroblastic proliferation; *liver, kidneys, and adrenals:* congestion, vacuolization, and edema

| Hackett and Sunderman, 1968 | Intravenous | Rat | 65 mg/kg | 0.5 hr to 8 days | *Lungs:* ultrastructural alterations, including edema of endothelial cells at 6 hr and massive hypertrophy of membranous and granular pneumocytes at 2–6 days |
| Hackett and Sunderman, 1969 | Intravenous | Rat | 65 mg/kg | 0.5 hr to 6 days | *Liver:* ultrastructural alterations of hepatocytes including nucleolar distortions at 2–24 hr, dilatation of rough endoplasmic reticulum at 1–4 days, and cytoplasmic inclusion bodies at 4–6 days |

[a] As tabulated by Sunderman et al. (1975).

TABLE 2   Clinical Manifestations of Nickel Carbonyl Poisoning in 25 Men[a]

Immediate symptoms	Dyspnea (80%), fatigue (80%), nausea (76%), vertigo (44%), headache (36%), odor of "soot" in exhaled breath (36%), vomiting (24%), and insomnia and irritability (24%)
Latent period	In half of subjects, an asymptomatic interval between recovery from initial symptoms and onset of delayed symptoms
Delayed symptoms	Dyspnea with painful inspiration (80%), nonproductive cough (64%), muscular weakness (44%), substernal pain (44%), chilling sensations (32%), muscular pain (28%), sweating (24%), visual disturbances (12%), diarrhea (12%), abdominal pain (4%), muscle cramps (4%), and hypoesthesia in legs (4%)
Physical and X-ray findings	Tachypnea and tachycardia (80%), interstitial pneumonitis on X-rays (60%), fever (40%), and cyanosis (36%)
Laboratory findings	Pulmonary function tests consistent with interstitial lung disease (40%), increased serum glutamic pyruvic transaminase (36%), increased serum glutamic oxaloacetic transaminase (32%), and low arterial pO_2 (32%)
Clinical course	Interval before hospitalization: median, 2 days; range, 0–7 days. Duration of hospitalization: median, 6 days; range, 0–27 days. Interval before recovery: median, 38 days; range, 1–88 days. Symptoms that persisted for more than 3 wk: fatigue (88%), exertional dyspnea (52%), muscular weakness (48%), headache (36%), abdominal pain (36%), muscular pain (32%), sweating (24%), visual disturbances (16%), and muscle cramps (8%)

[a]From Sunderman et al. (1975). Based on observations of Vuopala et al. (1970).

patients show uneventful recoveries. Sunderman (1971) reported that there were no deaths among 50 men with acute nickel carbonyl poisoning and that all returned to work within 3 wk following treatment with Dithiocarb. In comparison, of 31 acute nickel carbonyl-poisoned patients treated with dimercaprol (Sunderman and Kincaid, 1954), two died, and the period of convalescence for most of the others lasted several months.

Sunderman (1971) recommended the following therapeutic regimen for patients with known or suspected acute nickel carbonyl poisoning.

1. In cases where the extent or severity of exposure is unknown, give 0.2 g Dithiocarb and 0.2 g sodium bicarbonate with water every 2 min until 2 g Dithiocarb has been given. The divided doses and sodium bicarbonate are required to prevent nausea. If the symptoms of nickel carbonyl poisoning are minimal, further therapy

considerations may be deferred until the nickel content of the urine is known.

2. In mild cases (initial 8-hr specimen of urine contains less than 10 μg nickel/100 ml), it is probable that delayed symptoms will not develop or will be minimal. No therapy is required; however, an occasional patient may complain of fatigue and require rest. If severe delayed symptoms develop unexpectedly, the patient should be hospitalized and treated the same as patients with moderately severe nickel carbonyl poisoning.

3. In moderately severe cases (initial 8-hr specimen of urine contains 10–50 μg nickel/100 ml) a dosage of 25 mg Dithiocarb/lb body weight should be administered the first day. For example, a 160-lb man should receive an initial 2 g (in divided doses with sodium bicarbonate as described in 1, above); 4 hr later another gram; 0.6 g at the eighth hour; and 0.4 g 16 hr after exposure. Dithiocarb therapy should continue at a dosage of 0.4 g every 8 hr until the patient is free of symptoms and the urine nickel content is normal. The patient should be under close observation for at least 1 wk as delayed poisoning symptoms may develop.

4. In severe cases (initial 8-hr specimen of urine contains greater than 50 μg nickel/100 ml), patients are usually quite ill and require hospitalization. Unless the clinical condition is critical, the patient may be treated in the same way as the moderately severely poisoned patient. Parenterally administered Dithiocarb (12.5 mg/lb body weight) is suggested for critical cases. Further dosage is governed by clinical evaluation. The total amount given in the first 24 hr after exposure may be increased to as much as 50 mg/lb body weight.

Patients receiving Dithiocarb should abstain from alcoholic beverages for 1 wk following therapy or they may experience symptoms similar to those observed in people who ingest alcohol after an antabuse treatment. Sedatives, such as paraldehyde and chloral hydrate, tranquilizers, and other psycho-pharmacologic drugs are contraindicated during Dithiocarb therapy.

Studies of the effects of inhalation of nickel compounds other than nickel carbonyl are limited. Bingham et al. (1972) observed that rats exposed to an aerosol of nickel oxide exhibited an increased number of alveolar macrophages. They also noted a marked increase in mucus in rats exposed to an aerosol of nickel chloride. Both nickel compounds produced lung changes that were visible with the light microscope. After long exposure of the rats to nickel oxide, some thickening of the alveolar walls was visible. Inhalation of nickel chloride resulted in hyperplastic bronchial epithelium. Eliseev (1975) reported that rats inhaling 0.005–0.5 mg nickel chloride/m^3 air for 6 months showed suppressed iodine binding activity of the thyroid gland. Nickel introduced via the respiratory route had a greater inhibiting effect on the thyroid than nickel administered orally.

McConnell et al. (1973) reported a case of asthma associated with the inhalation of nickel sulfate. The patient, who worked in a nickel-plating company, also had nickel dermatitis. Immunologic studies showed circulating antibodies to the nickel salt; controlled inhalation exposure to a solution of nickel sulfate reproduced the illness. In cases of nickel toxicity of this type, the best therapy is removal from the source of nickel.

SUMMARY AND CONCLUSIONS

Nickel can be a toxic element. The toxicological manifestations of nickel depend on the form, level of exposure, and mode of entry into the body. Nickel or nickel salts are relatively nontoxic when taken orally. Abnormally high levels of dietary nickel are required to overcome homeostatic mechanisms that control nickel metabolism. Nickel toxicity in humans via the oral route occurs only in extreme and unusual circumstances. Nickel toxicity from parenteral administration has been observed only under experimental conditions and, therefore, is not a practical consideration for man. Nickel toxicity manifested by nickel dermatitis is relatively common. Since many items in everyday use contain nickel, this type of nickel toxicity will always be present unless new and better methods of prevention and therapy are found. The most serious type of nickel toxicity is that caused by the inhalation of nickel carbonyl. Fortunately, the general population is not exposed to air containing this compound; acute nickel carbonyl poisoning usually occurs as a consequence of an industrial accident.

REFERENCES

Amiraian, K., McKinney, J. A. and Duchna, L. 1974. Effect of zinc and cadmium on guinea-pig complement. *Immunology* 26:1135–1144.

Armit, H. W. 1908. The toxicology of nickel carbonyl. Part II. *J. Hyg.* 8:565–600.

Baer, R. L., Ramsey, D. L. and Biondi, E. 1973. The most common contact allergens 1968–1970. *Arch. Dermatol.* 108:74–78.

Barnes, J. M. and Denz, F. A. 1951. The effect of 2,3-dermercaptopropanol (BAL) on experimental nickel carbonyl poisoning. *Br. J. Ind. Med.* 8:117–126.

Barranco, V. P. and Soloman, H. 1972. Eczematous dermatitis from nickel. *J. Am. Med. Assoc.* 220:1244.

Bingham, E., Barkley, W., Zerwas, M., Stemmer, K. and Taylor, P. 1972. Responses of alveolar macrophages to metals. I. Inhalation of lead and nickel. *Arch. Environ. Health* 25:406–414.

Bryce, G. F., Roeske, R. W. and Gurd, F. R. N. 1966. L-histidine-containing peptides as models for the interaction of copper (II) and nickel (II) ions with sperm whale apomyoglobin. *J. Biol. Chem.* 241:1072–1080.

Callan, W. M. and Sunderman, F. W., Jr. 1973. Species variations in binding of

^{63}Ni (II) by serum albumin. *Res. Commun. Chem. Pathol. Pharm.* 5:459–472.

Clary, J. J. 1975. Nickel chloride-induced metabolic changes in the rat and guinea pig. *Toxicol. Appl. Pharmacol.* 31:55–65.

Cotton, D. W. K. 1964. Studies on the binding of protein by nickel. With special reference to its role in nickel sensitivity. *Br. J. Dermatol.* 76:99–109.

Dormer, R. L., Kerbey, A. L., McPherson, M., Manley, S., Ashcroft, S. J. H., Schofield, J. G. and Randle, P. J. 1973. The effect of nickel on secretory systems. Studies on the release of amylase, insulin, and growth hormone. *Biochem. J.* 140:135–142.

Eliseev, I. N. 1975. Data for comparative hygienic characteristics of nickel entering the body by oral and respiratory routes (in Russian). *Gig. Sanit.* 2:7–9.

Ferguson, A. B., Jr., Akahoshi, Y., Laing, P. G. and Hodge, E. S. 1962. Characteristics of trace ions released from embedded metal implants in the rabbit. *J. Bone Joint Surg.* 44A:323–336.

Fiedler, H. and Herrmann, I. 1971. Veränderungen der Thrombozyten-aggregation durch Verabreichung von Metallsalzen an Kaninchen. *Folia Haematol. (Leipzig)* 96:224–230.

Fiedler, H. and Hoffmann, H. D. 1970. Über die Wirkung von Nickel(II)–L-Glutamat und verschiedenen Kobaltkomplexen auf das Verhatten einiger Lipidkomponenten bei Kaninchen. *Acta Biol. Med. Ger.* 25: 389–398.

Fisher, A. A. 1964. Steroid aerosol spray in contact dermatitis. *Arch. Dermatol.* 89:841–843.

Fisher, A. A. 1967. Management of selected types of allergic contact dermatitis through the use of proper substitutes. *Cutis* 3:498–505.

Fisher, A. A. and Shapiro, A. 1956. Allergic eczematous contact dermatitis due to metallic nickel. *J. Am. Med. Assoc.* 161:717–721.

Freeman, B. M. and Langslow, D. R. 1973. Responses of plasma glucose, free fatty acids and glucagon to cobalt and nickel chlorides by *Gallus domesticus. Comp. Biochem. Physiol.* 46A:427–436.

Gmelin, C. G. 1826. Experiences sur l'action de la baryte, de la strontiane, du chrome, du molybdene, du tungstène, du tellure, de l'osmium, du platine, de l'iridium, du rhodium, du palladium, du nickel, du cobalt, de l'urane, du cérium, du feret du manganèse sur l'organisme animal. *Bull. Sci. Med.* 7:110–117.

Gordynya, R. I. 1969. Effect of a ration containing a nickel salt additive on carbohydrate metabolism in experimental animals (in Russian). *Vop. Ratsion. Pitan.* 5:167–170.

Hackett, R. L. and Sunderman, F. W., Jr. 1967. Acute pathological reactions to administration of nickel carbonyl. *Arch. Environ. Health* 14:604–613.

Hackett, R. L. and Sunderman, F. W., Jr. 1968. Pulmonary alveolar reaction to nickel carbonyl. Ultrastructural and histochemical studies. *Arch. Environ. Health* 16:349–362.

Hackett, R. L. and Sunderman, F. W., Jr. 1969. Nickel carbonyl. Effects upon the ultrastructure of hepatic parenchymal cells. *Arch. Environ. Health* 19:337–343.

Herxheimer, K. 1912. Ueber die gewerblichen Erkrankungen der Haut. *Dtsch. Med. Wochenschr.* 38:18–22.

Jasmin, G. 1974. Anaphylactoid edema induced in rats by nickel and cobalt salts. *Proc. Soc. Exp. Biol. Med.* 147:289–292.

Joó, F. 1968. Effect of inhibition of adenosine triphosphatase activity on the fine structural organization of the brain capillaries. *Nature* 219: 1378-1379.

Joó, F. 1969. Changes in the molecular organization of the basement membrane after inhibition of adenosine triphosphatase activity in the rat brain capillaries. *Cytobios* 3:289-301.

Kadlec, K. 1969. The role of chromium and nickel in occupational dermatology. *Prac. Lekar.* 21:18-23.

Kasprzak, K. S. and Sunderman, F. W., Jr. 1969. The metabolism of nickel carbonyl-[14]C. *Toxicol. Appl. Pharmacol.* 6:237-246.

Katz, S. A. and Samitz, M. H. 1975. Leaching of nickel from stainless steel consumer commodities. *Acta Dermatol. Venereol.* 55:113-115.

Kincaid, J. F., Strong, J. S. and Sunderman, F. W. 1953. Nickel poisoning. I. Experimental study of the effects of acute and subacute exposure to nickel carbonyl. *Arch. Ind. Hyg.* 8:48-60.

Kincaid, J. F., Stanley, E. L., Beckworth, C. H. and Sunderman, F. W. 1956. Nickel poisoning. III. Procedures for detection, prevention, and treatment of nickel carbonyl exposure including a method for the determination of nickel in biologic materials. *Am. J. Clin. Pathol.* 26:107-119.

Kolpakov, F. I. 1963. Permeability of skin to nickel compounds (in Russian). *Arkh. Patol.* 25:38-45.

Kurtin, A. and Orentreich, N. 1954. Chelation deactivation of nickel ion in allergic eczematous sensitivity. *J. Invest. Dermatol.* 22:441-445.

LaBella, F., Dular, R., Vivian, S. and Queen, G. 1973a. Pituitary hormone releasing or inhibiting activity of metal ions present in hypothalamic extracts. *Biochem. Biophys. Res. Commun.* 52:786-791.

LaBella, F. S., Dular, R., Lemon, P., Vivian, S. and Queen, G. 1973b. Prolactin secretion is specifically inhibited by nickel. *Nature* 245: 330-332.

Malyshev, V. V. 1971. Effect of nickel on heart beat (in Russian). *Nauchn. Tr. Irkutsk. Gos. Med. Inst.* 113:140-141.

McConnell, L. H., Fink, J. N., Schlueter, D. P. and Schmidt, M. G. 1973. Asthma caused by nickel sensitivity. *Ann. Intern. Med.* 78:888-890.

McKenzie, A. W., Aitken, C. V. E. and Ridsdill-Smith, R. 1967. Urticaria after insertion of Smith-Petersen vitallium nail. *Br. Med. J.* 4:36.

Mears, D. C. 1966. Electron-probe microanalysis of tissue and cells from implant areas. *J. Bone Joint Surg.* 48B:567-576.

Moiseeva, S. Z. 1973. Level of nickel in the organs and tissues of rabbits when its content in their rations is varied (in Russian). *Sb. Rab. Leningr. Vet. Inst.* 33:122-126.

Moseley, J. C. and Allen, H. J., Jr. 1971. Polyurethane coating in the prevention of nickel dermatitis. *Arch. Dermatol.* 103:58-60.

O'Dell, G. D., Miller, W. J., Moore, S. L. and King, W. A. 1970a. Effect of nickel as the chloride and the carbonate on palatability of cattle feed. *J. Dairy Sci.* 53:1266-1269.

O'Dell, G. D., Miller, W. J., King, W. A., Moore, S. L. and Blackmon, D. M. 1970b. Nickel toxicity in the young bovine. *J. Nutr.* 100:1447-1454.

O'Dell, G. D., Miller, W. J., King, W. A., Ellers, J. C. and Jurecek, H. 1971a. Effect of nickel supplementation on production and composition of milk. *J. Dairy Sci.* 53:1545-1548.

O'Dell, G. D., Miller, W. J., Moore, S. L., King, W. A., Ellers, J. C. and Jurecek, H. 1971b. Effect of dietary nickel level on excretion and nickel content of tissues in male calves. *J. Anim. Sci.* 32:767-773.

Onkelinx, C., Becker, J. and Sunderman, F. W., Jr. 1973. Compartmental analysis of the metabolism of ^{63}Ni(II) in rats and rabbits. *Res. Commun. Chem. Pathol. Pharm.* 6:663–676.

Phatak, S. S. and Patwardhan, V. N. 1950. Toxicity of nickel. *J. Sci. Ind. Res.* 9b:70–76.

Phatak, S. S. and Patwardhan, V. N. 1952. Toxicity of nickel-accumulation of nickel in rats fed on nickel-containing diets and its elimination. *J. Sci. Ind. Res.* 11b:173–176.

Rostenberg, A., Jr. and Perkins, A. J. 1951. Nickel and cobalt dermatitis. *J. Allergy* 22:466–474.

Samitz, M. H. and Katz, S. A. 1975. Nickel dermatitis hazards from prostheses. *In vivo* and *in vitro* solubilization studies. *Br. J. Dermatol.* 92:287–290.

Samitz, M. H. and Klein, A. 1973. Nickel dermatitis hazards from prostheses. *J. Am. Med. Assoc.* 223:1159.

Schroeder, H. A. and Mitchener, M. 1971. Toxic effects of trace elements on the reproduction of mice and rats. *Arch. Environ. Health* 23:102–106.

Schroeder, H. A. and Nason, A. P. 1974. Interactions of trace metals in rat tissues. Cadmium and nickel with zinc, chromium, copper, manganese. *J. Nutr.* 104:167–178.

Schroeder, H. A., Balassa, J. J. and Tipton, I. H. 1961. Abnormal trace metals in man—Nickel. *J. Chron. Dis.* 15:51–65.

Schroeder, H. A., Vinton, W. H., Jr. and Balassa, J. J. 1963. Effect of chromium, cadmium and other trace metals on the growth and survival of mice. *J. Nutr.* 80:39–47.

Schroeder, H. A., Balassa, J. J. and Vinton, W. H., Jr. 1964. Chromium, lead, cadmium, nickel and titanium in mice: effect on mortality, tumors and tissue levels. *J. Nutr.* 83:239–250.

Schroeder, H. A., Mitchener, M. and Nason, A. P. 1974. Life-term effects of nickel in rats: Survival, tumors, interactions with trace elements and tissue levels. *J. Nutr.* 104:239–243.

Spruit, D., Mali, J. W. H. and DeGroot, N. 1965. The interaction of nickel ions with human cadaverous dermis. *J. Invest. Derm.* 44:103–106.

Stoddart, J. C. 1960. Nickel sensitivity as a cause of infusion reactions. *Lancet* 2:741–742.

Sunderman, F. W. 1964. Nickel and copper mobilization by sodium diethyl-dithiocarbamate. *J. New Drugs* 4:154–161.

Sunderman, F. W. 1970. Nickel poisoning. In *Laboratory diagnosis of diseases caused by toxic agents,* ed. F. W. Sunderman and F. W. Sunderman, Jr., pp. 387–396. St. Louis: Warren H. Green.

Sunderman, F. W. 1971. The treatment of acute nickel carbonyl poisoning with sodium diethyldithiocarbamate. *Ann. Clin. Res.* 3:182–185.

Sunderman, F. W. and Kincaid, J. F. 1954. Nickel poisoning. II. Studies on patients suffering from acute exposure to vapors of nickel carbonyl. *J. Am. Med. Assoc.* 155:889–894.

Sunderman, F. W. and Sunderman, F. W., Jr. 1958. Nickel poisoning. VIII. Dithiocarb: A new therapeutic agent for persons exposed to nickel carbonyl. *Am. J. Med. Sci.* 236:26–31.

Sunderman, F. W., Range, C. L., Sunderman, F. W., Jr., Donnelly, A. J. and Lucyszn, G. W. 1961. Nickel poisoning. XII. Metabolic and pathologic changes in acute pneumonitis from nickel carbonyl. *Am. J. Clin. Pathol.* 36:477–491.

Sunderman, F. W., Jr. 1971. Effect of nickel carbonyl upon hepatic concentrations of adenosine triphosphate. *Res. Commun. Chem. Pathol. Pharm.* 2:545–551.

Sunderman, F. W., Jr., and Selin, C. E. 1968. The metabolism of nickel-63 carbonyl. *Toxicol. Appl. Pharmacol.* 12:207–218.

Sunderman, F. W., Jr., Roszel, N. O. and Clark, R. J. 1968. Gas chromatography of nickel carbonyl in blood and breath. *Arch. Environ. Health* 16:836–843.

Sunderman, F. W., Jr., Coulston, F., Eichhorn, G. L., Fellows, J. A., Mastromatteo, E., Reno, H. T., Samitz, M. H., Curtis, B. A., Vallee, B. L., West, P. W., McEwan, J. C., Shibko, S. I. and Boaz, T. D., Jr. 1975. *Nickel*, pp. 97–143. Washington, D.C.: National Academy of Sciences.

Taubman, S. B. and Malnick, J. W. 1975. Inability of Ni^{++} and Co^{++} to release histamine from rat peritoneal mast cells. *Res. Commun. Chem. Pathol. Pharm.* 10:383–386.

Tinckler, L. F. 1972. Nickel sensitivity to surgical skin clips. *Br. J. Surg.* 59:745–747.

Tsangaris, J. M., Chang, J. W. and Martin, R. B. 1968. Cupric and nickel ion interactions with proteins as studied by circular dichroism. *Arch. Biochem. Biophys.* 130:53–58.

Tseretili, M. N. and Mandzhavidze, R. P. 1969. Clinical observations of acute carbonyl nickel poisoning (in Russian). *Gig. Truda Prof. Zabol.* 13:46–47.

Vulcheva, Vl., Zlateva, M. and Mikhailov, Iv. 1970. Changes in the testes of white rats after chronic action of nickel sulfate (in Russian). *Khig. Zdraveopazvane* 13:558–564.

Vuopala, U., Huhti, E., Takkunen, J. and Huikko, M. 1970. Nickel carbonyl poisoning. Report of 25 cases. *Ann. Clin. Res.* 2:214–222.

Weber, C. W. and Reid, B. L. 1968. Nickel toxicity in growing chicks. *J. Nutr.* 95:612–616.

Weber, C. W. and Reid, B. L. 1969. Nickel toxicity in young growing mice. *J. Anim. Sci.* 28:620–623.

Wells, G. C. 1956. Effects of nickel on the skin. *Br. J. Dermatol.* 68:237–242.

Whanger, P. D. 1973. Effects of dietary nickel on enzyme activities and mineral content in rats. *Toxicol. Appl. Pharmacol.* 25:323–331.

Chapter 6

TOXICOLOGY OF VANADIUM

Michael D. Waters

Biochemistry Branch
Environmental Toxicology Division
Health Effects Research Laboratory
U.S. Environmental Protection Agency
Research Triangle Park, North Carolina

INTRODUCTION

Vanadium, a common constituent of the earth's crust, is found at an average concentration of about 150 μg/g. Although the metal is widely distributed in living things, it rarely occurs in high concentrations. Man's activities in petroleum and metallurgical refining have magnified naturally high concentrations of the metal in certain crude oils and ores. The ill effects of human exposure to the combustion products of vanadium-bearing residual oils

M. D. Waters was author of the Biomedical Effects section of the Scientific and Technical Assessment Report (STAR) on Vanadium, Office of Research and Development, U.S. Environmental Protection Agency, Washington, D.C. 20460. Major portions of this chapter are included in that report. A report prepared for the U.S. Environmental Protection Agency by the National Academy of Sciences' Panel on Vanadium of Committee on Medical and Biological Effects of Environmental Pollutants (NAS, 1974) served as a primary reference for the STAR document. This report has been reviewed by the Environmental Protection Agency and approved for publication. Approval does not signify that the contents necessarily reflect the views and policies of the Agency.

The author wishes to acknowledge the contributions and helpful criticism of Dr. Robert J. M. Horton, chairman, and the members of the Task Force on Vanadium (Environmental Research Center, Research Triangle Park, N.C., ERC/RTP) responsible for writing the Scientific and Technical Assessment Report (STAR) on Vanadium. The author further acknowledges the assistance of the Special Studies Staff, ERC/RTP, in compiling and editing the STAR document and the assistance of the Word Processing Center, HERL (ERC/RTP), in typing the manuscript for this chapter. Special gratitude is due to Dr. Joellen L. Huisingh for contributing the introductory section on vanadium chemistry and to Drs. K. V. Rajagopalan and F. William Sunderman, Jr., for their valuable review and criticism of the manuscript.

and to fumes and dusts in metallurgical refining have stimulated renewed interest in vanadium toxicology.

Studies in animals and man have shown that the toxicity of vanadium compounds is principally a function of the route of administration or exposure. Vanadium salts are considered highly toxic in a number of species when given by intravenous or subcutaneous injection. As a rule, vanadium compounds cause marked irritation to the respiratory tract in inhalation exposures. Soluble forms of the metal are readily absorbed by the respiratory route, and moderate toxicity is observed. Most vanadium compounds are poorly absorbed from the gastrointestinal tract and, hence, exhibit a low order of toxicity by the oral route.

Reviews on the toxic and irritative properties of vanadium compounds have been published by Symanski (1939), Sjφberg (1950), Faulkner Hudson (1964), Stokinger (1967), Athanassiadis (1969), and Schroeder (1970a). The National Academy of Sciences (1974) has recently published a review describing the industrial sources of vanadium in the environment and its biological effects. The present review will emphasize current information on the metabolism and toxicology of vanadium in man and in experimental animals.

Discovery

Andrés Manuel del Rio of Mexico City was apparently the first to recognize the new metal named, in 1801, erythronium after the red color of its salts. However, principal credit for the discovery of vanadium must be given to Nils Gabriel Sefström, a Swede. With the aid of Berzelius in Stockholm, Sefström succeeded in preparing in 1831 an oxide of a new element that Berzelius named vanadium (Weeks, 1956). It was nearly 100 yr later that Marden and Rich (1927) made the first pure metal beads so that the physical properties of the metal could be determined. Vanadium (V) is a bright white ductile metal with a melting point of $1,980 \pm 10°C$, a boiling point of $3,380°C$ at 1 atm, and a specific gravity of 6.11 at $18.7°C$ (Weast, 1970). Some of the physical properties of important vanadium compounds are shown in Table 1 (Weast, 1970).

Aspects of Vanadium Chemistry

Vanadium is a transition element of the first transition series and belongs to group Va, which includes in addition niobium and tantalum. Like the elements in group Vb (nitrogen, phosphorous, arsenic, antimony, and bismuth) each of the group V elements have five valence electrons giving rise to a maximum oxidation state of +5. Although compounds of vanadium are known that contain vanadium in oxidation states of 0, +2, +3, +4, and +5, vanadium in the +4 and +5 valence states are most stable. Discrete V^{4+} and V^{5+} ions are not known to exist, and vanadium is usually found bound to oxygen as a negatively charged polymeric oxyanion that tends to complex

TABLE 1 Some Physical Properties of
Vanadium Compounds[a]

Compound	Melting point (°C)	Boiling point (°C)	Solubility in water (g/100 cm³)	
			Cold	Hot
Vanadium pentoxide	690	1,750	0.8	No data
Vanadium trioxide	1,970	No data	Slightly soluble	Soluble
Sodium metavanadate	630	No data	21.1	38.8
Vanadium tetrachloride	28 ± 2	148.5	Decomposes	No data
Vanadium oxychloride	No data	126.7	Decomposes	No data
Ammonium vanadate	200[b]	No data	0.52	6.95[b]

[a]Data from Weast (1970).
[b]Decomposes.

to polarizable ligands such as phosphorous and sulfur (Buckingham, 1973).

Vanadium pentoxide, V_2O_5, is the most common commercial form of vanadium and is the form commonly found in industrial exposure situations. It dissolves in water ($0.8g/100$ cm³), giving a pale yellow acidic solution containing vanadium species that are moderately strong oxidizing agents (Cotton and Wilkinson, 1962). The distribution of ionic species in such a solution is dependent on both pH and vanadium concentration. It is only in very dilute solutions that the vanadium ion remains monomeric. For example, at a pH between 5 and 9 and at a concentration of vanadium of less than 10^{-3} M, more than 50% of the vanadium ions will be present as $H_2VO_4^-$ (often written as VO_3^-); this form will be in equilibrium with the isopolymeric trimer, $V_3O_9^{3-}$. As the concentration of vanadium in the solution increases, the oxyanions condense to form the tetramer, $V_4O_{12}^{4-}$. Only in strongly acidic solutions (below pH 4) do cationic species such as VO_2^+ exist (Kepert, 1972).

Vanadium in the +5 oxidation state is reduced to vanadium +4 by relatively mild reducing agents. This would suggest that an organism exposed to vanadium in the +5 oxidation state would have an environment conducive to either nonenzymic or enzymic reduction of +5 vanadium to +4. Experiments by Johnson et al. (1974), discussed below, demonstrated that rats injected with vanadium in the +5 oxidation state contained vanadium in their tissues in the +4 oxidation state.

The +4 state is the most stable oxidation state for vanadium. Nearly all of the complexes of V^{4+} are derived from the vanadyl ion (VO^{2+}). Most of these complexes are anionic and a few are nonelectrolytes. Vanadium in this oxidation state forms a large number of five or six coordinate complexes such as vanadyl acetylacetonate, which exhibits a square-pyramidal structure (Earnshaw and Harrington, 1973), and vanadyl porphyrins found in crude petroleum.

Vanadium in the +3 oxidation state, as in V_2O_3, is completely basic and dissolves in acid to give the green hexaaquo ion $[V(H_2O)_6]^{3+}$. This is the species that probably accounts for the "green tongue" that is symptomatic of acute exposure to vanadium. Several bright green complexes of vanadium +4 are known (Cotton and Wilkinson, 1962; Durrant and Durrant, 1970), which might also account for the green tongue symptom. It should be noted that vanadium in the +3 state is a strong reducing agent that slowly attacks water with the liberation of hydrogen ions and the production of vanadium +4. Since the hexaaquo ion of vanadium III is easily oxidized to the +4 state, it would appear unlikely to be a long-lived species within biological organisms unless stabilized by complexation with organic ligands. Such is the case in certain marine organisms as noted in the following section.

Distribution in Nature

Vanadium is one of the more abundant trace elements in nature. Its geochemical and biochemical behavior is primarily a function of oxidation state (Goren, 1966). In the earth's crust relatively insoluble salts in the trivalent state are common. Natural sources of airborne vanadium are believed to be marine aerosols and continental dust. The contribution of volcanic action in producing airborne vanadium is believed to be small (Zoller et al., 1973).

Presence in soil, water, plant, and animal tissues. There are several reviews on vanadium in rocks, soils, and sediments (Bertrand, 1950; Vinogradov, 1959; Cannon, 1963; Pratt, 1966; NAS, 1974). The principal source of vanadium in soil is the parent rocks from which the soils are formed. Vanadium levels for soils reported in the literature range from 3 to 310 $\mu g/g$; the highest concentrations are found in shales and clays.

The concentration of vanadium in fresh water is largely dependent on the extent of leaching from soil and rocks. In the process vanadium is oxidized from the trivalent to the more soluble pentavalent state. Microorganisms are thought to play a significant role in the process of solubilization. In a spectrographic analysis by Kopp and Kroner (1968) of trace metals in rivers and lakes of the United States, the average for 54 positive samples was 40 $\mu g/liter$ with a range from 2 to 300 $\mu g/liter$. In drinking water a study by Durfor and Becker (1963) found that 91% of samples analyzed were below 10 $\mu g/liter$. Vanadium was undetectable in many samples; the maximum concentration found was 70 $\mu g/liter$ and the average was about 4.3 $\mu g/liter$. A large proportion of the soluble vanadium in rivers is removed by precipitation as the rivers reach the sea. Bowen (1966) has estimated that only about 0.0001% of the vanadium entering the oceans is retained in soluble form. Concentrations of vanadium in seawater range from 0.2 to 29 µg/liter. A biological mechanism as postulated by Krauskopf (1956) may be responsible for the removal of vanadium to the bottom sediments of the sea.

Estimates of vanadium concentrations in plants and animals are provided in Table 2 (Vinogradov, 1959; Schroeder, 1970a). In most cases the vanadium content of plants, animals, and humans is directly related to the physical environment. Roots of plants, for example, contain nearly the same content as the soil in which they are grown. Aerial portions of most plants are lowest in vanadium exhibiting concentrations unrelated to soil levels. Bertrand (1950) found vanadium in each of 62 plant species analyzed. The mean concentration in higher plants was 0.16 µg/g fresh weight, 1.0 µg/g dry weight, and 7 µg/g in ash. Cowgill (1973) determined that the vanadium concentration in fresh water plants ranged from 0.4 to 80.0 µg/g—the latter value in the pickerel weed (*Pontedaris cordata*), a probable accumulator of vanadium. The fly agaric mushroom (*Amanita muscaria*) is another known accumulator of vanadium (Bertrand, 1950).

Vanadium is found in both terrestrial and aquatic animals. However, in reviewing the literature prior to 1950 pertaining to animal exposure to vanadium, Bertrand (1950) considered it probable that vanadium was present in all animals, with the liver and skeletal tissues showing the highest amounts. Wild animals generally have higher amounts of vanadium in their tissues than humans; however, in contrast to humans, very little is found in their lungs. Limited data for several tissues of wild animals are provided in Table 3 (Schroeder 1970a). Fatty tissues of animals and humans appear to have a strong affinity for vanadium. Higher concentrations in hair, keratin, and gelatin probably result from ionic bonding to the many sulfhydryl groups present in these materials.

Using neutron activation analysis, Fukai and Meinke (1962) determined that the concentration of vanadium in soft tissues of fish was 1,000 times that

TABLE 2 Vanadium in Plants and Animals[a]

Plants	Vanadium concentration (µg/g dry weight)	Animals	Vanadium concentration (µg/g dry weight)
Plankton	5	Coelenterates	2.3
Brown algae	2	Annelids	1.2
Bryophytes	2.3	Mollusks	0.7
Ferns	0.13	Echinoderms	1.9
Gymosperms	0.69	Crustaceans	0.4
Angiosperms	1.6	Insects	0.15
Bacteria	+	Fish	0.14
Fungi	0.67	Mammals	<0.4

[a]Data from Schroeder (1970) and Vinogradov (1959).

TABLE 3 Vanadium in Tissues of Wild Animals[a]

Tissue[b]	Number of samples	Vanadium concentration (μg/g wet weight)	
		Mean	Range
Kidney	4	0.94	0.0-2.07
Liver	4	0.25	0.0-0.94
Heart	4	1.16	0.0-3.40
Spleen	1	1.16	

[a]Data from Schroeder (1970).
[b]Beaver, deer, woodchuck, rabbit, muskrat, and fox.

in sea weed and mollusks. The highest concentrations of vanadium in marine organisms have been found in certain sea squirts (e.g., *Phallusia mamillata*, 1,900 μg/g) certain sea cucumbers (e.g., *Sticopus mobii*, 1,200 μg/g), and a mollusk (*Pleurobranchus plumula*, 150 μg/g). Trivalent vanadium in certain ascidians is present as a chromoprotein called hemovanadin with sulfuric acid in green cells termed vanodocytes; in other forms, the hemovanadin is present free in the plasma (Faulkner-Hudson, 1964). It is of interest that vanadium in crude oil is largely present as an organometallic porphyrin compound. For this reason the metal is not readily removed with other mineral impurities during the refining process.

Essentiality of Vanadium

Vanadium is known to be essential for the mold *Aspergillus niger* (Bertrand, 1942) and the green alga *Scenedesmus obliguus* (Arnon and Wessel, 1953; Arnon, 1958). Vanadium may play a role in photosynthesis in the latter organism. A requirement for vanadium by the yeast *Candida slooffii* is demonstrable at high temperatures (Roitman et al., 1969). The growth effects of vanadium in *Azotobacter* and other bacteria have been related to the ability of vanadium in lieu of molybdenum to catalyze nitrogen fixation reactions (Horner et al., 1942; Takahashi and Nason, 1957; Nason, 1958). Vanadium does not appear to be essential in higher plants.

Trace amounts of vanadium, corresponding to those quantities normally found in nutrients and tissues, are apparently essential for the chick and rat. Several investigators (Hopkins and Mohr, 1971; Schwarz and Milne, 1971; Strasia, 1971; Nielsen and Ollerich, 1973) have demonstrated vanadium deficiency in the chick and rat. Diets containing less than 10 μg/kg resulted in significant reduction in feather growth in chickens but did not affect their growth rate (Hopkins and Mohr, 1971). A dietary concentration of 100 μg/kg diet was found to be nearly optimal to maintain normal growth in rats (Schwartz and Milne, 1971). Rats fed a diet deficient in vanadium produced fewer offspring in the third or fourth generation, compared with vanadium-supplemented controls (Hopkins and Mohr, 1974).

In a recent review by Hopkins and Mohr (1974) 100 μg/kg vanadium is suggested as a conservative minimum dietary vanadium requirement for animals fed purified diets. The requirement for animals consuming natural diets is thought to be significantly greater. Since many feed and food ingredients contain concentrations of vanadium far below 100 μg/kg, Hopkins and Mohr (1974) have concluded that marginal vanadium deficiency could be a nutritional problem. This is in contrast to the absence of demonstrable vanadium deficiency in the normal animal except where efforts have been made to prevent metal contamination.

HUMAN EXPOSURE

Industrial Exposure

Most of the information on human exposure to vanadium is of occupational origin. Vanadium and its compounds possess many valuable chemical and physical properties that have resulted in increased production and use in recent years. From the point of view of industrial hygiene, the most important vanadium compounds are vanadium pentoxide, vanadium trioxide, ferrovanadium, vanadium carbide, and vanadium salts such as sodium and ammonium vanadate. The oxides and salts are commonly used in industry in the powder form, giving rise to the possibility of dust and aerosol formation when the substances are crushed or ground. Many metallurgical processes involve production of vapors containing vanadium pentoxide that condense to form respirable aerosols. Boiler-cleaning operations generate dusts containing the pentoxide and trioxide compounds. Combustion of residual fuels having a high vanadium content is likely to produce aerosols of the pentoxide, as well as oxide complexes of vanadium with other metals. (Details on concentration and duration of industrial exposure are given in the section Respiratory Effects of Human Industrial Exposure.)

The portal of entry for vanadium in most industrial situations is the respiratory tract. The gastrointestinal tract is also an important portal of entry in the case of dusts and larger particles that are coughed up and swallowed. Soluble forms of vanadium can penetrate the skin (Stockinger, 1967), although this route of exposure appears to be of lesser significance industrially.

Community Exposure

The major route of nonindustrial exposure to vanadium is via the gastrointestinal tract from food and water. The oral intake of vanadium approximates that of copper. Tipton et al. (1969) have determined the intake of vanadium from food and water by three adults for 70 to 347 days (Table

TABLE 4 Balance of Vanadium in Humans[a]

	Mean	Range
Daily intake		
Food (μg)	100	6.1–170
Water (μg)	4	0–7
Air (μg)	3.1	0–10
Total	107.1	
Daily output		
Feces (μg)	102	37–150
Urine (μg)	18	12–23
Sweat, hair, etc.	?	
Balance (μg)	−18	
Absorbed (%)	16.8	14–30
Total in body (μg)	25.0	

[a]Data calculated from Tipton et al. (1969) with portions of table from Schroeder (1970a).

4). The calculated intake of vanadium from water is 4.0% of that from food, whereas the intake from air is about 3.1% of that from food. The proportion of vanadium supplied in water varies from 0 to about 7% depending on the water supply. Similarly, the proportion of the total vanadium intake attributable to inhaled air varies from 0 in most parts of the country to around 10% in New York City.

According to Schroeder (1970b) the average daily intake of vanadium from food and water is 116 μg. On the basis of 1966 National Air Sampling Network (NASN, 1965-1969) data, Schroeder has calculated (Table 5) that the maximum intake of vanadium from air in 44 cities of the United States

TABLE 5 Vanadium Intake in Urban and Nonurban Areas[a]

Cities studied	44
Nonurban areas studied	29
Range, urban (μg/m^3)	0.001–0.458
Range, nonurban (μg/m^3)	0.0005–0.024
In motor vehicle exhausts	+
Intake (μg/day)	
Air, maximum	9.16
Food and water	116
Total intake (% from air)	7.9
Retained in lung (μg/yr)	1.3
Maximum found in lung (μg)	680
Minimum found in lung (μg)	<10
Average total body content (μg)	22
Soil (μg/g)	100
Principal source	Petroleum combustion products

[a]Data from Schroeder (1970b).

and Puerto Rico is 9.16 μg. Thus, in this study 7.9% of the daily intake of vanadium came from air at maximum urban levels. Among the 27 metals listed by Schroeder, only mercury and lead showed a higher percent of intake from air than vanadium.

The concentration of vanadium in ambient air varies considerably throughout the United States, but the highest levels—annual averages exceeding 100 ng/m³—occur in metropolitan areas along the eastern seaboard (NASN, 1965-1969). The primary source of airborne vanadium in these areas is the fuel oil (much of which comes from the vanadium-rich crude oil of Venezuela) consumed in the generation of electricity and in space heating. Figure 1 represents the fate of vanadium in the urban air shed (Tullar and Suffet, 1973).

Airborne vanadium concentrations have increased in recent years, probably in response to the increasing use of oil as a fuel. Also, there is a

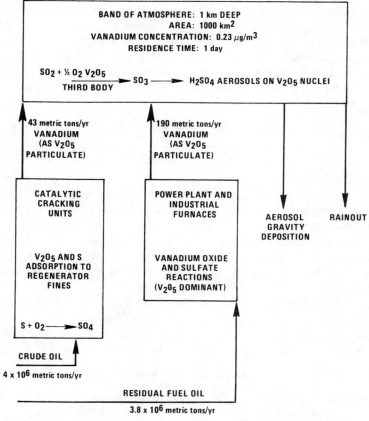

FIGURE 1 The fate of vanadium in the urban air shed (Tullar and Suffet, 1973).

strong seasonal pattern in vanadium levels (winter averages are more than triple summer averages) that undoubtedly reflects the seasonal pattern in fuel oil consumption for heating purposes (NASN, 1965-1969).

Maximum vanadium concentrations occur in areas of greatest population density, during the coldest part of the year, and during the late evening hours. Particle size distribution studies (Lee et al., 1972) have shown that most vanadium-bearing particulate matter is very small—well within the respirable range for human beings.

Medicinal Use

In the past vanadium compounds have been prescribed infrequently for medicinal purposes. In the early 1900s, vanadium was used as a therapeutic agent for anemia, chlorosis, tuberculosis, and diabetes. It was also used as an antiseptic, a spirochetocide, and a tonic to improve appetite, nutrition, and general health. Sodium metavanadate was given by mouth in doses of 1-8 mg, and the sodium tartrate was injected intramuscularly at levels as high as 150 mg. More recent studies on the effect of vanadium on blood cholesterol levels (Curran et al., 1959; Somerville and Davies, 1962) employed oral doses of soluble diammonium oxytartratovanadate for long periods to determine whether cholesterol levels could be lowered. Additional material on this subject will be found in the section Other Effects of Human Exposure.

METABOLISM

Gastrointestinal Absorption

As previously mentioned, vanadium salts are, in general, poorly absorbed from the human gastrointestinal tract. According to Curran et al. (1959), from 0.1 to 1% of 100 mg vanadium in the form of the very soluble diammonium oxytartarovanadate was absorbed from the human gut. Sixty percent of the absorbed vanadium was excreted via the kidneys within 24 hr in their study. The remainder of vanadium was retained in the liver and bone. It was mobilized rapidly from the liver and slowly from the bone following cessation of oral therapy.

In an unpublished study by Mountain (cited by Stokinger, 1967) vanadyl sulfate was fed to adult male rats in daily doses ranging from 650 to 1,250 μg (160-310 μg vanadium). The mean absorption in this case was about 0.5%, with considerable variation as judged by urinary values.

Respiratory Absorption

Specific information on the deposition of vanadium compounds in the respiratory tract following inhalation is not readily available. As with other particulate matter, deposition of vanadium would be expected to be greatest in the submicron particle size fraction. Soluble vanadium compounds are readily absorbed following inhalation and deposition in the lung. However, the rates of pulmonary absorption have not been quantitatively determined, and no estimate has been made of the amount of vanadium that is coughed up, swallowed, and then possibly absorbed through the gastrointestinal tract. Lewis (1959b) reported that workers exposed to airborne vanadium excreted four times as much vanadium in their urine as did controls.

Animal experiments have generally shown complete clearance of the relatively soluble vanadium pentoxide from the lung within 1–3 days following acute exposure (Sjøberg, 1950; Levina, 1972). Stokinger (1967) has demonstrated, however, that some metal is present for a month or more following cessation of chronic exposure.

Levina (1972) has reported that vanadium trioxide is cleared from the lung more rapidly than the pentoxide or ammonium vanadate following intratracheal instillation in rats. The fact that the pentoxide remains longest, she suggests, may be related to its "aggressiveness" or damage to the lung tissue.

In an autopsy study Tipton and Shafer (1964) have shown that vanadium accumulates in the lungs of the general population with age, reaching approximately 6.5 $\mu g/g$ wet weight of tissue in the greater-than-65 age group. The accumulation of vanadium with age was not observed in organs other than the lung. This is not surprising since vanadium is poorly absorbed from the gastrointestinal tract (Curran and Burch, 1967) and that which is absorbed is rapidly excreted via the urine. However, accumulation of vanadium in the lung suggests that some of the vanadium compounds deposited in the lung represent relatively insoluble forms (Schroeder, 1970a). Evidence of prolonged residence in the lung and hilar lymph nodes of complexed forms of vanadium is also provided by the work of Carlberg et al. (1971) with coal miners.

Percutaneous Absorption

Stokinger (1967) has reported that skin absorption occurs when a nearly saturated solution (20%) of sodium metavanadate is applied to rabbit skin. The application of this compound to the skin causes irritation in the rabbit. Human skin comes into contact with vanadium salts in the industrial environment, and skin sensitivity does develop (Sjøberg, 1950; Troppens, 1969). However, the skin appears to be a minor route of uptake of the metal.

Transport

Although vanadium is present in the serum, the carrier or "transvanadin" has never been isolated. Serum values for vanadium in 13 normal individuals ranged from 0.35 to 0.48 $\mu g/ml$ (mean 0.42 $\mu g/ml$) in a study by Schroeder et al. (1963). The vanadium was carried in the lipid fraction with none in the serum proteins. Schroeder et al. estimated that a total of 1,380 μg vanadium was circulating in the normal serum when samples were taken. They pointed out that this was approximately equivalent to the daily intake of patients on an institutional diet. This institutional diet apparently provided about 10 times the usual dietary intake of vanadium. Vanadium was not detected in washed red cells from 19 control subjects, but five individuals taking diammonium oxytartarovanadate by mouth (4.5 mg V/day) had elevated vanadium levels in red cells (0.37–0.81 ppm; mean 0.48 ppm) with equivalent amounts in serum (mean 0.47 μg/ml). Thus it appears that excess vanadium in the serum spills over to be adsorbed or absorbed by the red cells. The data of Schroeder et al. (1963) suggest that the serum carrier, if one exists, may be saturated or near saturated at levels (approximately 0.5 μg/ml) that may occur in the serum of individuals on institutional diets.

The influence of the oxidation state of intravenously injected compounds of vanadium-48 on uptake and distribution to selected organs and subcellular elements of liver in rats was studied by Hopkins and Tilton (1967). These investigators reported no significant differences in rate or amount of uptake of nanogram quantities of vanadium of three different oxidation states ($VOCl_3$, $VOCl_2$, and VCl_3). Hence, it appears either that the oxidation state is not critical to transport or that vanadium is somehow converted to a common oxidation state *in vivo*. Roshchin et al. (1965) have found some evidence for partial conversion of vanadium trioxide to the pentavalent form in blood serum and in weakly acidic and basic solutions *in vitro*. More recently, Johnson et al. (1974) have described the *in vivo* conversion of vanadium pentoxide to a tetravalent state (see below).

Distribution and Storage

Studies by Hopkins and Tilton (1967) in rats demonstrated that liver, kidney, spleen, and testes continued to accumulate intravenously injected vanadium-48 up to 4 hr and retained most of the radioactivity for up to 96 hr. Most others organs retained only 14–84% of their 10-min uptake when examined after 96 hr. After 96 hr, 46% of the vanadium-48 had been excreted in the urine and 9% in the feces. Differential centrifugation of homogenized liver showed that over the first 96 hr vanadium-48 in the mitochondrial and nuclear fractions increased from 14 to 40% of the total. Radioactivity in the microsomal fraction remained relatively constant.

Söremark et al. (1962) reported highest uptake of vanadium-48 from $^{48}V_2O_5$ in young rats in rapidly mineralizing areas of dentin and bone. In another study Söremark and Üllberg (1962) demonstrated high uptakes in mice in mammary glands, liver, renal cortex, and lung. In pregnant female rats Söremark et al. (1962) noted a concentration of radiovanadium in the fetus as well as in maternal bones and teeth. Thus it is important to note the potential exposure of the unborn offspring to vanadium via the maternal–fetal circulation. However, it should be mentioned that Schroeder et al. (1963) failed to find vanadium in the human fetus and detected the metal in only one infant of less than 6 wk old.

In $^{48}VOCl_3$ retention studies in rats Strain et al. (1964) found that hair retention correlates with that in aorta, bone, and liver, although levels in hair are much lower than in the three other tissues. Hair retention did not correlate with blood retention. Blood levels were undetectable after 20 days. Bone retention was highest in young rats of both sexes.

According to Schroeder et al. (1963) storage of available vanadium in man is mainly in fat and serum lipids. Large amounts of vanadium are also reported in crude fat from beef, pork, and lamb. Keratin has been suggested as a minor storage depot (Schroeder et al., 1963). As previously mentioned, animal tissues are generally low in vanadium. None was found in lungs of several animals despite the known accumulation in the lungs of man. Söremark (1967) has reported tissue levels ranging from less than 1 to several parts per billion (wet weight).

Excretion

Vanadium that is absorbed is excreted mainly in urine but also in feces. In long-term balance studies Tipton et al. (1969) demonstrated that approximately 17% of dietary vanadium is excreted in the urine (Table 6). Vanadium was not detected by two different methods of analysis in urines of 36 normal subjects (Tipton et al., 1969). However, when 4.5–9.0 mg vanadium as diammonium oxytartarovanadate was fed daily to 16 elderly persons, urinary excretion, although quite variable, amounted to a mean of 5.2% of the amount ingested (Tipton et al., 1969).

A study by Dimond et al. (1963) in which ammonium vanadyl tartrate was fed to young and middle-aged patients demonstrated that urinary excretion was unpredictable relative to oral dosage. These investigators suggested that variable absorption was the reason for wide fluctuations in urinary excretion. Jaraczewska and Jakubowski (1964) also concluded that concentrations of vanadium pentoxide in air in industrial exposures could not, with

TABLE 6 Excretion of Vanadium by Three Human Subjects[a,b]

Excretion route	Subject C		Subject D		Subject E	
	μg	% of total	μg	% of total	μg	% of total
Feces	120 ± 3.0	86	150 ± 2.0	83	37 ± 0.5	75
Urine	18 ± 0.2	13	23 ± 0.2	13	12 ± 0.2	24
Sweat, hair, etc.[c]	–	1	–	4	–	1
Total	140 ± 3.0	100	180 ± 2.0	100	49 ± 0.5	100

[a]Data from Tipton et al. (1969).
[b]Mean values given. Subjects C and D were studied for 347 days; subject E for 70 days.
[c]Inferred values to yield 100%.

certainty, be correlated with vanadium concentrations in urine. However, the latter authors pointed out the similarity in urinary clearance of intravenously injected and tracheobronchially administered sodium vanadate in guinea pigs.

Since most of the ingested vanadium is not absorbed, the preponderance of vanadium elimination is via the feces. Data from Tipton et al. (1969) on three normal individuals are shown in Table 6. An average of 81% of ingested vanadium was excreted via the feces, compared with 17% excreted via the urine.

Homeostasis

The relatively high serum levels (0.35–0.48 ppm) in normal individuals and low or undetectable urine levels reported by Schroeder et al. (1963) suggest that vanadium in serum is not filtered by the renal glomeruli. Alternatively, renal tubular reabsorption may be very efficient. The exception is when high levels of soluble vanadium are administered. Filtration or reduced reabsorption may function to eliminate vanadium when the ability of serum lipids or lipoproteins to bind vanadium is exceeded. Estimates of intestinal absorption following the administration of soluble vanadium compounds seem to indicate a proportional reduction in absorption with increasing dose. In any case, it appears that homeostatic mechanisms, possibly involving a low affinity transport system, are operative to maintain consistently low tissue levels of vanadium.

TOXICOLOGICAL EFFECTS OF HUMAN EXPOSURE

Respiratory Effects of Human Industrial Exposure

Most of the clinical symptoms observed following industrial exposures to vanadium reflect its irritative effects on the respiratory tract. Dutton (1911) is believed to be the first to have described the health effects of industrial exposure to vanadium-bearing ores. Dutton reported a dry, paroxysmal cough with hemoptysis and irritation of the eyes, nose, and throat. A temporary increase in hemoglobin and red cells was followed by a reduction in both and the onset of anemia. Vanadium was recovered in all bodily secretions. Postmortem examination revealed highly congested lungs with destruction of the alveolar epithelium and congested kidneys with evidence of hemorrhagic nephritis. Unfortunately, the workers studied frequently suffered from pulmonary tuberculosis, which undoubtedly produced many symptoms that were aggravated by vanadium exposure. Also, Dutton did not provide details as to the number of workers examined or the incidence of the symptoms he described.

A subsequent observation by Symanski (1939) of relatively healthy metal workers exposed to vanadium pentoxide dust revealed severe conjunctivitis, rhinitis, pharyngitis, chronic productive cough, and tightness of the chest. X-rays demonstrated bronchitis and Symanski expected bronchiectasis with longer exposure. Symanski's report differed from Dutton's (1911) in that the former found no evidence for a generalized systemic action of vanadium.

Rundberg (1939) observed bronchitis with purulent sputum, general weakness, and skin irritation of the face and hands of 20 men handling vanadium pentoxide in a metallurgical works. Balestra and Molfino (1942) reported productive cough, bronchitis, and shortness of breath in 25 workers exposed to vanadium pentoxide dust from petroleum ash. Other substances were involved, but chest X-rays showed definite lung markings suggesting pneumonoconiosis. Bronchiectasis was suspected in two cases.

Wyers (1946, 1948) described his supervision of between 50 and 90 workers exposed to vanadium pentoxide as an oil combustion residue and to slag from production of ferrovanadium. Findings included bronchospasm, often with elevated blood pressure and an accentuated pulmonary second sound, a proxysmal cough, dyspnea, skin pallor, tremor of fingers, palpitation, chest pains, and reticulation of the lungs. Thus Wyers emphasized the irritant effects of vanadium pentoxide on the respiratory tract but also found some evidence of systemic toxicity. An indication of exposure to vanadium first reported by Wyers was green tongue, believed to result from reduction of vanadium pentoxide to the trioxide in the mouth. Data in Wyers' report are limited to 10 case histories (that is, occurrence in one or two of 10 men observed), and no control group was examined.

An extensive report with data on the dust content of the air in a metallurgical plant producing vanadium pentoxide was published by Sjøberg (1950). The dust particles were relatively large in size—39% less than 12 μm, 22% less than 8 μm—and it was estimated that 6.5 mg V_2O_5/m^3 represented the worst exposure condition. Thirty-six men between age 20 and 50 yr had been employed in the plant since 1946: 22 had a dry cough; wheezing sounds could be detected in 31; and 27 were short of breath. One man developed acute pneumonitis, and four others developed bronchopneumonia. There was no concrete evidence of systemic toxicity.

A dry eczematous dermatitis developed in nine men in Sjøberg's (1950) study, but only one man showed a positive patch test. Sjøberg (1951) and Sjøberg and Ringer (1956) believed that allergy might play a role in the development of eczema and pneumonitis following vanadium exposure. Zenz et al. (1962) also considered this an explanation for the more severe symptoms found on reexposure in their study.

Six years later in a follow-up to his 1950 study, Sjøberg and Ringer (1956) reported that the 16 men most severely affected still complained of shortness of breath, coughing, fatigue, and wheezing. Bronchitis was present in two of the men. However, spirometric measurements, cardiac function tests, electrocardiograms, hematological tests, and urinalyses were essentially normal.

Bronchitis and conjunctivitis resulting from exposure to soot (containing 6-11% vanadium) in cleaning the stacks of oil-fired boilers were first

TABLE 7   Thermal-Precipitator Samples Taken from
Superheater during Cleaning Operation[a]

Diameter of particles (μm)	No. particles per ml air	Concentration of particles		
		No. %	mg/m^3 of air	wt. %
0.15–1.0	3,300	93.6	0.36	2.9
1.1–5.5	217	6.14	4.09	33.3
5.5–11	9	0.26	7.85	63.8

[a] Data from Williams (1952).

recognized by Frost (1951). Frost reported no systemic symptoms, but a subsequent report of a boiler-cleaning operation by Williams (1952) noted secondary symptoms of lassitude and depression. Within 0.5–12 hr of exposure primary symptoms were sneezing, nasal discharge, lacrimation, sore throat, and substernal pain. Within 6–24 hr secondary symptoms developed, consisting of dry cough, wheezing, labored breathing, lassitude, and depression. In some cases the cough became paroxysmal and productive. Symptoms lessened only after removal from the working environment for 3 days. Air sampling showed that most dust particles were smaller than 1 μm. The vanadium concentration ranged from 17.2 mg/m^3 in a superheater chamber to 58.6 mg/m^3 in a combustion chamber (Tables 7 and 8).

Other observations of boiler-cleaning operations were made by Fallentin and Frost (1954), Sjøberg (1955), Thomas and Stiebris (1956), Hickling (1958), Roshchin (1962), and Troppens (1969). Troppens describes the symptoms as "a slight cold condition (usually recognized as the flu) [which] ended in bronchitis. . . . the workers following recovery were tired, debilitated, irritable, without any appetite, and complained of watery eyes such as would come from a slight case of conjunctivitis." Troppens (1969) describes the first symptoms as swelling of face and eyes as early as 20 min after entering the boiler area. Removal from exposure for 2–3 wk resulted in disappearance of symptoms. Skin blemishes described as allergic dermatoses were attributed to

TABLE 8   Analysis of Dust from Boiler during Cleaning[a]

	Dust concentration (mg/m^3 air)	Vanadium concentration (% of dust)	Vanadium concentration (mg/m^3 air)
Superheater chamber	659	6.1	40.2
Superheater chamber	239	7.2	17.2
Combustion chamber	489	12.7	58.6

[a] Data from Williams (1952).

absorption of vanadium through sensitive skin. Vanadium in the urine was elevated 1.5- to threefold. Mention was made of increased susceptibility of the vanadium worker to asthmatic bronchitis and emphysema (twice as high as other workers according to data derived from Koelsch, reference not given).

With regard to all reports of respiratory symptoms relating to boiler-cleaning operations, it should be noted that sulfates and sulfuric acid are also present in boiler soot and may in part be responsible for irritative effects. Faulkner Hudson (1964) has suggested that quick onset of symptoms (lacrimation with nose and throat irritation) with rapid recovery following removal from exposure is characteristic of exposure to acid sulfates. Response to vanadium exposure is characterized by some delay in the onset of irritative symptoms (a few hours to several days) with persistence of symptoms following removal from exposure (Faulkner Hudson, 1964).

Additional reports have appeared relating to health effects of occupational handling of pure vanadium pentoxide or vanadate dusts. Among them are reports by Pielsticker (1954), Gulko (1956), Matantseva (1960), and Zenz et al. (1962). The report by Zenz et al. (1962) described a uniform acute illness that occurred in 18 workers pelletizing pure vanadium pentoxide; it was characterized by a rapidly developing mild conjunctivitis, severe pharyngeal irritation, a nonproductive persistent cough, diffuse rales, and bronchospasm. With severe exposure, four men complained of itching skin and a sensation of heat in the face and forearms. The symptoms became more severe after each exposure, suggesting a sensitivity reaction; however, duration of symptoms was not prolonged by the subsequent exposures.

Studies concerned primarily with mining, milling, and smelting operations have been published by Vintinner et al. (1955), Lewis (1959b), Rajner (1960), and Roshchin (1964). The investigation of Lewis (1959b) is particularly significant in that the maximum exposure was only 0.925 mg V (as V_2O_5)/m^3 of air, and in most cases 0.3 mg V/m^3 was the exposure level. More than 92% of dust particles were smaller than 0.5 μm in every process area sampled. Symptoms of cough with sputum production, eye, nose, and throat irritation, and wheezing were related to physical findings of wheezes, rales, or rhonchi, injected pharynx, and green tongue. All of these symptoms and physical findings were statistically significant, compared with controls (Tables 9 and 10). The report by Rajner (1960) on 30 vanadium workers in a metallurgical plant in Czechoslovakia describes particularly severe symptoms, but gives no estimates of exposure except in conjunction with urinary vanadium levels. In acutely affected workers vanadium values were about 4,000 μg/liter of urine. The average values among permanent employees was 45 μg/liter, but among vanadium pentoxide smelter workers, the average was 400 μg/liter. When a new production process was introduced, symptoms of acute vanadium intoxication occurring in these workers included severe respiratory difficulties, headaches, dejection, and loss of appetite after 16 hr

TABLE 9 Symptoms in Vanadium Workers[a]

Symptom	Incidence (%)		
	Control	Exposed	χ^2 value
Cough	33.3	83.4	13.71[b]
Sputum	13.3	41.5	5.55[c]
Exertional dyspnea	24.4	12.5	0.592
Eye, nose, throat, irritation	6.6	62.5	23.17[b]
Headache	20.0	12.5	0.124
Palpitations	11.1	20.8	0.538
Epistaxis	0	4.2	0.148
Wheezing	0	16.6	5.20[c]

[a]Data from Lewis (1959b).
[b]Significant beyond $p = 0.01$.
[c]Significant at $p = 0.02$.

of work. Acute inflammatory changes of the upper respiratory tract with copious mucus production, edema of the vocal cords, and profuse nose bleeding were reported. Workers who had been exposed up to 22 yr (27 of the workers), mostly of ferrovanadium and vanadium pentoxide smelting operations, complained of coughing and eye, nose, and throat irritation (all 27 workers), breathing difficulties during physical exertion (more than 14 cases), and headaches (12 cases). Clinical findings included intensive hyperemia of the mucosa of the nasal septum in 20 workers and perforation of the nasal septum in four workers. Intensive hyperemia of the mucosa of the throat and larynx with dilated fine capillaries was found in half the workers. Bronchoscopy indicated the presence of chronic bronchitis, and bronchial smears revealed sloughed epithelium. "Slight (pulmonary) functional disorders and beginning emphysema" were reported in five cases. Pneumoconiosis was not

TABLE 10 Physical Findings in Vanadium Workers[a]

Physical finding	Incidence (%)		
	Control	Exposed	χ^2 value
Tremors of hands	4.5	4.2	0.320
Hypertension	13.3	16.6	0.0002
Wheezes, rales, or rhonchi	0.0	20.8	6.93[b]
Hepatomegaly	8.9	12.5	0.003
Eye irritation	2.2	16.6	2.94
Injected pharynx	4.4	41.5	12.62[b]
Green tongue	0.0	37.5	14.53[b]

[a]Data from Lewis (1959b).
[b]Significant beyond $p = 0.01$.

detected on X-ray, and no heart changes or alterations in blood chemistry were reported.

Other occupational areas in which respiratory effects of vanadium exposure have been reported include operations connected with the gasification of fuel oil (Fear and Tyrer, 1958), with the maintenance of gas turbines (Browne, 1955), and in the manufacture of phosphor for television picture tubes (Tebrock and Machle, 1968). Elevated blood pressure was noted in the latter study in men exposed to vanadium pentoxide.

A recurrent inadequacy of all of these reports on industrial exposure to vanadium is the failure to consider or evaluate the influence of smoking on clinical findings.

In summary, the consensus of industrial hygienists is that there is insufficient evidence to support the view that vanadium, except at extremely high concentrations, causes generalized systemic toxicity. However, extensive evidence exists that vanadium dust (usually the pentoxide) is severely irritating to the mucous membranes of the eyes, nose, throat, and respiratory tract. Bronchitis and bronchospasm are characteristic symptoms, and pneumonia occasionally develops. Chronic productive cough and wheezing persist even after the subject is removed from exposure. The observation has been made by many reporters that vanadium workers are more susceptible to colds and other respiratory illnesses than others (Roshchin, 1962, 1967; Schümann-Vogt, 1969; Troppens, 1969). Recent studies by Waters et al. (1974, 1975) have demonstrated that vanadium salts are quite toxic for alveolar macrophages *in vitro*. Toxicity was related to the extent of dissolution of particles of V_2O_5, V_2O_3, and VO_2. In view of the important role of the alveolar macrophage in pulmonary defense, these studies suggest that exposure to vanadium may impair the lung's resistance to respiratory infection. Thus an attractive hypothesis to account for these reports and for many of the chronic respiratory symptoms is that vanadium exposure predisposes the individual to respiratory infection. While there is no evidence to the contrary, a chemico-bacterial basis for the respiratory symptoms observed in industrial workers remains to be demonstrated.

Maximum permissible concentrations. As defined by the American Conference of Governmental Industrial Hygienists (1961), threshold limit values (TLVs) "refer to air borne concentrations of substances and represent conditions under which it is believed that nearly all workers may be repeatedly exposed day after day without adverse effect.... Threshold limit values refer to time-weighted concentrations for a 7- or 8-hour workday and 40-hour workweek."

In 1961 the American Conference of Governmental Industrial Hygienists (ACGIH, 1961) adopted TLVs for vanadium compounds. In 1971 on the basis of additional data, new TLVs were established (ACGIH, 1971a). The TLVs were revised upward in 1972 in that concentrations were expressed as vanadium rather than vanadium pentoxide; however, the value for vanadium

M. D. Waters

pentoxide fume was designated a ceiling concentration. (A ceiling value should not be exceeded and is, in effect, a maximum allowable concentration.) Ceiling values were redefined in 1974 under the standards issued by the Occupational Safety and Health Administration (OSHA, 1974), Department of Labor. Table 11 lists TLVs for vanadium compounds for 1961, 1971, and 1972 and the OSHA standards issued in 1974.

An outline of the industrial hygiene literature cited in the *Documentation of Threshold Limit Values*, 3d ed. (ACGIH, 1971b), is given in Table 12. These data along with the experimental studies of Roshchin (1952), Stokinger (1973), and Zenz and Berg (1967) constitute the scientific evidence upon which the TLVs were based.

The 1974 OSHA standard for vanadium pentoxide dust and fume is compared with standards for some other metallic oxides in Table 13.

The TLVs for vanadium established mainly on the basis of the irritative effects of vanadium compounds on the respiratory mucosae. Schroeder (1970a), after Stokinger (1967), has ascribed the irritative effects of vanadium pentoxide to the acidity of its aqueous solutions. Vanadium trichloride is perhaps the most potent irritant, followed by vanadium pentoxide, ammonium and sodium vanadate, vanadium trioxide, and vanadium alloys, in that order. These irritative effects are tolerated by workers since they are not accompanied by pain and, except in severe exposure, do not develop rapidly.

TABLE 11   Threshold Limit Values[a] and Occupational Safety
and Health Administration Standards[b] for
Vanadium Compounds

Compound	Threshold limit values (mg/m^3)
1961	
Vanadium pentoxide (dust)	0.5
Vanadium pentoxide (fume)	0.1
Ferrovanadium (dust)	1.0
1971	
Vanadium pentoxide (dust) as V_2O_5	0.5
Vanadium pentoxide (fume) as V_2O_5	0.05
1972	
Vanadium pentoxide (dust) as V	0.5
Vanadium pentoxide (fume) as V	0.05 (ceiling)
1974	
Vanadium pentoxide (dust) as V_2O_5	0.5 (ceiling)[b]
Vanadium pentoxide (fume) as V_2O_5	0.1 (ceiling)[b]

[a]Data from ACGIH (1961, 1971a,b).
[b]Data from OSHA (1974).

TABLE 12 Industrial Exposures to Vanadium

Industry	Investigator	Vanadium compound	Concentration (mg/m^3)	Symptoms
Vanadium refinery	Sjøberg (1950)	V_2O_5	≤12	Mild respiratory irritation
Boiler cleaning	Sjøberg (1955)	$V_2O_5(V_2O_3)$	2–85	Respiratory irritation
Boiler cleaning	Williams (1952)	$V_2O_5(V_2O_3)$	30–104	Intoxication (used respirators to some extent)
Vanadium ore mining and processing	Vintinner et al. (1955)	V_2O_5	3–100	Local respiratory effects; no systemic poisoning
Boiler cleaning	McTurk et al. (1956)	V_2O_5	99	No intoxication (gauze filters worn)
V_2O_5 processing	Gulko (1956)	(V_2O_5)	0.5–2.2	Eye and bronchial irritation
Vanadium refinery	Lewis (1959)	(V_2O_5)	0.2–0.5	Respiratory irritation
Phosphor plant	Tebrock and Machle (1968)	Europium-activated yttrium orthovanadate YVO_4 : Eu	1.5 (as V_2O_5)	Conjunctivitis, tracheobronchitis, and dermatitis
Vanadium refinery	Faulkner-Hudson (1964)	V_2O_5 and NH_4VO_3	0.25	Green tongue, metallic taste, throat irritation, and cough

167

TABLE 13   Comparison of OSHA (1974) Standards
for Vanadium Pentoxide Dust and Fume with
Standards for Other Metallic Oxides

Compounds	TLV (mg/m^3)
Boron oxide	15
Calcium oxide	5
Iron oxide (fume)	10
Magnesium oxide (fume)	15
Osmium tetroxide	0.002
Vanadium pentoxide (dust)	0.5 (ceiling)
Vanadium pentoxide (fume)	0.1 (ceiling)
Zinc oxide (fume)	5

Respiratory Effects of Human Experimental Exposure

The previously cited study of Zenz and Berg (1967) involved nine healthy volunteers, aged 27–44, who had previously submitted to lung function tests for the purpose of developing base-line data. Two of the volunteers, exposed to vanadium pentoxide dust at 1 mg/m^3 for 8 hr, developed sporadic coughing after 5 hr and frequent coughing near 7 hr. Coughing lasted 8 days, but lungs remained clear and there were no other signs of irritation. Lung function tests, complete blood counts, urinalyses, and nasal smears were normal up to 3 wk. After this time the same two volunteers were accidently exposed to a "heavy cloud" of vanadium pentoxide dust for 5 min. A productive cough developed within 16 hr, and within 24 hr rales and expiratory wheezes developed throughout the lung. Pulmonary function remained normal. Isoproterenol (1:2,000) relieved the symptoms for about 1 hr. Then coughing began again and continued for 7 days. There were no symptoms. Eosinophils were not present in nasal mucus.

In the next test the exposure concentration was reduced to 0.2 mg/m^3 (± 0.05 SD) for 8 hr. Light microscopy indicated that 98% of the particles were smaller than 5 μm. Again, by the following day all five men exposed had developed a loose cough. Coughing, without other systemic symptoms, persisted 7-10 days. Pulmonary function tests and differential white blood counts remained normal. Vanadium in the urine was highest (0.013 mg/100 ml or 0.13 μg/ml) on the third day, with none detectable after 7 days. Maximal fecal vanadium was 3 μg/g, with none detectable after 14 days.

Finally, two volunteers were exposed to 0.1 mg/m^3 of vanadium pentoxide (0.056 mg V/m^3) for 8 hr. Within 24 hr considerable mucus had formed. The mucus was cleared by slight coughing, which became more severe after 48 hr, subsided after 72 hr, and disappeared after 96 hr. Zenz and Berg (1967) described the symptoms as a "distinct clinical picture of pulmonary irritation," despite the fact that pulmonary function tests and other clinical findings remained normal. All of the individuals exposed in the study by Zenz and Berg developed acute symptoms of marked pulmonary irritation upon initial exposure to vanadium pentoxide dust. Two individuals exposed a second time to the dust at 0.1 mg/m^3 (0.056 mg V/m^3) developed the irritative symptoms. Hence, the TLV for vanadium pentoxide dust and fume was lowered to 0.05 mg V/m^3. It is likely, however, that even at the new TLVs, healthy individuals will suffer symptoms of respiratory irritation. Furthermore, individuals with a history of respiratory disorders may be expected to experience distress before TLVs or OSHA standards are reached.

Diagnosis of Human Exposure

The respiratory symptoms resulting from vanadium exposure are so similar to those of acute infection of the respiratory tract that unequivocal diagnosis is difficult. Certain biochemical indices, along with evidence of probable exposure, can assist in diagnosis, but no specific test is recommended. Determination of the vanadium content in blood and urine provides definitive qualitative documentation of exposure. In view of the work of Schroeder (1963), it would seem desirable to measure the vanadium content of the serum separately from that of the cellular elements since the concentration of vanadium in the latter may be more indicative of exposure levels. A decreased urinary output of ascorbic acid may be one characteristic of vanadium exposure as reported by Watanabe et al. (1966) although differences from controls do not appear great enough to make the test useful clinically.

The most sensitive test developed to date involves measurement of the cystine content of the fingernails. This test was correlated to vanadium exposure in workers and has provided a useful clinical tool in the management of health of vanadium workers (Mountain et al., 1955). A decrease in cystine in fingernails was demonstrated when urinary vanadium levels were only 0.02–0.03 μg/ml. A similar reduction in the cystine content of rat hair was also reported (Mountain et al., 1953) when vanadium in the diet ranged from 25 to 1,000 ppm. There is some evidence to suggest that vanadium may directly inhibit the synthesis of cystine or cysteine (Mountain et al., 1953, 1955). Also, these reductions in cystine content in nails and hair may be related to excretion of cystine in the urine (cystinuria) since an increased neutral sulfur fraction, which is indicative of a cystinuria, is observed in the urine of rats fed vanadium (Mountain, 1963). Cystinuria is known to be associated with Wilson's disease, in which there is a genetically determined decrease in serum levels of the copper protein ceruloplasmin (ferroxidase I), a plasma globulin implicated in iron metabolism. A possible implication of increased cystine excretion in the urine would be that vanadium somehow interferes with copper metabolism. Keenan (1963) noted that in spectrographic analyses of livers of animals exposed to vanadium, the intensity of copper lines diminished as vanadium intensity increased. Conversely, de Jorge et al. (1970) have demonstrated decreased serum levels of vanadium and increased serum levels of copper and ceruloplasmin in Charcot-Marie muscular atrophy. These changes were proportional to symptom severity.

Health Effects of Community Exposure

Stocks (1960) reported the results of a statistical study in which airborne concentrations of 13 trace elements were correlated with mortality from lung cancer, pneumonia, and bronchitis in 23 localities in Great Britain. Beryllium and molybdenum proved to correlate best, whereas arsenic, zinc, and vanadium showed weak associations with mortality from lung cancer. After eliminating beryllium, molybdenum, and social index (taking into consideration population density, sex, and age), vanadium retained a coefficient of correlation with lung cancer of 0.347. With regard to mortality from pneumonia in the localities studied, beryllium was an important element in both sexes. Vanadium was also correlated with mortality from pneumonia, but only in males. When social index and beryllium were held constant, a significant coefficient of correlation with mortality from pneumonia (0.443) remained for vanadium. Molybdenum was strongly correlated with mortality involving bronchitis in both sexes. When social index and molybdenum were eliminated, vanadium gave a coefficient of correlation with mortality involving bronchitis of 0.563. Beryllium, molybdenum, and vanadium also showed associations with mortality from cancers other than of the lungs in males, but not in females.

Another statistical study by Hickey et al. (1967) considered 10 metals in the air, including vanadium, in 25 communities in the United States. Various techniques, including canonical analysis, were used to correlate airborne metal concentrations with mortality indices (1962 and 1963) involving eight disease categories. The incidence of several diseases, including "diseases of the heart," nephritis, and "arteriosclerotic heart," could be correlated reasonably well with air levels of vanadium, cadmium, zinc, tin, and nickel. The addition of vanadium to cadmium produced a reduction of more than 10% (the greatest reduction) in the error of variance. A high intercorrelation between vanadium and nickel was unexplained. Parenthetically, it is of interest that the consumption of hard water containing vanadium has statistically associated with a lower incidence of cardiovascular disease (Strain, 1961). Furthermore, Schroeder (1966) reported a significant negative correlation between the vanadium content of municipal waters and death rates due to arteriosclerotic heart disease.

These studies of Stocks (1960) and Hickey et al. (1967) are exploratory in nature. The relationships disclosed cannot be considered to be causal in nature without further study. The Hickey study is of a very preliminary nature, with no adjustments for the basic pertinent variables normally employed. The more extensive Stocks investigation considers a number of important adjustments (e.g., age, sex) and various interactions such as other metals and social factors. Other pollutants—smoking, for example—have not been examined. Intercorrelations were not fully explored.

An additional multivariate analysis of air vanadium levels in relation to selected white male mortality levels is contained in an unpublished Environmental Protection Agency staff study by Pinkerton et al. (1972). Several categories of cardiovascular disease were used, and also influenza-pneumonia. Vanadium was not correlated with the latter, but was correlated with the cardiovascular categories. However, adjustments for population density produced considerable reduction in some of these relationships. The author commented that "These results suggest that the observed statistical associations of air manganese and air vanadium were not causal associations, and represented either a reflection of other more directly associated causes or statistical artifacts."

Another difficulty with interpretation of these observations lies in their differences from observations on health effects seen in vanadium workers. Additional occupational and population studies on chronic illness in relation to vanadium exposure are needed to determine whether some consistency of findings and evidence of a dose–response relationship exists.

Other Effects of Human Exposure

As evidenced by the very high levels of vanadium prescribed medicinally at the turn of the century (see the Medicinal Use section), it is well established that the metal is not very toxic for humans when ingested. However, Faulkner Hudson (1964) has reported that lethal dose for a 70-kg (154 lb) person would be only 30 mg V_2O_5 (0.43 mg V_2O_5/kg) if introduced directly into the circulation in soluble form. No adverse health effects have been reported from ingestion of vanadium at any levels normally found in food or water. But Barannik et al. (1969) in a study of goiter in two areas in the Kiev region of the Soviet Union found food levels of vanadium and chromium to be higher in the goitrous area than in the comparison area.

Vanadium has been prescribed in recent years only in experimental investigations of its effects on circulating cholesterol levels. In 1959 Curran et al. conducted a clinical study in which five normal adult males were fed 150-200 mg/day soluble diammonium oxytartarovanadate (21-30 mg V/day) for 6 wk. At the end of this time plasma cholesterol was significantly reduced. Lewis (1959a) conducted a study of 32 vanadium workers and 45 controls; ages were matched in the two groups, and all persons were over age 40. The vanadium workers were observed to excrete greater than normal amounts of vanadium and exhibited slightly lower serum cholesterol levels than controls. Mean control cholesterol values were 230.9 and 226.7 mg/100 ml. Levels in the vanadium workers were less by 26 and 20 mg/100 ml ($p < 0.05$). A clinical study of Somerville and Davies (1962) of 12 patients (nine of whom were hypercholesterolemic) given diammonium vanadotartrate showed no significant changes in serum cholesterol levels over 5.5 months. The mean pretreatment control level of serum cholesterol was 411 mg/100 ml, and the mean age was 49.2 yr. Hence, in addition to being hypercholesterolemic, these individuals were older than those originally studied by Curran et al. (1959). Dimond et al. (1963) observed temporary drops (not statistically significant) in cholesterol in two of six patients given ammonium vanadyl tartrate for several weeks at levels between 50 and 100 mg/day. No statistically significant changes were observed in blood lipids, phospholipids, triglycerides, 17-ketosteroids, or 17-hydroxycorticosteroids. The subjective symptoms of fatigue and lethargy were present in two patients while they were taking vanadium. All complained of cramps and loosened stools, and all developed green tongue. Schroeder et al. (1963) reported that their findings were similar to those of Dimond et al. (1963) and expressed the view that the slight effects of vanadium on serum cholesterol were pharmacological and not caused by correction of a physiological deficiency. They pointed out that dietary regimens based on the consumption of unsaturated fats, which lower plasma cholesterol in humans, are associated with the intake of 1–4 mg V/day and that the feeding of vanadium-poor, saturated fats raises cholesterol. Hence, they note the interesting possibility that the ratio of vanadium to fat may be a factor in the homeostasis of circulating cholesterol.

TOXICOLOGICAL EFFECTS ON EXPERIMENTAL ANIMALS

General Toxicity

The toxicity of vanadium salts for several experimental animals and for humans by different routes of administration is illustrated in Table 14 (Browning, 1962; Schroeder et al., 1963; Schroeder, 1970a). The smaller animals, including the rat and mouse, tolerate the metal fairly well. The rabbit, horse, and human are more sensitive (Faulkner Hudson, 1964). In general, toxicity is low when the metal is given by the oral route, moderate by the respiratory route, and high by injection. Lethal doses for various vanadium salts injected intravenously in rabbits and subcutaneously in guinea pigs, rats, and mice are shown in Table 15 (Faulkner Hudson, 1964). The toxicity of vanadium also varies considerably with the nature of the compound, although vanadium is toxic both as a cation and as an anion. As a general rule, toxicity increases as valency increases, pentavalent vanadium being the most toxic. Among the oxides of vanadium, the pentavalent vanadium pentoxide is more soluble and more toxic than the less common trioxide or dioxide.

According to Roshchin (1967) the toxic effects of vanadium compounds in experimental animals are highly specific; however, at present the mechanisms of these effects are incompletely defined.

TABLE 14 Toxicity of Vanadium Salts to Mammals[a]

Method and duration of administration	Animal	Valence	Amount	Effect
Oral (in diet), months	Rat	5	1,000 ppm	Severe
Oral (in diet), life	Rat	5	100 ppm	Moderate
Oral (in water), life	Rat	5	8 ppm	None
Oral (in diet), months	Man	5	13.5–18 ppm	Slight
Inhalation, 2 hr daily, months	Rat	3	40–70 mg/m^3	Severe
	Rabbit	3	40–70 mg/m^3	Severe
	Rat	5	18 mg/m^3	Severe
	Rabbit	5	18 mg/m^3	Severe
	Man	–	1–50 mg/m^3	Slight
	Man	–	33–66 mg/m^3	Lethal (estimated)
Parenteral intraperitoneal	Rat	3	4–5 mg/m^3	Lethal
Subcutaneous	Rat	5	6–100 mg/kg	Lethal
Intravenous acute	Rat	5	6 mg/kg	Lethal
	Rabbit	5	1–20 mg/kg	Lethal
	Cat	5	7.18 mg/kg	Lethal
	Man	5	16.8 mg/km	Lethal (estimated)

[a]Data from Schroeder et al. (1965, 1970a) and Browning (1962).

Respiratory Effects and Histopathological Alterations

A number of studies that have dealt with respiratory exposure to vanadium pentoxide are outlined in Table 16 (Sjøberg, 1950; Gulko, 1956; Roshchin 1963; Pazynich, 1966). Sjøberg has reported in great detail experiments in which rabbits were exposed to vanadium pentoxide dust particles, nearly all of which were smaller than 10 μm in diameter. High concentrations over short periods of time were relatively toxic; 205 mg V_2O_5/m^3 (or 115 mg V/m^3) was lethal in 7 hr. At these levels tracheitis was marked and was accompanied by pulmonary edema and bronchopneumonia.

TABLE 15 Lethal Doses (mg V_2O_5/kg)[a]

Compound	Rabbit[b]	Guinea pig	Rat	Mouse
Colloidal vanadium pentoxide	1–2	20–28	–	87.5–117.5
Ammonium metavanadate	1.5–2.0	1–2	20–30	25–30
Sodium orthovanadate	2–3	1–2	50–60	50–100
Sodium pyrovanadate	3–4	1–2	40–50	50–100
Sodium tetravanadate	6–8	18–20	30–40	25–50
Sodium hexavanadate	30–40	40–50	40–50	100–150
Vanadyl sulfate	18–20	34–45	158–190	125–150
Sodium vanadate	–	30–40	10–20	100–150

[a]Data from Faulkner Hudson (1964).
[b]Rabbits were injected intravenously, other animals subcutaneously.

TABLE 16 Respiratory Effects of Vanadium Pentoxide in Experimental Animals[a]

Investigator	Animal	Form	Concentration (mg/m^3)	Time	Pathological findings
Sjøberg (1950)	Rabbit	Dust	205	7 hr	Conjunctivitis, tracheitis, pulmonary edema, broncho-pneumonia, enteritis, fatty liver, death
Sjøberg (1950)	Rabbit	Dust	20–40	1 hr/day, several months	Chronic rhinitis, tracheitis, emphysema, atelectasis, bronchopneumonia, pyelonephritis
Gulko (1956)	Rabbit	Dust	10–30	Continuous, acute	Bronchitis, pneumonia, weight loss, bloody diarrhea
Roshchin (1963)	Rat	Dust, fume	80–700 10–70	Continuous, acute	Hemorrhagic inflammation in lungs, hemorrhage in internal organs, paralysis, respiratory failure, death
Roshchin (1963)	Rat	Dust, fume	10–30 3–5	2 hr/day, several months	Hemorrhagic inflammation in lungs, purulent bronchitis, pneumonia
Pazynich (1966)	Rat	Fume	0.027	Continuous, 70 days	Hemorrhagic inflammation in lungs, vascular conges-tion, and hemorrhage in liver, kidneys, and heart

[a]Data from Sjøberg (1950), Gulko (1956), Roshchin (1963), and Pazynich (1966).

174

Conjunctivitis, enteritis, and fatty infiltration of the liver were also observed. Vanadium was detected in ashed lung, liver, kidney, and intestine.

Sjøberg (1950) also carried out long-term studies in which rabbits were exposed to 20-40 mg V_2O_5/m^3 (or 11-22 mg V/m^3) intermittently for 1 hr each day for several months. Upon sacrifice of the animals, pathological changes observed included chronic rhinitis and tracheitis, emphysema, patches of lung atelectasis, and bronchopneumonia. Pyelonephritis was seen in some cases. Vanadium was detected in ashed lung, liver, and kidney but not, as with heavy exposure, in the intestine. There were no fibrotic changes or specific chronic lesions in the lungs, nor was there a visible accumulation of particles. These findings, plus the presence of vanadium in the liver and kidney, were evidence of rapid clearance and/or absorption from the lung.

Gulko (1956) showed that continuous exposure of rabbits to 10-30 mg V_2O_5/m^3 (5.6-16.8 mg V/m^3) was toxic to the animals and caused bronchitis, pneumonia, loss of weight, and bloody diarrhea.

Roshchin (1963) described the results of acute inhalation studies using rats to which vanadium pentoxide was administered at 10-70 mg/m^3 as the condensation aerosol (fume) or 80-700 mg/m^3 as the grinding aerosol (dust), where ammonium vanadate was given (presumably as the grinding aerosol) at 1,000 mg/m^3, and where ferrovanadium was given as the grinding aerosol at 1,000-10,000 mg/m^3. The minimum concentration of vanadium pentoxide, in the form of a condensation aerosol, that gave rise to mild signs of acute poisoning was 10 mg/m^3 air. The absolute lethal concentration for the condensation aerosol was 70 mg/m^3. Grinding aerosols (containing large particles) were only one-fifth as toxic as condensation aerosols. Inhalation of grinding aerosols of ferrovanadium did not produce acute toxicity, perhaps because particles were too large. Acute toxic effects, however, were observed following intratracheal installation of ferrovanadium, which may be related to its biological solubility and degree of absorption.

Acute inhalation toxicity was characterized by irritation of the respiratory mucosa, with nasal discharge, sometimes containing blood. Animals breathed with difficulty and with crepitations. Behavior was passive; the animals refused to eat and lost weight. Dysentery, paralysis of the hind limbs, respiratory failure, and death ensued in cases of severe poisoning. Pulmonary abscesses were found frequently in animals that recovered. Animals that died or were killed at various times after exposure showed severe congestion, particularly in the capillaries, and tiny hemorrhages in all internal organs. There was evidence of increased intracranial pressure. Livers and kidneys showed fatty degeneration. Lungs showed capillary congestion, tiny hemorrhages, perivascular and focal edema, bronchitis, and focal interstitial pneumonia. The bronchitis and bronchopneumonia were often purulent and the small bronchi were constricted. The severity of pathological changes could be related to vanadium content in the air; in the cases of slight toxicity the pathological changes were mainly observed in the lungs.

When rats were exposed intermittently to a condensation aerosol of vanadium pentoxide 2 hr every other day for 3 months at 3-5 mg/m^3 (or a grinding aerosol of V_2O_5 at 10-30 mg/m^3 for 4 months), pathological changes were seen only in the lungs. Blood vessels of the lungs were engorged with blood and showed a swollen endothelium; capillary congestion, perivascular edema, lymphostasis, and tiny hemorrhages indicated altered vascular permeability and disturbances of pulmonary blood and lymph circulation. Foci of edema were sometimes seen, and in some cases there was desquamative bronchitis. Small bronchi were often constricted. Interstitial tissue was infiltrated by lymphocytes and histiocytes. Connective tissue proliferation was sometimes seen in the zone of lymphocytic infiltration. Some animals showed purulent bronchitis or pneumonia, and occasionally lung abscess developed.

Similar effects were observed with vanadium trioxide and vanadium trichloride (Roshchin, 1967); the latter compound, being more soluble, showed more marked histopathological effects on internal organs. Pentavalent compounds of vanadium were three to five times more toxic than those of trivalent vanadium (in terms of median lethal concentration). Grinding aerosols of vanadium metal, vanadium carbide, and ferrovanadium were not highly toxic; however, chronic exposure to them at high concentrations produced many of the same symptoms as described above for vanadium pentoxide.

In summary the basic manifestations of vanadium exposure of the experimental animals included the following, according to Roshchin (1963):

Marked irritation of the respiratory mucosa;

Vascular injury that resulted in capillary stasis, perivascular edema, and tiny hemorrhages, i.e., a hemorrhagic inflammation process;

A spastic effect on the smooth muscle of the bronchi that resulted in asthmatic-type bronchitis and expiratory difficulty on acute exposure;

Vascular changes—resulting from absorption—in internal organs and brain, which in turn cause neurological symptoms, toxic nephritis, and disorders of protein metabolism (see below).

With respect to the respiratory tract in experimental animals, the major differences between acute and chronic effects of vanadium relate to the development, with prolonged exposure, of chronic inflammation in bronchi, accompanied by greater tendency to septic bronchopneumonia. Atelectasis, interstitial infiltration and proliferation, and emphysema were also noted.

The histopathological changes observed in kidney and liver following acute exposure to vanadium at high concentrations (tens of milligrams per cubic meter) are not usually seen with intermittent low-level exposure.

Neurophysiological Effects

Other physiological effects have been mentioned in cases of severe exposure of animals to oxides and salts (Roshchin, 1967). These include disturbances of the central nervous system (impaired conditioned reflexes and neuromuscular excitability) and cardiovascular changes (occurrence of arrhythmias and extrasystole and prolongation of the Q–RST interval and decrease in the height of the P and T waves of the EKG). The significance of these findings with respect to human environmental exposure, if any, is not clearly defined at present.

Two studies have reported changes in the functional state of mice and rats following exposure to vanadium. In a series of experiments by Selyankina (1961) dissolved vanadium pentoxide or ammonium vanadate was administered orally to rats or mice in doses of 1–0.005 mg V/kg body weight-day for periods ranging from 21 days at the higher levels to 6 months at the lower levels. The threshold dose causing functional disturbances of conditioned reflex activities in mice and rats was 0.05 mg V/kg body weight. A dose of 0.005 mg V/kg body weight proved to be inactive. On the basis of these experiments, a vanadium concentration of 0.1 mg/liter was recommended as a maximum permissible concentration for water basins in the Soviet Union. This level was calculated to permit an intake of 0.005 mg V/kg body weight-day by a 60-kg person consuming 3 liters of water.

A report by Pazynich (1966) involved continuous inhalation exposure of rats for 70 days to condensation aerosols of vanadium pentoxide at levels of 0.027 ± 0.002 mg/m^3 and 0.002 ± 0.00013 mg/m^3. Rats in both groups experienced normal gains in weight as did the controls. After 30 days the motor chronaxy of the extensor muscles of the tibia in the rats exposed at the higher level (0.027 mg/m^3) decreased by an average of 0.8 μsec ($p < 0.01$), and the chronaxy of the corresponding flexor muscles increased by an average of 4 μsec ($p < 0.001$). Thus the chronaxy ratio of antagonistic muscles had fallen from 1.5 at the beginning of the experiments to 1.0 on the 20th day ($p < 0.02$), to 0.5 on the 30th day ($p < 0.01$), and continued to decrease to a level of about 0.25. The chronaxy ratio returned to normal (1.5) about 18 days following cessation of exposure on the 70th day. No changes were observed in motor chronaxy in controls or in rats exposed at the lower level (0.002 mg/m^3). Statistically significant changes in other parameters observed at the high-level exposure (0.027 mg/m^3), but not at the lower level, included depressed whole blood cholinesterase, decreased total serum protein, depressed serum β-globulins, and decreased oxyhemoglobin content of venous blood. Also observed in the high level exposure group were elevated serum γ-globulins; increased number of blood leukocytes showing yellow, orange, and red nuclear fluorescence with acridine orange; and increased oxygen consumption as indicated by the minced livers. The pattern of leukocyte nuclear fluorescence returned to normal 20 days following cessation of exposure. Histopathological changes observed following high-level exposure included marked lung congestion, focal lung hemorrhages, and extensive bronchitis. Liver changes included central vein congestion with scattered small hemorrhages, scattered infiltration between lobes, and granular degeneration of hepatocytes. The kidneys showed marked granular degeneration of the epithelium of the convoluted tubules. In the heart myocardial vascular congestion was observed with focal perivascular hemorrhages.

Because of the large differences in concentrations employed in the first two experimental groups (0.027 and 0.002 mg/m^3), a second experiment was performed in which rats were exposed continuously to vanadium pentoxide at 0.006 ± 0.00056 mg/m^3 for 40 days. During the first month of exposure no changes as compared with controls were observed in the parameters investigated; that is, chronaxy of antagonistic muscles of the tibia and blood leukocyte nuclear fluorescence. After 30 days, there was a statistically significant decrease in chronaxy ratio. During the sixth week of exposure animals were stressed by receiving only water and no food. After 3.5 days of this treatment, chronaxy ratios decreased to 0.92, compared with 1.5 in controls, and the number of leukocytes displaying altered nuclear fluorescence increased by a factor of 4.83. The overall results of the study led Pazynich (1966) to recommend the no-effect level of 0.002 mg/m^3 as the mean daily maximum permissible concentration of vanadium pentoxide in the atmosphere of the Soviet Union.

Metabolic Alterations

In the studies described above by Roshchin (1963) and in his subsequent investigations (Roshchin et al., 1965) a number of metabolic alterations were observed. After exposure of rabbits to a grinding aerosol of vanadium trioxide (40–75 mg/m^3, 2 hr/day for up to 12 months), several changes were reported: the test animals exhibited a progressive weight loss amounting to an average of 4.6% at the termination of the experiment; controls gained weight by 12.3%; the number of blood leukocytes declined after the fifth month from between 7,000 and 9,000/mm^3 to 5,000/mm^3 by the end of the test; no change was noted in controls. Hemoglobin levels in the test animals decreased from 75 to 68%.[1] Serum ascorbate levels in the blood progressively decreased to about 20% of control between 7 and 8 months. Protein sulfhydryl levels in the serum of exposed animals decreased by 30% as compared with controls. Respiration in liver and brain tissue of test animals was reduced by one-half by the end of the experiment as compared with controls, but the respiratory quotient was unchanged. Blood cholinesterase in exposed rabbits increased by an average of 25% after the fifth month.

---

[1] A normal rabbit hemoglobin is 8–15 g/100 ml and a normal rabbit hematocrit is 30–50%.

In these studies the weight loss along with the depressions in the levels of white cells, hemoglobin, and protein sulfhydryl groups in the blood and the decreased tissue respiration were taken as indicators of the "general toxic effect" of vanadium. Increased cholinesterase activity was held to be indicative of sensitization. Bronchial asthma was taken as a clinical symptom of sensitization.

Chronic poisoning from the inhalation of trivalent vanadium (V_2O_3 and VCl_3) resulted in blood changes by the end of the second and third month. These changes were characterized by decreased albumin and increased globulins (mainly γ-globulins), such that the albumin–globulin ratio was halved, and by an increase in serum concentrations of three amino acids—cystine, arginine, and histidine. There was also a 10% increase in nucleic acid in the blood and a "considerable" increase in blood chloride. Roshchin (1967) has said that "the effect of vanadium on the metabolism of proteins and nucleic acids is responsible for the immunological and allergic reactions which constitute important manifestations of vanadium poisoning."

In an attempt to explain the mechanism of the initial nonspecific hematopoetic effect of vanadium and the subsequent anemia, Roshchin (1967) hypothesized that

> ... the redox system of hydrogen carriers is inhibited or blocked, and in response to the resulting hypoxia, there is increased regeneration of the formed elements of the blood ... Possible vanadium interferes with tissue respiration at the stage of dehydrogenation effected by coenzyme I[2] belonging to the group of dehydrases [sic]. By inhibiting this coenzyme, vanadium (similarly to lead) interferes with the incorporation of iron in the porphyrin complexes, thereby retarding the synthesis of hemoglobin. The anti-vitamin C effect of vanadium is closely related to the inhibition of hemoglobin synthesis. Vitamin C deficiency in the body likewise inhibits the utilization of iron for hemoglobin synthesis, iron becoming accumulated in the reticuloendothelial tissue. . . .

Roshchin also points out that vanadium is known to inhibit the activity of monoamine oxidase, which catalyzes the conversion of serotonin to 5-hydroxyindoleacetic acid. After a 3-month chronic exposure of rabbits to vanadium pentoxide dust, Roshchin observed that urine levels of 5-hydroxyindoleacetic acid had fallen to 33% below control values. He suggests, therefore, that inhibition of monoamine oxidase may result in accumulation of serotonin in the central nervous system, leading to functional disturbances. The sensitivity of smooth muscle to accumulation of serotonin, he notes, could result in bronchospasm, diarrhea, and urinary incontinence. The dystrophic and necrotic process in the kidneys and the high permeability of the blood vessels could, in his opinion, also be explained by elevated serotonin levels.

[2] Coenzyme I was previously used to denote nicotinamide adenine dinucleotide (NAD).

Roshchin also suggests an interaction of vanadium with an unspecified enzyme to account for the observed decrease in sulfhydryl groups in blood proteins and to explain the reduced cystine content of keratinized tissues.

Many of the clinical findings observed and interactions hypothesized by Roshchin and others can be accounted for or amplified by examination of the rather fragmentary knowledge of biochemical effects of vanadium exposure in experimental animals and *in vitro*.

Bergel et al. (1958) reported that the catabolism of cystine and cysteine is increased by exposure to vanadium. In studies *in vitro*, Anbar and Inbar (1962) demonstrated that in the presence of VO^{2+} pyridoxal 5-phosphate induces the catabolism of sulfhydryl amino acids. They pointed out that "the activation of pyridoxal phosphate by vanadyl ions is rather specific to α-, β-elimination and strongly suggests a decrease of $-SH$ groups in the organism." This observation provides a corollary to the observed lowering of cystine levels in keratinized tissues (Mountain et al., 1953, 1955). Decreased synthesis of cysteine and cystine was thought to account for reduced levels of cystine in hair and nails. The essential point, however, is that metabolic processes that depend on either of these amino acids may be depressed in the presence of vanadium.

Cysteine is required in the biosynthesis of coenzyme A, being added to 4-phosphopantothenic acid in the presence of adenosine triphosphate (ATP) to form the intermediate 4'-phosphopantothenyl cystine. Coenzyme A plays a central role in many biosynthetic and oxidative pathways (Mahler and Cordes, 1966) as will be discussed. Mascitelli-Coriandoli and Citterio (1959a,b) have demonstrated that treatment with sodium metavanadate lowers the content of coenzyme A in rat liver. The administration of an antimetabolite of pantothenic acid, ω-methylpantothenic acid, to humans results in a syndrome consisting of postural hypotension, dizziness, tachycardia, fatigue, drowsiness, epigastric distress, anorexia, numbness and tingling of hands and feet, and hyperactive deep reflexes. It is not known whether these symptoms reflect an induced deficiency of pantothenic acid or the toxicity of the antimetabolite. The symptoms, however, are not unlike those resulting from exposure to high concentrations of vanadium. The common denominator in both cases may be reduced hepatic coenzyme A levels.

The requirement for coenzyme A in biochemical pathways where acetate is a starting material suggests that these processes will be impaired by excessive exposure to vanadium. In 1954 Curran (1954) demonstrated that the synthesis of cholesterol from [^{14}C]acetate in rat liver was diminished in the presence of vanadium. Subsequently, Azarnoff and Curran (1957) and Azarnoff et al. (1961) showed that one site of inhibitory action of vanadium in the cholesterol biosynthetic pathway was at the level of squalene synthetase—the enzyme that catalyzes the conversion of farnesyl pyrophosphate to squalene. Vanadium was also shown to mobilize aortic cholesterol in atherosclerotic rabbits more rapidly than was the case in control (Curran and

Costello, 1956). With respect to the observed effect of vanadium in lowering circulating cholesterol levels, Curran and Burch (1967) have suggested that a regulatory enzyme, for the biosynthesis of cholesterol, acetoacetyl coenzyme A deacylase, is activated by vanadium in young animals but inhibited by vanadium in older ones. This suggestion could explain the fact that cholesterol levels appear to be lowered by vanadium in younger animals, including humans, and not in older ones; however, no such enzyme has been demonstrated.

Because acetyl coenzyme A is a precursor of fatty acids, it has been suggested that vanadium may depress the synthesis of triglycerides and phospholipids. Levels of triglycerides were decreased in livers of rats given vanadium (Curran, 1954); however, serum triglycerides in humans were increased following ingestion of vanadium (Curran et al., 1959). The incorporation of labeled phosphate into liver phospholipids was decreased following injection of vanadyl sulfate (Snyder and Cornatzer, 1958). This finding could have resulted from inhibition of phospholipid biosynthesis or from increased oxidative degradation as suggested by the Bernheims (Bernheim and Bernheim, 1938, 1939).

Coenzyme A is also required in the synthesis of coenzyme Q, or ubiquinone, of the mitochondrial electron transport chain. Aiyar and Sreenivasan (1961) demonstrated that ubiquinone synthesis in isolated mitochondria was reduced in the presence of vanadium. When cysteine was given with vanadium, the effect on ubiquinone synthesis was partially reversed. Further addition of ATP and coenzyme A completely prevented the inhibition of ubiquinone synthesis.

Coenzyme A is required in the biosynthesis of many other biochemicals; however, the effect of vanadium on these biosynthetic processes has not been investigated.

Wright et al. (1960) have suggested that vanadium uncouples mitochondrial oxidative phosphorylation in liver homogenates *in vitro*, resulting in depletion of the ATP energy stores. The addition of ammonium metavanadate to the diet at a level to supply 25 μg/g vanadium also was reported to uncouple oxidative phosphorylation in liver mitochondria of young chicks (Hathcock et al., 1966). Hathcock et al. (1966) suggested that vanadate may replace the phosphate ion in the reactions leading to the synthesis of ATP such that a high-energy vanadyl intermediate (V $\sim$ X) or an ADP $\sim$ V is formed and hydrolyzed in a manner analogous to that proposed for arsenate induced arsenolysis. Recent studies by DeMaster and Mitchell (1973) support this mechanism for vanadium uncoupling of glyceraldehyde-3-phosphate dehydrogenase. In this same report the vanadium (V) oxyanion was found to inhibit oxidative phosphorylation in intact rat liver mitochondria but did not act as an uncoupler.

Aiyar and Sreenivasan (1961) have shown that vanadium salts inhibit succinic dehydrogenase, which would also reduce ATP synthesis. Succinic dehydrogenase, a key enzyme of the citric acid cycle and the electron transport system, is activated by sulfhydryl groups. Vanadium could inhibit this enzyme by mediating a decrease in available –SH groups. Sulfhydryl groups are also involved in the regulation of the deiodination of thyroxine at the cellular level (Anbar and Inbar, 1962). Thyroxine along with Ca^{2+}, and perhaps other agents, act through a so-called "U-factor" to cause swelling of mitochondria and uncoupling of oxidative phosphorylation (Mahler and Cordes, 1966).

Perry et al. (1955, 1969) have reported that in the presence of vanadium the oxidation of tryptamine by monoamine oxidase from guinea pig liver and kidney was accelerated by 125%. Studies by Lewis (1959c), however, indicated that vanadium inhibited monoamine oxidase because the urinary output of 5-hydroxyindoleacetic acid was reduced in dogs injected with sodium metavanadate. Decreased output of 5-hydroxyindoleacetic acid suggests the possibility of accumulation of serotonin as previously found by Roshchin (1967).

A recent study by Johnson et al. (1974) demonstrated that weight loss of rats receiving daily injections of sodium metavanadate (1.25–2.5 mg/kg) was correlated with the accumulation of the metal in the liver. The activities of two hepatic enzymes having molybdenum prosthetic groups, xanthine oxidase and sulfite oxidase, were not influenced by the presence of vanadium. Total liver concentrations of molybdenum also were not affected. Hence, it was concluded that vanadium toxicity in rats was unrelated to the antagonism of molybdenum utilization. However, it was noted that although vanadium was administered in the diamagnetic pentavalent state, the livers from these animals displayed an electron paramagnetic resonance (EPR) spectrum characteristic of vanadium (IV). The spectrum obtained was slightly broadened, compared with that observed in a standard solution of sodium metavanadate reduced with dithionite. Hence, the authors suggested that vanadium is in a protein-bound form in the liver of rats. The authors further demonstrated the presence of paramagnetic vanadium in the kidneys and to a limited extent in lungs, but none in hearts of rats injected with sodium metavanadate. Since colorimetric quantitation revealed the presence of vanadium in these organs in proportion to the EPR signals, the authors speculated that the ability of liver and kidney to reduce the vanadate by one electron might be related to a specific detoxification mechanism present in these tissues.

Other health-related effects of vanadium have been noted in the literature. Vanadium has been reported to decrease the incidence of dental caries when added to the diet of hamsters (Geyer, 1953). Subsequent studies (Hein and Wisotzky, 1955; Muhler, 1957; Hadjimarkos, 1966, 1968; McLundie et al., 1968) have failed to demonstrate a clearly beneficial effect with regard to dental caries in humans, however.

Vanadium in high concentration has been reported to reduce hemoglobin content and to produce anemia in experimental animals (Roshchin,

1967). However, when low levels of vanadium were administered to humans or animals with a normal hemoglobin level, little effect was observed (Franke and Moxon, 1937; Lewis, 1959a). Further reports suggest beneficial effects of low-level vanadium in treating nutritional anemia (Beard et al., 1931; Myers and Beard, 1931; Lewis, 1959a; Hadjimarkos, 1966).

An older report by Califino and Caselli (1949) suggested that the ability of hemoglobin to transport oxygen is reduced in the presence of vanadium. Nitrite can oxidize ferrous iron of hemoglobin to the ferric state producing methemoglobinemia in nitrite poisoning. Shakman (1974) recently questioned whether vanadium could have such an effect.

Mutagenic, Carcinogenic, and Teratogenic Effects

No concrete data were found on carcinogenic, mutagenic, or teratogenic effects of vanadium exposure in humans or experimental animals [however, see Stockinger (1967) and Hickey et al. (1967)]. Life-term studies in mice given small amounts of vanadyl ions (5 μg/ml) as the sulfate in drinking water showed no greater incidence of spontaneous tumors than in controls (Kanisawa and Schroeder, 1967).

SUMMARY AND SYNTHESIS

Vanadium is widely distributed in the physical and biological environment. It is usually present in small or very small quantities in all media and living forms. Vanadium soil concentrations vary according to the type of rock or soil and have been found to be highest in shales and clays. The estimated average concentrations in the earth's crust is 150 μg/g. The concentration of vanadium in fresh water is dependent on the soil and rocks in the area of the water source. All plants contain vanadium. Levels tend to be higher and related to soil level in roots, lower and independent of soil levels in leaves. The metal is essential for the production of soil and plant ammonia by the bacteria of the root nodules of legumes and is accumulated by certain agaric mushrooms. Small amounts of vanadium are present in most animal tissues. Accumulation of vanadium to very high levels has been noted in sea squirts.

Vanadium has been shown to be an essential nutrient for chicks and rats. As mentioned previously, it is also essential for soil nitrogen-fixing microorganisms. Nutritional requirements have not been widely studied. It is possible that the element is essential to many plants and animals. Spontaneous deficiency states have not been observed, nor has the mechanism of essentiality been determined.

Most information on human vanadium exposure is of occupational origin. Vanadium exposure in industry stems principally from dusts in metallurgical work and boiler-cleaning operations. Nonindustrial exposure is from foodstuffs, with small amounts coming from water and air. Gastrointestinal absorption of vanadium is poor, but respiratory absorption may be very efficient. The percentage of vanadium that enters the body through the lungs is high as compared with other metals, and it appears that vanadium from contaminated air accumulates in the lungs with age. Absorbed vanadium is transported in serum lipid fractions to the highly vascular organs and is excreted in the urine and to a lesser extent in the feces. Retained vanadium is stored mainly in fat. Keratin may be a minor storage depot.

Oxides and salts of vanadium are respiratory irritants at levels lower than those resulting in systemic toxicity. Clinical symptoms of acute exposure to vanadium pentoxide aerosol at 0.1 to several mg/m^3 consist of conjunctivitis, pharyngitis, persistent cough, and tightness of the chest. The onset of symptoms is often delayed by several hours, depending on the severity of the exposure; however, symptoms may persist for 1 wk or more. Green tongue results from the oral reduction of the pentoxide to a trivalent form. Some individuals may become sensitized to vanadium so that they exhibit more severe dermal and respiratory symptoms on subsequent exposures at the same concentrations.

Chronic inhalation of vanadium pentoxide dusts in industry has resulted in rhinitis, pharyngitis, bronchitis, chronic productive cough, wheezing, shortness of breath, and fatigue. Pneumonitis and bronchopneumonia have also been observed. Vanadium workers are probably predisposed to secondary respiratory infections, which may account for certain chronic histopathological alterations. Exposure to vanadium may be detected by elevated vanadium levels in urine and depressed cystine content of fingernails.

Statistical studies of an exploratory nature have shown positive correlations of vanadium content in urban air with mortality involving bronchitis, pneumonia, and cancer (in males) in one investigation, and with "diseases of the heart," nephritis, and "arteriosclerotic heart" in another. These studies are incompletely adjusted for important, probably relevant factors. In addition, most of these causes of excess mortality have not been reported in vanadium workers, who are exposed to higher levels of the metal and its compounds than is the public.

Vanadium salts may reduce serum cholesterol levels in young men and animals but are not very toxic when given orally. They are highly toxic when given intravenously; lethal doses range from 1 to 20 mg/kg in lower animals and man.

The irritative respiratory effects of vanadium oxides and salts in experimental animals (rats, mice, and rabbits) are, in general, similar to those reported in humans. However, the higher concentrations employed in acute animals exposures have produced marked vascular injury in the lungs and in internal organs as well. It is believed that absorption of vanadium in severe acute exposure (tens of mg/m^3) is responsible for damage to the liver, kidney, heart, and brain by causing vascular constriction, congestion, and hemorrhage, Intermittent exposure (for 1-2 hr daily or every other day) at lower levels (3-5 mg/m^3 vanadium pentoxide as condensation aerosol) produces a hemorrhagic inflammation process only in the lungs. However, continuous exposure at very low levels (0.005–0.027 mg V_2O_5 fume/m^3) has produced hemorrhagic respiratory inflammation as well as vascular congestion and hemorrhage in other internal organs (one study only).

Biochemical alterations resulting from vanadium exposure include depression of synthesis or increased catabolism of cystine and cysteine with an overall lowering of serum protein sulfhydryl groups. Serum ascorbic acid levels are lowered (perhaps because of insufficient cysteine as glutathione to maintain the oxidized state) such that the removal of iron from ferritin is no longer facilitated by ascorbate. Hemoglobin synthesis is probably, therefore, reduced, and the lowered hemoglobin levels and anemia that accompany severe chronic exposure to vanadium are manifest clinically.

The reduced hepatic levels of coenzyme A following exposure to vanadium may reflect the reduced availability of cysteine—a precursor in the biosynthesis of coenzyme A. Coenzyme A, in turn, plays a central role in many biosynthetic and oxidative pathways. Several other enzyme disturbances have also been observed in animal studies. Continued investigation may be expected to elucidate the interrelated biochemical, physiological, and histopathological effects of vanadium exposure. Whether a mutagenic, carcinogenic, or teratogenic potential exists for vanadium remains to be determined.

REFERENCES

American Conference of Governmental Industrial Hygienists. 1961. *Threshold limit values for chemical substances and physical agents in the workroom environment.* Cincinnati, Ohio.

American Conference of Governmental Industrial Hygienists. 1971a. *Threshold limit values for chemical substances and physical agents in the workroom environment with intended changes for 1972.* Cincinnati, Ohio.

American Conference of Governmental Industrial Hygienists. 1971b. *Documentation of the threshold limit values for substances in workroom air,* 3d ed. Cincinnati, Ohio.

Aiyar, A. S. and Sreenivasan, A. 1961. *Proc. Soc. Exp. Biol. Med.* 107:914.

Anbar, M. and Inbar, M. 1962. *Nature* 196:1213.

Arnon, D. I. 1958. In *Trace elements,* ed. C. A. Lamb, O. G. Bentley, and J. M. Beattie, pp. 1–32. New York: Academic Press.

Arnon, D. I. and Wessel, G. 1953. *Nature* 172:1039.

Athanassiadis, Y. C. 1969. *Preliminary air pollution survey of vanadium and its compounds: A literature review,* Publ. No. APTD 69-48. Raleigh: National Air Pollution Control Administration.

Azarnoff, D. L. and Curran, G. L. 1957. *J. Am. Chem. Soc.* 79:2968.

Azarnoff, D. L., Brock, F. E. and Curran, G. L. 1961. *Biochem. Biophys. Acta* 51:397.

Balestra, G. and Molfino, F. 1942. *Rass. Med. Ind. (Turin)* 13:5. (Abst. in *J. Ind. Hyg.* 26:7, 1944).

Barannik, P. I., Mikhalyuk, I. A. and Motuzkov, I. N. 1969. *Gig. Sanit.* (English ed.) 35:289.

Beard, H. H., Baker, R. W. and Myers, V. C. 1931. *J. Biol. Chem.* 94:123.

Bergel, F., Bray, R. C. and Harrap, Y. R. 1958. *Nature* 181:1654.
Bernheim, F. and Bernheim, M. L. C. 1938. *Science* 88:481.
Bernheim, F. and Bernheim, M. L. C. 1939. *J. Biol. Chem.* 127:353.
Bertrand, D. 1942. *Ann. Inst. Pasteur* 68:226.
Bertrand, D. 1950. *Bull. Am. Mus. Natl. Hist.* 94:403.
Bowen, H. J. M. 1966. *Trace elements in biochemistry*. London: Academic Press.
Browne, R. C. 1955. *Br. J. Ind. Med.* 12:57.
Browning, E. 1962. *Toxicity of industrial metals*. London: Butterworths.
Buckingham, D. A. 1973. In *Inorganic biochemistry*, ed. G. L. Eichhorn, vol. 1, p. 18. Amsterdam: Elsevier.
Califano, L. and Caselli, P. 1949. *Pubbl. Stn. Zool. Napoli* 21:261.
Cannon, H. L. 1963. *Soil Sci.* 96:196.
Carlberg, J. R., Crable, J. V., Lintiaca, L. P., Norris, H. B., Holtz, J. L., Mauer, P. and Wolowicz, F. R. 1971. *Am. Ind. Hyg. Assoc. J.* 32:432.
Cotton, F. A. and Wilkinson, G. 1962. *Advanced inorganic chemistry*, pp. 673–680. New York: Interscience.
Cowgill, U. M. 1973. *Appl. Spectrum* 27:5.
Curran, G. L. 1954. *J. Biol. Chem.* 210:765.
Curran, G. L. and Burch, R. E. 1967. *Proceedings, 1st annual conference on trace substances in environmental health*. Columbia: University of Missouri.
Curran, G. L. and R. L. Costello, R. L. 1956. *J. Exp. Med.* 103:49.
Curran, G. L., Azarnoff, D. L. and Bolinger, R. E. 1959. *J. Clin. Invest.* 38:1251.
de Jorge, F. B., Ferreira, L. C. G. and Tilbery, C. P. 1970. *Rev. Bras. Med.* 27:303.
DeMaster, E. G. and Mitchell, R. A. 1973. *Biochemistry* 12:3616.
Dimond, E. G., Caravaca, J. and Benchimol, A. 1963. *Am. J. Clin. Nutr.* 12:49.
Durfor, C. N. and Becker, E. 1963. Public water supplies of the 100 largest cities in the U.S., 1962. Water supply paper No. 1812. Washington, D.C.: Geological Survey.
Durrant, P. J. and Durrant, B. 1970. *Introduction to advanced inorganic chemistry*, 2d ed., pp. 996–1007. New York: Wiley.
Dutton, W. F. 1911. *J. Am. Med. Assoc.* 56:1948.
Earnshaw, A. and Harrington, T. J. 1973. *The chemistry of the transition elements*, pp. 54–57. Oxford: Clarendon Press.
Fallentin, B. and Frost, J. 1954. *Nord. Hyg. Tidskr. (Stockholm)* 3:58.
Faulkner Hudson, T. G. 1964. *Vanadium toxicology and biological significance*. Amsterdam: Elsevier.
Fear, E. C. and Tyrer, F. H. 1958. *Trans. Assoc. Ind. Med. Off.* 8:153.
Franke, K. W. and Moxon, A. L. 1937. *J. Pharmacol. Exp. Ther.* 61:89.
Frost, J. 1951. *Ugeskr. Laeger (Copenhagen)* 113:1309.
Fukai, R. and Meinke, W. W. 1962. *Limnol. Oceanogr.* 7:186.
Geyer, C. F. 1953. *J. Dent. Res.* 35:590.
Goren, M. 1966. U.S. patent 3, 252, 756.
Gulko, A. G. 1956. *Gig. Sanit. (Moscow)* 21:24.
Hadjimarkos, D. M. 1966. *Nature* 209:1137.
Hadjimarkos, D. M. 1968. *Adv. Oral Biol.* 3:253.
Hathcock, J. N., Hill, C. H. and Tove, S. B. 1966. *Can. J. Biochem.* 44:983.
Hein, J. W. and Wisotzky, J. 1955. *J. Dent. Res.* 34:756.
Hickey, R. J., Schoff, E. P. and Clelland, R. C. 1967. *Arch. Environ. Health* 15:728.

Hickling, S. 1958. *N.Z. Med. J.* 57:607.

Hopkins, L. L., Jr., and Mohr, H. E. 1971. In *Newer trace elements in nutrition*, ed. W. Mertz and W. E. Cornatzer, pp. 195–213. New York: Marcel Dekker.

Hopkins, L. L., Jr., and Mohr, H. E. 1974. *Fed. Proc.* 33:1773.

Hopkins, L. L., Jr., and Tilton, B. E. 1967. *Am. J. Physiol.* 211:169.

Horner, C. K., Burck, D., Allison, F. and Sherman, M. S. 1942. *J. Agric. Res.* 65:173.

Jaraczewska, W. and Jakubowski, M. 1964. *Med. Pracy* 15:375.

Johnson, J. L., Cohen, H. L. and Rajagopalan, K. V. 1974. *Biochem. Biophys. Res. Commun.* 56:940.

Kanisawa, M. and Schroeder, H. A. 1967. *Cancer Res.* 27:1192.

Kennan, R. G. 1963. Unpublished data in: Mountain, J. T. *Arch. Environ. Health* 6:357.

Kepert, D. L. 1972. *The early transition metals*, pp. 181–190. New York: Academic Press.

Kopp, J. F. and Kroner, R. C. 1968. *Trace metals in water of the United States.* Cincinnati, Ohio: Federal Water Pollution Control Administration.

Krauskopf, K. B. 1956. *Geochim. Cosmochim. Acta* 9:1B.

Lee, R. E., Goranson, S. S., Enroine, R. E. and Morgan, G. B. 1972. *Environ. Sci. Technol.* 6:1025.

Levina, E. N. 1972. *Gig. Truda Prof. Zabol. (Moscow)* 16:40.

Lewis, C. E. 1959a. *Am. Med. Assoc. Arch. Ind. Health* 19:419.

Lewis, C. E. 1959b. *Am. Med. Assoc. Arch. Ind. Health* 19:497.

Lewis, C. E. 1959c. *Am. Med. Assoc. Arch. Ind. Health* 20:455.

Mahler, H. R. and Cordes, E. H. 1966. *Biological chemistry.* New York: Harper & Row.

Marden, J. W. and Rich, M. N. 1927. *Ind. Eng. Chem.* 19:786.

Mascitelli-Coriandoli, E. and Citterio, C. 1959a. *Nature* 183:1527.

Mascitelli-Coriandoli, E. and Citterio, C. 1959b. *Nature* 184:1641.

Matantseva, E. I. 1960. *Gig. Truda Prof. Zabol. (Moscow)* 7:41.

McLundie, A. C., Shepherd, J. B. and Mobbs, D. R. A. 1968. *Arch. Oral Biol.* 13:1321.

McTurk, L. C., Hirs, C. H. W. and Eckardt, R. E. 1956. *Ind. Med. Surg.* 25:29.

Mountain, J. T. 1963. *Arch. Environ. Health* 6:357.

Mountain, J. T., Delker, L. L. and Stokinger, H. E. 1953. *Am. Med. Assoc. Arch. Ind. Hyg. Occup. Med.* 8:406.

Mountain, J. T., Stockell, F. R., Jr., and Stokinger, H. E. 1955. *Am. Med. Assoc. Arch. Ind. Health* 12:494.

Muhler, J. C. 1957. *J. Dent. Res.* 36:787.

Myers, V. C. and Beard, H. H. 1931. *J. Biol. Chem.* 94:89.

Nason, A. 1958. In *Trace elements*, ed. C. A. Lamb, O. G. Bently, and J. M. Beattie, pp. 269–296. New York: Academic Press.

National Academy of Sciences. 1974. *Vanadium.* Washington, D.C.

National Air Sampling Network. 1965–1969. Vanadium data stored in the National Aerometric Data Bank. Unpublished data. U.S. Environmental Protection Agency, National Environmental Research Center, Research Triangle Park, N.C.

Neilsen, F. H. and Ollerich, D. A. 1973. *Fed. Proc.* 32:929.

Occupational Safety and Health Administration. 1974. *Fed. Reg.* 39:23540.

Pazynich, V. M. 1966. *Gig. Sanit. (Moscow)* (English ed.) 31:6.

Perry, H. M., Jr., Teitlebaum, S. and Schwartz, P. L. 1955. *Fed. Proc.* 14:113.
Perry, H. M., Jr., Schwartz, P. L. and Sahagian, B. M. 1969. *Proc. Soc. Exp. Biol. Med.* 130:273.
Pielsticker, F. 1954. *Arch. Gewerbepathol. Gewerbehyg. (Berlin)* 13:73.
Pinkerton, C., Hammer, D. I., McClain, K., Williams, M. E., Bridbord, K. and Riggins, W. B. 1972. Relationship of manganese and vanadium in the ambient air to heart disease and influenza–pneumonia mortality rates. Unpublished staff study. U.S. Environmental Protection Agency, National Environmental Research Center. Research Triangle Park, N.C.
Pratt, P. F. 1966. In *Diagnostic criteria for plants and soils*, ed. H. D. Chapman, Riverside: University of California Press.
Rajner, V. 1960. *Cesk. Otolaryngol.* 9:202.
Roitman, I., Travassas, L. R., Azevedo, H. P. and Curry, A. 1969. *Sabouraudia J. Int. Soc. Hum. Anim. Mycol.* 7:15.
Roshchin, I. V. 1952. *Gig. Sanit. (Moscow)* 11:49.
Roshchin, I. V. 1962. *Gig. Truda Prof. Zabol. (Moscow)* 6:17.
Roshchin, I. V. 1963. In *Toxicology of the rare metals*, ed. Z. I. Izraelson, AEC-tr-6710. Washington, D.C.: Atomic Energy Commission.
Roshchin, I. V. 1964. *Gig. Truda Prof. Zabol. (Moscow)* 9:3.
Roshchin, I. V. 1967. *Gig. Sanit. (Moscow)* 32:26.
Roshchin, I. V., Il'nitskaya, A. V., Lutsenko, L. A. and Zhidkova, L. A. 1965. *Fed. Proc. Trans. Suppl.* 24:611.
Rundberg, G. 1939. *Nord. Med. (Stockholm)* 2:1845.
Schroeder, H. A. 1966. *J. Am. Med. Assoc.* 195:81.
Schroeder, H. A. 1970a. *Vanadium.* Air quality monographs, monograph No. 70-13. Washington, D.C.: American Petroleum Institute.
Schroeder, H. A. 1970b. *Arch. Environ. Health* 21:798.
Schroeder, H. A., Balassa, J. J., and Tipton, I. H. 1963. *J. Chron. Dis.* 16:1047.
Schümann-Vogt, B. 1969. *Zentralbl. Arbeitsmed. Arbeitsschutz (Darmstadt)* 19:33.
Schwarz, K. and Milne, D. B. 1971. *Science* 174:426.
Selyankina, K. P. 1961. *Gig. Sanit. (Moscow)* 26:6.
Shakman, R. A. 1974. *Arch. Environ. Health* 28:105.
Sjøberg, S. G. 1950. *Acta. Med. Scand.* 138:1. (Suppl. 238)
Sjøberg, S. G. 1951. *Acta Allerg.* 4:357.
Sjøberg, S. G. 1955. *Arch. Ind. Health* 11:505.
Sjøberg, S. G. and Ringer, K. G. 1956. *Nord. Hyg. Tidskr. (Stockholm)* 37:217.
Snyder, F. and Cornatzer, W. E. 1958. *Nature* 182:462.
Somerville, J. and Davies, B. 1962. *Am. Heart J.* 54:54.
Söremark, R. 1967. *J. Nutr.* 92:183.
Söremark, R. and Üllberg, S. 1962. In *Use of radioisotopes in animal biology and the medical sciences*, vol. 2, pp. 103–114. New York: Academic Press.
Söremark, R., Üllberg, S. and Appelgren, L. E. 1962. *Acta Odontol. Scand.* 20:225.
Stocks, P. 1960. *Br. J. Cancer* 14:397.
Stokinger, H. E. 1967. In *Industrial hygiene and toxicology*, 2d ed., ed. F. A. Patty, vol. 2, pp. 1171–1182. New York: Interscience.
Stokinger, H. E. 1973. Unpublished laboratory data. U.S. Department of Health, Education and Welfare, National Institute of Occupational Safety and Health, Cincinnati, Ohio.

Strain, W. H. 1961. Unpublished data in: Masironi, R. *Bull. WHO* 40:305, 1969.

Strain, W. H., Berliner, W. P., Lankau, C. A., McEvoy, R. K., Pories, W. J. and Greenlaw, R. H. 1964. *J. Nucl. Med.* 5:664.

Strasia, C. A. 1971. Vanadium: Essentiality and toxicity in the laboratory rat. Ph.D. thesis. Purdue University. *Diss. Abstr.* 32:646B.

Symanski, J. 1939. *Arch. Gewerbepathol. Gewerbehyg. (Berlin)* 9:295.

Takahashi, H. and Nason, A. 1957. *Biochim. Biophys. Acta* 23:433.

Tebrock, H. E. and Machle, W. 1968. *J. Occup. Med.* 10:692.

Thomas, D. L. G. and Stiebris, K. 1956. *Med. J. Aust.* 1:607.

Tipton, I. H. and Shafer, J. J. 1964. *Arch. Environ. Health* 8:58.

Tipton, I. H., Stewart, P. L. and Dickson, J. 1969. *Health Phys.* 16:455.

Troppens, H. 1969. *Dtsche. Gesundheitswes. (Berlin)* 24:1089.

Tullar, I. V. and Suffet, I. H. 1973. The fate of vanadium in an urban air shed, the lower Delaware River Valley. Department of Chemistry, Environmental Engineering and Science, Drexel University, Philadelphia.

Vinogradov, A. P. 1959. *The geochemistry of rare and dispersed chemical elements in soils*, 2d ed. Translated from Russian by Consultants Bureau, Inc., New York.

Vintinner, F. J., Vallenas, R., Carlin, C. E., Weiss, R., Macher, C. and Ochoa, R. 1955. *Arch. Ind. Health* 12:635.

Watanabe, H., Murayama, H. and Yamaoka, S. 1966. *Jap. J. Ind. Health* 8:23.

Waters, M. D., Gardner, D. E. and Coffin, D. L. 1974. *Toxicol. Appl. Pharmacol.* 28:253.

Waters, M. D., Gardner, D. E., Aranyi, C. and Coffin, D. L. 1975. *Environ. Res.* 9:32.

Weast, R. C. (ed.) 1970. *CRC handbook of chemistry and physics*, 51st ed. Cleveland: Chemical Rubber Co.

Weeks, M. E. 1956. *Discovery of the elements*, 6th ed. Easton, Pa.: Journal of Chemical Education.

Williams, N. 1952. *Br. J. Ind. Med.* 9:50.

Wright, L. D., Li, L. F. and Trager, R. 1960. *Biochem. Biophys. Res. Commun.* 3:264.

Wyers, H. 1946. *Br. J. Ind. Med.* 3:177.

Wyers, H. 1948. *Proceedings of the 9th International Congress of Industrial Medicine*, p. 900. Bristol: John Wright.

Zenz, C. and Berg, B. A. 1967. *Arch. Environ. Health* 5:542.

Zenz, C., Bartlett, J. P. and Thiede, W. H. 1962. *Arch. Environ. Health* 5:542.

Zoller, W. H., Gordon, G. E., Gladney, E. S. and Jones, A. G. 1973. In *Trace elements in the environment*, ed. E. L. Kothny, pp. 31–47. Advances in chemistry series 123. Washington, D.C.: American Chemical Society.

Chapter 7

TOXICOLOGY OF SELENIUM AND TELLURIUM

Lawrence Fishbein
National Center for Toxicological Research
Jefferson, Arkansas

INTRODUCTION

The elements selenium and tellurium (atomic numbers 32 and 52, respectively) are in proximity in the periodic table of elements (e.g., the VIA group) and exhibit a chemical similarity in a number of areas. Numerous aspects, however—such as utility, environmental occurrence (air, water, soil, and plants), effects on nutrition of domestic animals, fowl, and humans, interrelationships with other metals and vitamins, toxicity, and biochemical effects—are so disparate as to warrant individual consideration.

Salient features of the chemical properties (Bagnall, 1966; Elkin and Margrave, 1968; Painter, 1941), analysis (Green and Turley, 1961; Watkinson, 1967; Shendrikar, 1974), metabolic and biochemical effects (Ganther et al., 1966; Paulson et al., 1968; Byard, 1969; Ganther and Corcoran, 1969; Jenkins, et al., 1969; Palmer et al., 1969, 1970; Ganther, 1970, 1973; Levander, 1972; Rotruck et al., 1972), interrelationships with vitamin E (Diplock, 1970; Levander and Morris, 1970; Diplock et al., 1971), and arsenic and heavy metals (Kar and Das, 1963; Parizek and Ostadalova, 1967; Levander and Argrett, 1969; Dennis, 1971; McConnell and Carpenter, 1971; Parizek, 1971; Koeman et al., 1973; Kosta et al., 1975), natural occurrence (Rosenfeld and Beath, 1964b; Sindeeva, 1964; Mason 1952; Lansche, 1967), occurrence in human diets (Oelschlager and Menke, 1969; Morris and Levander, 1970; Schroeder et al., 1970; Frost, 1971; Arthur, 1972), levels in human tissues (Bowen and Cawse, 1963; Brune et al., 1966; Burk et al., 1967; Dickson and Tomlinson, 1967; Fuller et al., 1967; Allaway et al., 1968), industrial production and utilization patterns (Lakin and Davidson, 1967; Lansche, 1967; Elkin and Margrave, 1968; FDA, 1974), and environmental exposure levels (West, 1967; Pillay et al., 1969; Stahl, 1969; Hashimoto et al., 1970; Johnson, 1970; Olson and Frost, 1970; Davis, 1972; Shendrikar and West, 1973) of selenium have all been reported.

191

The major objective of this chapter is to highlight the toxicological, carcinogenic, teratogenic, and mutagenic aspects of selenium and tellurium that are of greatest relevance to humans. In any consideration of the potential toxicological (as well as environmental) hazards of selenium and tellurium, as with other elements, it is essential to designate the form of the element (valence state), as well as to delineate the dosage or degree of exposure and route of administration in order to permit a more rational assessment of the potential hazard to humans.

The chemical form of the metalloid selenium, the animal species, and the presence of other metallic salts with similar electronic configurations (e.g., arsenic) can influence the intestinal absorption of Se derivatives. The gastrointestinal absorption of Se, particularly in the form of selenites, selenates, and seleniferous compounds by mammals, is rapid and efficient. The rate of absorption of selenide and elemental Se is comparatively low, and some organic selenium compounds are poorly absorbed. Absorbed selenium is complexed with plasma proteins and is distributed to all tissues with its intracellular distribution, varying with the tissue and selenium concentration.

Excretion of selenium is by feces, urine, and exhaled air, depending on the intake and chemical form of selenium, the presence of metals such as As, Hg, Cd, and Tl, and the route of administration.

TOXICITY OF SELENIUM

Toxic Effects in Laboratory Animals

Acute toxicity. Acute selenium poisoning in laboratory animals has been produced by a toxic dose of selenium compound administered orally, subcutaneously, intraperitoneally, or intravenously. The most commonly tested selenium salts are sodium selenite or selenate. The lethal dose has varied in the hands of different observers, owing probably to species differences, age of the animals, mode of administration, and the purity of the salts. In general, however, the selenate is less toxic than the selenite. The toxicity of selenium in different forms has been determined by Franke and Moxon (1936, 1937).

Smith et al. (1937a), Painter (1941), and Heinrich and MacCanon (1957) cited the differences of the minimal lethal dose for various animals. Table 1 illustrates the minimum lethal dose of several selenium compounds. The results from different laboratories are not exactly comparable because the minimum lethal dose is not rigidly defined. The minimum lethal dose of sodium selenate expressed in mg/kg for different laboratory animals was reported by Moxon and Rhian (1943) as follows: 3.0-5.7, rat; 0.9-1.5, rabbit; and about 2.0, dog. (Their values apply to subcutaneous, intraperitoneal, or intravenous injections.) Smith et al. (1937a) reported the minimum lethal dose of selenium as sodium selenate or selenite in rabbits, rats, and cats as 1.5 to 3.0 mg/kg, regardless of the route of administration.

TABLE 1 Minimum Lethal Doses of Selenium Compounds

Selenium compound	Mode of administration	Experimental animals	Number of animals used	mg Se/kg body weight	References
Na₂SeO₃	Intraperitoneal	Rat	155	3.25–3.5	Franke and Moxon, 1936
Na₂SeO₄	Intraperitoneal	Rat	90	5.25–5.75	Franke and Moxon, 1937
Na₂SeO₃	Intravenous	Rat	45	3.0	Smith et al., 1937a
Na₂SeO₄	Intravenous	Rat	37	3.0	Smith et al., 1937a
Na₂SeO₃	Intravenous	Rabbit	9	1.5	Smith et al., 1937a
Na₂SeO₄	Intravenous	Rabbit	16	2.0–2.5	Smith et al., 1937a
Colloidial selenium	Intravenous	Rat		6.0	Muehlberger and Schrenk, 1928
Selenocystine	Intraperitoneal	Rat	18	4.0	Moxon, 1940
Selenomethionine	Intraperitoneal	Rat	16	4.25	Moxon, 1940

193

Marked differences in the acute toxicity of selenium sulfide and sodium selenite have been reported by Henschler and Kirschner (1969). (The acute oral LD50s of the compounds were 3,700 and 48 mg/kg, respectively, in the mouse.) In this species, daily doses of up to 125 mg Se sulfide/kg for 7.5 wk were well tolerated.

The toxicity of sodium selenite and sodium selenate in the rat has been compared with that of compounds of arsenic, molybdenum, tellurium, and vanadium by Franke and Moxon (1937). Table 2 illustrates the highest toxicity of both sodium tellurite and sodium selenite.

The acute toxic effects of hydrogen selenide in guinea pigs with a single exposure of 10, 30, or 60 min to graded concentrations of the gas ranging from 0.002 to 0.57 mg/liter was investigated by Dudley and Miller (1937). All animals exposed to 0.02 mg/liter for 60 min died within 25 days; 93% of those exposed to 0.043 mg/liter for 30 min died within 30 days, and all exposed to 0.57 mg/liter for 10 min died within 5 days. Decreasing the concentrations of H_2Se and increasing the time of exposure to 2, 4, or 8 hr produced death in 50% of the guinea pigs within 8 hr.

The acute toxicity of selenium dust (average mass median particle diameter was 1.2 μm) to rats, guinea pigs, and rabbits was described by Hall et al. (1951). Exposure of these animals for 16 hr to an atmosphere containing approximately 30 mg selenium dust/m^3 produced mild interstitial pneumonitis in the animals. Rats exposed to selenium fumes by vacuum cooperation developed acute toxicity. It was suggested that the fumes may have contained some selenium dioxide.

The acute toxic dose of selenic anhydride was 5 mg/kg when animals were exposed to air containing 0.09 mg selenium/liter of air (Filatova, 1951). Concentrations of 0.15–0.6 mg/liter produced convulsions and death in 4 hr. Conjunctival irritation and respiratory distress preceded death and the lungs

TABLE 2   Minimum Doses Fatal to Rats among Compounds of
Tellurium, Selenium, Arsenic, Vanadium, and Molybdenum[a]

Compound	Formula	Minimum fatal dose[b]
Sodium tellurite	Na_2TeO_3	2.25–2.50 mg Te/kg
Sodium tellurate	Na_2TeO_4	20.0–30.0 mg Te/kg
Sodium selenite	Na_2SeO_3	3.25–3.50 mg Se/kg
Sodium selenate	Na_2SeO_4	5.25–5.75 mg Se/kg
Sodium arsenite	Na_2HAsO_3	4.25–4.75 mg As/kg
Sodium arsenate	Na_2HAsO_4	14.0–18.0 mg As/kg
Sodium vanadate	$NaVO_3$	4.0–5.0 mg V/kg
Ammonium molybdate	$(NH_4)_6Mo_7O_{24}$	Above 160.0 mg Mo/kg

[a]From Franke and Moxon (1936).
[b]Minimum doses fatal to at least 75% of young rats within 48 hr after substances were injected intraperitoneally.

showed edema. Degenerative changes were also noted in the liver, kidneys, spleen, and heart.

Although elemental selenium has been stated to be relatively nontoxic because of its relative insolubility, it should be noted that colloidal selenium has proved fatal to rats when injected intravenously (Muehlberger and Schrenk, 1928).

The acute toxic effects of dimethyl selenide indicate that this compound has a low degree of toxicity in rats and mice. For example, the LD50 by intraperitoneal injection was 1.3 and 1.6 g selenium (as dimethyl selenide) for mice and rats, respectively (McConnell and Portman, 1952). The low toxicity of dimethyl selenide suggests that detoxification of selenium in the animal proceeds from a highly toxic form to a less toxic compound that is eliminated through the respiratory tract.

Experimental studies of the toxic effects of organic selenium compounds are comparatively limited owing to the toxicity hazards involved in their preparation and, hence, to their general unavailability. Klug et al. (1953), Moxon (1940), and Moxon et al. (1938) tested a number of selenoamino acids for their acute oral toxic effects. The minimum fatal dose of selenomethionine by intraperitoneal injection in rats was 4.25 mg selenium (as seleno-methionine)/kg LD50 of selenium compounds expressed as mg/kg by intra-peritoneal administration were: selenohomocystine, 3.5 ± 0.4; 3-, 1-, *dl-*, or *meso*-selenocystine, 4.0 ± 0.2; selenomethionine, 4.5 ± 0.3; selenotetragluta-thione, $6.0 \pm .3$. The symptoms of acute selenosis were produced with each compound except selenomethionine, which produced a comatose state in which the animal became hypersensitive to external stimuli (Klug et al., 1953).

Acute effects including symptoms similar to those of "blind staggers" have been described by Franke and Moxon (1936) in dogs and rats given massive doses of selenium (and in horses, mules, and cattle by Miller and Williams, 1940). The symptoms, appearing 5 min after subcutaneous and 15 min after oral administration, include vomiting, garlic odor of the breath, syspnea, tetanic spasms, and death from respiratory failure. Pathological changes include congestion of the liver with areas of focal necrosis, congestion of the kidney, endocarditis, myocarditis, petechral hemorrhages of the epi-cardium, atony of the smooth muscles of the gastrointestinal tract, gall bladder, and bladder, and erosion of the long bones, especially the tibia in about two-thirds of the cases. It is clear that the organ most affected by selenium in animals is the liver. In general, the liver undergoes a reversible fatty degeneration (in laboratory animals) provided the duration of exposure to selenium is not too long (Smith, 1939); if, however, the exposure continued over a longer period of time, liver cirrhosis was likely to occur. The spleen also becomes enlarged and the stomach and intestinal tract show hemorrhages. In the kidney there is usually only mild tubular degeneration. Early rise in blood hemoglobin, produced by naturally toxic grain, were found to be identical with those produced by sodium selenite (Cerwenka and Cooper, 1961). Rats, guinea pigs, and rabbits exposed to selenium fumes gave no evidence of injury except in the lungs, which were hemorrhagic. Animals killed 1 month after exposure to the fumes had a mild interstitial pneumonitis (Hall et al., 1951).

Rats exposed to high concentrations of selenic anhydride exhibited toxic edema of lung and parenchymatous organs (Felatova, 1951). The most toxic selenium derivative, hydrogen selenide (1–4 μg/liter air) produces irritation of the upper respiratory tract, failure of the olfactory sense, pulmonary edema, pneumonia, and death (Rosenfeld and Beath, 1964b). Subacute selenosis may be produced by selenium salts or selenoamino acids (LD50 = 10–15 μg/kg in the rat). Effects include increased pancreatic weight and metabolic rate (DeVasconcellos and Hampe, 1963; Halverson et al., 1966) and hemolysis (hypochronic anemia and increased serum bilirubin (Halverson et al., 1970).

Cummins and Kimura (1971) recently described comparative oral toxicity studies in Sprague-Dawley rats and dogs of the following selenium compounds: sodium selenate, selenourea, biphenyl selenium, selenium sulfide (1–30μm particle size), and elemental selenium (1–30μm particle size), as well as various active antidandruff compounds. The oral LD50 values in rats for a number of selenium compounds, shown in Table 3, demonstrate the large variations in LD50 values that occur depending on both the oxidation state and its aqueous solubility. The aqueous solubilities were carried out in 0.01 N HCl to more closely simulate acidic conditions in the stomach. The least toxic selenium compound was the insoluble elemental selenium within LD50 of 6.7 g/kg. Signs of drug action include pilomotor activity, decreased body activity, dyspnea, diarrhea, anorexia, and cachexia. Fatalities occurred within 18–72 hr; survivors appeared outwardly normal at the end of the 7-day observation period. The most toxic of the selenium compounds tested was the

TABLE 3   Solubilities and LD50 Values for Tested
Selenium Compounds

Compound	Solubility in 0.01 N HCl (mg/ml)	Rat oral LD50[a] (mg/kg)
Na_2SeO_3	700	7 (4.4–11.2)
$H_2N-C-NH_2$ $\overset{\parallel}{Se}$	30	50 (35.7–70.0)
SeS_2	Insoluble (< 1)	138 (110–172)
⟨benzene⟩–Se–⟨benzene⟩	5	360 (308–421)
Elemental Se	Insoluble (< 1)	6,700 (6,000–7,300)

[a]Values in parentheses represent 95% confidence limit.

highly soluble sodium selenite with an oral LD50 of 7 mg/kg. The signs of drug action were similar to those seen with high doses of the elemental selenium. Selenium sulfide (a component in shampoos) was about 20 times less toxic than sodium selenite (e.g., 138 mg/kg vs. 7 mg/kg). It was also found in this study that with the exception of biphenyl selenium, toxicity could be correlated with blood selenium levels. For example, sodium selenite, being the most toxic, gave the highest blood level followed in order by selenourea, selenium sulfide, and elemental selenium.

It was suggested that the high blood selenium level produced by the relatively nontoxic biphenyl selenium compound can be attributed to the covalently bound selenium compound, which is the most lipophilic compound leading to better absorption, resisting catabolism, and apparently circulating as parent compound. This appears to support the popular theory that selenium compounds are only highly toxic when the selenium is metabolized to the Se^{4+} oxidation state (Potter and Elvehjem, 1937; Rosenfeld and Beath, 1945). Once the selenium is metabolized to this ionic species, it has the ability to bind very tenaciously to proteins (Cummins and Martin, 1967), hence probably acting as a toxicant by some means of enzyme inhibition. The above study of Cummins and Kimura (1971) also showed that the oral LD50 values for 1 and 2.5% SeS_2 shampoos in rodents were about 10 times the emetic doses, suggesting that a toxic amount of shampoo could not be ingested without emesis occurring. The acute oral LD50 values in mice and rats of selenium sulfide detergent 'suspensions (Selsun Blue antidandruff lotion shampoo contains 1% SeS_2; Selsun Suspension contains 2.5% SeS_2) expressed in ml/kg (14.2 and 5.3 ml/kg for mice and 7.8 and 4.9 ml/kg for rats) are roughly equivalent in selenium sulfide content to 142, 133, 78, and 123 mg/kg, respectively.

Acute toxicity studies of selenium hexafluoride (SeF_6), carried out in rabbit, guinea pig, rat, and mouse at 100, 50, 25, 10, 5, and 1 ppm for 4 hr, showed that exposures down to and including 10 ppm (Ct, 40 ppm-hr) were uniformly fatal (Kimmerle, 1960). Five parts per million (Ct, 20 ppm-hr) resulted in pulmonary edema from which the animals survived and 1 ppm was without grossly observable effects. However, in daily repeated exposures of 1 hr each for 5 days at 5 ppm, definite signs of pulmonary injury were observed. The acute toxicity of SeF_6 from these studies would appear to be about half that of ozone (Ct, 25 ppm-hr) and represents an LD50 for small laboratory rodents.

Chronic toxicity. The concentrations of selenium in the diet, or given orally, leading to selenosis, depend on the chemical form of the selenium and other dietary components. In general, however, the concentration necessary to produce chronic selenium poisoning has been observed in rats and dogs at dietary levels of 5-10 ppm (Anspaug and Robison, 1971). Chronic effects from prolonged feeding with diets containing added selenium in amounts of 5-15 ppm include liver injury in the form of atrophy, necrosis, cirrhosis, and hemorrhage (Lillie and Smith, 1940; Cameron, 1947), marked and progressive anemia with hemoglobin values as low as 2 g % in some species. Changes in the endocrine glands, especially the ovaries, pituitary, and adrenals, have been noted by Vesce (1947) following oral administration of 5-12.5 mg sodium selenide to guinea pigs over two periods of 20 days. Vesce (1947) suggested that the pituitary lesions (edema and hemorrhage) may be the primary cause of the widespread disturbance of the other endocrine organs. Selenium anhydride, by repeated inhalation of 0.01-0.03 mg/liter, caused loss of weight, anemia, conjunctival irritation, and drowsiness, with lesions of the organs similar to those of acute intoxication with the anhydride (Filatova, 1951).

Laboratory animals chronically poisoned with small doses of inorganic selenium given over a long period of time show a wide distribution of selenium in the tissues (Smith et al., 1937b). Liver, kidney, spleen, pancreas, heart, and lungs contain the largest amounts of selenium. If the feeding of inorganic selenium is discontinued, most of the selenium is excreted within 2 wk, although small amounts are found for about a month or longer (Smith et al., 1937b).

At concentrations that will produce toxic effects, selenium is observed to be readily accumulated in tissue with the liver and kidney usually having the highest concentrations (Rosenfeld and Beath, 1945). Under toxic conditions, the liver and kidney can reach selenium concentrations of 20 or 30 ppm, while most of the other tissues range from 4 to 7 ppm. When these concentrations are reached, apparently further intake is balanced by excretion. In acute poisoning, blood and spleen concentrations reach 25-30 ppm; in chronic poisoning these levels are only 1-2 ppm (Rosenfeld and Beath, 1964). The concentration in tissues depends not only on the quantity of selenium ingested but also on the chemical form at the time of intake. Selenium in the form of organic compounds was found to be retained longer and concentrated to higher levels than in inorganic forms (Smith et al., 1938). In chronic selenium poisoning the order of toxicity for equivalent amounts of selenium from various sources was wheat > corn > barley > selenite > selenate (Rosenfeld and Beath, 1964b).

A striking early toxicity was originally reported by Schroeder (1967) of selenite given to male rats in drinking water from the time of weaning, with no apparent toxicity from the same dose of selenate. Female rats also showed signs of toxicity, but of less severity. In a more recent study, Schroeder and Mitchener (1971a,b) evaluated the toxic effects of selenate, selenite, and tellurite on 313 rats (Long-Evans strain; BLU:LE) in a controlled environment given 2 ppm selenium or tellurium or these compounds in drinking water from the time of weaning until natural death. After 1 yr the doses of selenium were increased to 3 ppm. Selenite was extremely toxic, suppressing growth and causing high mortality, especially in males. The experiments were ended at 20-23 months. Selenate and tellurite were not toxic in terms of growth, survival, and longevity. Selenium accumulated in five organs (kidney, liver, heart, lung, and spleen) of control rats, although the diet contained only 0.05 $\mu g/g$ wet weight. Concentrations were more than doubled in selenium-fed animals; the kidney exhibited the greatest such concentration.

Although tolerable for growth and longevity, feeding selenate was accompanied by two potentially adverse effects not found in the controls. There was an increase in aortic plaques and lipids in both sexes, with significant increases in serum cholesterol levels. There was also some slight disturbance in glucose metabolism in older males and younger females, although it was not accompanied by glycosuria. Of added potential significance is the considerable increase in incidence of benign and malignant tumors in the older animals, indicating that selenium in the hexavalent form was both tumorigenic and carcinogenic (16.9% malignant tumors were found in the controls, 41.7% in the selenate group, and 18.2% in the tellurite group).

Pletnikova (1970) described the biological action and tolerable level of selenium in drinking water ingested by rats, guinea pigs, and rabbits. Selenites were found to be more toxic than selenates. The following LD50 values were found for Na_2SeO_3: male rats, 7.08-7.75; female rats, 10.50-13.19; guinea pigs, 5.06; and rabbits, 2.25 mg/kg. Rats receiving 0.0005 mg/kg selenite daily for 6 months showed decreases in liver function and in the activity of several enzymes. The concentration of glutathione and of its oxidized form increased. No physiological effects were observed in the experimental animals receiving daily doses of 0.00005 mg/kg selenite.

Harr et al. (1967b) fed semipurified rations that contained up to 16 ppm of added selenium as selenite or selenate. Lesions included toxic hepatitis, myocarditis, nephritis, and pancreatitis. The minimal toxic level of exposure under these conditions was 0.25 ppm based on liver lesions and 0.75 ppm estimated from longevity and lesions of the heart, kidney, and spleen. Rats fed rations containing 6-16 ppm selenium grew very slowly and died before 90 days of exposure; rats fed 4 ppm selenium grew slowly and 10% lived 15-24 months. Those fed the 0.5 ppm rations grew as well and lived as long (25% for 2 yr) as the controls. Livers from rats fed the 0.5 ppm rations were hyperplastic and livers from those fed 2.0 ppm selenium were cirrhotic. Although Harr and co-workers found median survival age greater in rats receiving diets with increasing levels than in those receiving a constant level, their data were admittedly inconclusive concerning selenium tolerance. However, Jaffe and Mondragon (1969) recently studied the changes of liver selenium after intake of organic selenium at different levels and for various periods in 236 young Sprague-Dawley rats and suggested the existence of an adaptation mechanism that allows rats exposed to chronic selenium ingestion to store less of this element than previously unexposed controls. In this study selenium (in seleniferous sesame press) cake was used in the diet of the rats. On a diet containing 4.5 ppm selenium exhibited a slower, more continuous rise of liver selenium than when the high-selenium diet was fed.

Toxicity in Livestock and Poultry

Selenium has long been considered an element toxic to farm animals and poultry. Since 1957 when Schwarz and Foltz (1957) demonstrated its effectiveness in preventing liver necrosis in the rat, it has been studied as an essential element as well (Proctor et al., 1958; Hogue et al., 1962; Mathias et al., 1965; Ewan et al., 1968; Muth et al., 1968).

Selenium excess. Organic compounds of selenium have been shown to produce several syndromes, depending on the species of animal, selenium concentration of the forage, the amount eaten, and possibly concomitant phytotoxins (Rosenfeld and Beath, 1964a,b). Selenium has been found in toxic amount in wheat and other plants in North and South America, Australia, New Zealand, South Africa, Algeria, Morocco, Spain, Bulgaria, France, the Soviet Union, Germany, Ireland, and Israel (Underwood, 1971). Although all animals are susceptible to selenium toxicosis, poisoning is most common in forage-eating animals such as cattle, horses, and sheep that may graze selenium-containing plants. All degrees of selenium poisoning exist from a mild, chronic condition to an acute form resulting in death of the animal.

Cattle and sheep that graze on primary indicator plants (100–10,000 ppm selenium) develop a clinially acute syndrome known as blind staggers, which is characterized by anorexia, emaciation, and collapse. Lesions include hepatic necrosis, nephritis, and hyperemia, and ulceration of the upper intestinal tract (Radeleff, 1964; Rosenfeld and Beath, 1964a,b).

Alkali disease, a subacute form of organic selenosis of cattle, sheep, and horses, is caused by feed that contains 25–50 ppm selenium. Lesions include alopecia, elongated, weak, and cracked hooves in horses, and erosion of the articular cartilages in cattle. Animals become lame; the heart, liver, and kidney are degenerate and fibrotic; conception is reduced, feti are reabsorbed; and the tissues contain 5–50 ppm selenium (Radeleff, 1964; Rosenfeld and Beath, 1964a,b).

Cattle, hogs, and horses, consuming seleniferous grains and plants in which selenium is bound to protein, suffer loss of hair and hooves, lassitude, anemia, and joint stiffness. It is likely that selenium displaces sulfur in keratin, resulting in structural changes in hair, nails, and hooves.

Chronic poisoning by selenium salts occurs from dietary exposures (e.g., 44 ppm selenium for horses, or a daily intake of 0.5–1.0 mg Se/kg for ruminants) (Muth and Binns, 1964). Effects of chronic inorganic selenosis include follicular skin, hemolytic anemia and serum bilirubin, increased weight of the spleen, pancreas, and liver, fatigue, lassitude, and dizziness.

Cattle (250 kg) fed 0.5 mg Se/kg three times a week became inappetent and depressed (Maag and Glenn, 1964). Concentration of selenium in blood was 3 ppm and lesions included gastroenteritis and polioencephalomalacia. Sheep (45 kg) fed up to 75 mg selenite/day developed myocardial degeneration and fibrosis, pulmonary congestion, and edema.

It is thus apparent that excess selenium is toxic with consumption of plant materials containing 400–800 ppm of selenium shown to be fatal to sheep, hogs, and calves. A chronic selenium toxicity in livestock occurs when animals consume seleniferous plants containing 3–30 ppm of selenium over a prolonged period of time. The toxicity of selenium to the various species of animals differs with the diet being consumed; for example, a high-protein diet may protect an animal from selenium poisoning. The following factors appear to be the most important in regard to selenium toxicity (although it is recognized that many factors may prevail): size and frequency of doses; characteristics of the compounds (organically bound selenium as found in forages and grains may be more toxic than inorganic selenium salts and metallic selenium); presence of synergistic, potentiating, or antagonistic substances; inherent susceptibility of the animals; efficiency of elimination after absorption; and dietary protein level. Although it is not apparently possible to state with any degree of absolute accuracy what constitutes the minimum toxic dose of selenium in each of its forms for the different kinds of livestock, the single acute oral minimum lethal dose of selenium for most animals generally ranges from 1 to 5 mg/kg body weight. The toxicity may be decreased by adding inorganic or organic arsenic to the drinking water or diet. Conversely, the toxicity may be increased with concomitant cobalt deficiency. Levels in the total diet of 25 ppm and greater of selenium may be sufficient to produce acute poisoning in most animals, while levels of 5 ppm or greater in the diet may produce chronic or subacute poisoning characterized by hemolytic anemia, hepatic necrosis, enlarged spleen, and reduced weight gain. Natural cases of chronic selenosis involving cattle, swine, and horses, but not sheep, have been reported (Dinkel et al., 1957). Table 4 summarizes the selenium requirements, tolerance levels, and toxic levels in livestock and poultry feeds (Patrias, 1969).

Pathological tissue changes attributed to selenium toxicoses vary with both the species of animal and the duration of the disease. Such major changes may include the following: swelling and congestion of the liver with varying degrees of degeneration progressing to focal necrosis and fibrosis in some animals; congestion, degeneration, and necrosis of renal tubular

TABLE 4 Selenium Requirements and Toxicity in Livestock and Poultry Feeds

Type of livestock	Requirement level (ppm of ration)	Tolerance level (ppm of ration)	Toxic level (ppm of ration)
Cattle	0.10	2.0	8.0
Sheep	0.10	?	10.0
Swine	0.10	2.5	7.0
Poultry	0.15	5.0	15.0
Turkey	0.20	?	?

epithelium and fibrosis of the kidneys; congestion of the spleen with hyperplasia of splenic nodules; hemorrhagic enteritis ranging from mild to severe; ulceration of the stomach; subendocardial and subserous hemorrhages in the heart with myocardial congestion, necrosis, cellular infiltration, and fibrosis; edema and congestion of the brain with neuronal degeneration in the cerebral and cerebellar cortices; articular erosions and deformed hooves; various developmental anomalies of the embryo (Glenn et al., 1964; Morrow, 1968; Caravaggi et al., 1970; Underwood, 1971; Buck and Ewan, 1973); and various chemical changes, including decreased vitamin A, ascorbic acid, and serum protein levels and an increase in nonprotein nitrogen levels. It is important to note that the organs affected by excessive levels of selenium are also those affected by selenium deficiency. For example, skeletal and heart muscle deficiency and the liver and other organs may exhibit degenerative changes and necrosis as well.

In regard to tissue concentrations in selenium toxicoses, the kidney and liver usually contain the largest amounts and the brain the least amount. The concentration in these tissues depends both on the length of time during which selenium was ingested and on the quantity ingested. Organic selenium also accumulates in higher quantities in these tissues than does inorganic selenium. Levels of Se from 4 to 25 ppm may be seen in liver and kidney in both chronic and acute cases of poisoning, and the urine from animals suffering from selenium toxicosis may contain from 0.1 to 8.0 ppm selenium (Maag and Glenn, 1964; Muth and Binns, 1964; Rosenfeld and Beath, 1964a,b; Underwood, 1971; Buck and Ewan, 1973).

Selenium deficiency. It is generally acknowledged that the selenium requirement of animals is influenced by the selenium balance of the organism as well as other factors. It has been amply shown, however, that if the selenium content of forage is low enough, deficiency diseases may occur even when the content of other factors in the forage is apparently normal.

Natural selenium deficiency in animals and birds has been observed in many countries including the major livestock areas of the world (Underwood, 1971). It is also evident that selenium-deficient areas are becoming increasingly recognized throughout the world as more analyses of feeds and diets are conducted with highly sensitive chemical and instrumental techniques that permit determination of selenium levels as low as 0.005 ppm in feeds and tissues. In newborn animals it takes the form of a degenerative disease of the striated muscles, or "white muscle disease" of calves, lambs, foals, and rabbits, and as a necrosis of the liver in pigs and as exudative diathesis in birds. Underwood (1971) suggested that there were basically three groups of deficiency diseases: those caused by vitamin E deficiency alone (encephalomalacia in chicks and sterility in rats); those due to deficiency in vitamin E and selenium, such as liver necrosis in rats and pigs, exudative diathesis in birds, and muscular dystrophy in several species, all of which are controlled by either selenium or vitamin E; and those due to selenium deficiency alone such as retarded growth and infertility in chicks and ruminants.

The best known and most investigated disease caused by selenium deficiency is nutritional muscular degeneration (NMD)—a degenerative disease of the striated muscles that occurs without neural involvement in a wide range of animal species. The essential importance of selenium deficiency in the etiology of NMD is supported by investigations in which it could be shown that the organs of animals suffering from NMD contained remarkably less selenium than the corresponding organs of healthy animals (Hartley and Grant, 1961; Lindberg and Siren, 1965).

NMD in farm animals is primarily a disease of young individuals. In sheep it may be divided into three different types—congenital, delayed, and a type occurring in hoggets (Gardiner et al., 1962).

The pathogenesis of muscular dystrophy is still not clearly understood. While it has been shown that naturally occurring myopathies of sheep and cattle are preventable by low levels of selenium, large doses of vitamin E are either ineffectual or afford only partial protection (Blaxter, 1962).

Selenium deficiency in animals generally appears as the selenium content of forages becomes less than 0.1 ppm. The critical level for dietary selenium below which deficiency symptoms are observed is apparently 0.02 ppm for ruminants and 0.03-0.05 ppm for poultry. Experimental evidence to date would indicate that 0.1 ppm of selenium in the ration of all classes of livestock (Allaway, 1969) and 0.1-2 ppm for chickens (Thompson and Scott, 1969) and 0.2 ppm in the diet of turkeys is the requirement for selenium. Selenium toxicity, from whatever source, produces loss of fertility, congenital malformations, defects in the eyes, small litters, and emaciated young. It is also of interest to note that both deficiency and toxicity of selenium cause retarded growth, muscular weakness, infertility, and focal necrosis of the liver.

A summary of various selenium-vitamin E deficiency effects is given in Table 5, and Table 6 illustrates the biological potency of selenium.

Carcinogenic, Teratogenic, Cytogenic, and Mutagenic Effects

Carcinogenic effects. The earliest reported apparent carcinogenic effect of selenium was that of Nelson et al. (1943), who were able to produce 11 hepatic neoplasms of low malignancy and four microscopic adenomatoid hyperplasias in seven groups of 18 male Osborne-Mendel rats fed either seleniferous grain or a solution of ammonium potassium sulfide and ammonium potassium selenide at a dietary level of 5-10 ppm selenium. Fifty-three rats survived for 18 months and liver tumors developed in 11, while in four others pronounced hepatic cell hyperplasia was observed. The tumors, which were either single or multiple, developed in cirrhotic livers, but a direct relationship to the degree of cirrhosis was not apparent, except that no tumors existed in animals developing cirrhosis but not surviving for 18

TABLE 5 Selenium–Vitamin E Deficiency Effects

Symptom	Species	Symptom	Species
Hepatic necrosis (hepatosis dietetica)	Pig	Infertility	Cow, sheep, pig
Mulberry heart	Pig	Yellow fat disease	Horse, pig
Muscle dystrophy	Sheep, cattle, pig, chick	Gizzard myopathy	Turkey, chick
		Paradentosis	Sheep, rat
Liver necrosis	Pig, rat	Pancreatic lesions	Chick
Muscle calcification	Sheep, cattle, pig	Cardiac muscle degeneration	Sheep, pig, cattle
Exudative diathesis	Chick	Sudden death	Sheep, pig
Growth depression	Rat	Alopecia	Rat, monkey
Hemorrhage	Rabbit	Cataracts	Rat
Elevated SGOT	Sheep, cattle	Skeletal muscle degeneration (white muscle disease)	Pig, sheep
Elevated aryl sulfatase and β-glucuronidase	Sheep		
Changes in serum protein levels	Pig	Encephalomalacia	Cattle, sheep, chick
Elevated serum ornithine carbamyl transferase (SOCT)	Pig	"Ill-thrift"	Sheep, cattle

months. The hepatic tumors were not completely encapsulated and were composed of fairly regular to irregular cords of large hepatic cells with cellular atypism apparently not prominent. Slight to moderate numbers of mitoses were seen and no metastic lesions occurred. The more disordered lesions were termed "low-grade carcinoma" despite an equivocal histological picture and the possibility that these might have represented one phase of hepatic regeneration rather than neoplasia (Shapiro, 1972). Fitzhugh et al. (1944) subsequently reported a low incidence of spontaneous tumors, less than 1%, was found in control rats. Seifter and co-workers (1946) observed multiple thyroid adenomas and adenomatous hyperplasia of liver in rats fed bis-4-acetaminophenyl selenium dehydroxide (0.05%) for 105 days. Shapiro (1972)

TABLE 6 Biological Potency of Selenium

	Concentration of selenium
Toxic dietary levels	5 ppm
Usual dietary level	0.1–0.5 ppm
Forage levels in selenium-deficiency areas	0.01–0.05 ppm
Lowest level attained in purified diets	About 0.005 ppm
Lowest level of supplemental selenium giving nutritional response	0.01 ppm or less
Lowest tissue level (lamb muscle) without signs of deficiency	About 0.05 ppm
Corresponding g atoms selenium/kg tissue	6.35×10^{-7}

has suggested that this compound probably has inherent carcinogenic activity per se (aside from its selenium content). It is generally acknowledged that on the basis of the above two studies (Nelson et al., 1943; Seifter et al., 1946) selenium was listed as a carcinogen in the food additive amendment of 1958 (Delaney amendment).

Tscherkes et al. (1961) fed 40 heterozygous male rats 12% protein diets containing 4.3 ppm selenium as Na_2SeO_4 (0.43 mg Se/100 g diet) for periods up to 32 months. Ten of 23 rats, surviving 18 months, developed tumors; three rats had hepatic cancer with lung metastases in two; four had sarcomas in which the site was not well determined, and three had hepatic adenomas. In a second group of 40 animals fed 0.86 mg Se/100 g diet, three rats developed sarcomas of mediastinum and retroperitoneal lymph nodes that appeared at 14–19 months. The criteria of neoplasia in these studies included uneven basophilia, preserved architecture, variation in size of cells and nuclei, and fatty degeneration. Tscherkes et al. (1961) concluded that neoplastic changes were caused by selenium and were due to an antagonistic relationship between selenium and methionine. The following should be noted, however, in the studies of Tscherkes et al. (1961): No non-selenium-fed controls were mentioned; no citation was made as to whether the animals developed cirrhosis; and sarcomatous lesions are known to develop spontaneously in some rat colonies.

Volgarev and Tscherkes (1967) subsequently noted that elevating dietary protein from 12 to 30% appeared to lessen the incidence of "carcinoma," although selenium in the diet was increased concomitantly, suggesting a significant protective role for protein. In the above study (Volgarev and Tscherkes, 1967), an overall tumor incidence of 8.5% in 200 rats was reported. The cancer incidence was 35% in the series of animals fed 0.43 mg selenium plus 12% casein, 8.5% in rats fed 0.86 mg Se/100 g diet plus 30% casein, and 0% in the third group fed 0.43 mg Se, 12% casein plus additives such as L-tocopherol, cystine, nicotinic acid, or choline.

In those animals with neoplasms, two of the four with hepatoma had pulmonary metastasis. Of the seven sarcomas, six were of the mesenteric lymph nodes, and only one (in series I) involved the liver. Four rats had hepatocellular changes that were classified as adenomas, and four more had sufficient cellular damage to be classified as precancerous. The control rats of series III, although fed the same rations as the series I animals, failed to develop tumors. This difference can possibly be explained by the fact that the experiments of series III were started 2 yr later than series I and involved animals obtained from a different source. Since the controls in series III developed no tumors, the effectiveness of the other dietary additives in the prevention of carcinogenesis could not be evaluated.

Harr et al. (1967a) described the effects of low levels of sodium selenate in semipurified and commercial diets of female Wistar rats over a 4-yr span. Sodium selenate was added in amounts of 2 ppm (with 12% casein), 2, 6, and 8 ppm (with 22% casein), and 4, 6, and 8 ppm (with methionine). In addition, soluble oxytetracycline was added to the drinking water at a level of 75 ppm and an additional group received the hepatocarcinogen acetylamino-fluorene (2-AAF) for comparison. The experimental animals were grouped into 272 statistical cells on the basis of 34 diets and eight logarithmically equal divisions of the animal's life-span from 28 to 1,150 days. The rats on control rations showed no evidence of malnutrition. The hepatic changes of the older animals on these diets included accentuated lobular pattern, hyperemia, cellular degeneration, mildly proliferative hepatocytes, and double and multiple nuclei. Sixty-three neoplasms were found in this study: 43 occurred in the 88 rats fed 2-AAF; the other 20 (of which 11 occurred in the controls) were randomly distributed through the experimental diets and included no hepatic neoplasms. *In situ* and hyperplastic lesions were most numerous in animals that had been on a control ratio for 50–250 days following an 84-day selenium feeding and in animals alternately receiving added selenium 1 wk and a control diet the next, apparently pointing to the major role played by other components of the diet in addition to selenium. The significant role played by the particular diet employed is illustrated by the fact that selenium added to a commercial diet produced less toxic and hyperplastic changes than selenium added to semipurified diets. Forty *in situ* lesions occurred in 727 selenium-fed animals, and these did not regress when selenium-free diets were instituted. Harr et al. (1967a) concluded that selenium was not related to the occurrence of neoplasia and further suggested that proliferating foci were related to hepatic necrosis. Other areas of major interest in the study of Harr et al. should be noted and include the following: (1) inflammations—there were 74 inflammatory lesions with the incidence of these lesions directly correlated with increasing dietary levels of selenium. The most frequent sites of inflammation were the uterus and intestines; less often involved were the heart and kidney. Associated lesions included myocardial sclerosis and renal amyloidosis. (2) Degeneration—there were 168 degenerative lesions that included defects of the epiphyseal plates, ascites, hydrothorax, icterus, degeneration of the kidney tubules, and eosinophilic infiltration. Of the 168 affected animals, 136 died or were moribund between 28 and 177 days of age.

The most salient features of the study of Harr et al. (1967a) were these: (1) hyperplasia was most marked and *in situ* lesions were most numerous in animals that had been on the control ration for 50–250 days following an 84-day selenium-feeding regimen and in animals alternately receiving added selenium 1 wk and a control diet the next; (2) two- to threefold resistance to selenium toxicity occurred in rats fed commercial rations when compared with semipurified diets; (3) of 21 parameters studied, selenium levels employed affected all except erythrocyte concentration and spleen weight; (4) the selenite form of added selenium was associated with increased myocardial inflammation and ischemia and reduced incidence of pneumonia and nephritis when compared to selenate; (5) hyperplastic liver lesions occurring in animals fed selenium did not regress when the added selenium was removed from the diet; (6) 63 neoplasms were observed, none of which could be attributed to the addition of selenium.

Tsuzuki et al. (1960) reported the occurrence of cylindromatous ("basalioma") subcutaneous dorsal neck tumors in mice exposed to metallic selenium vapor, oral metallic selenium, or rubbed with 10% metallic selenium ointment. No tumors occurred with similarly treated rabbits, and only four of a large group of animals were affected.

Innes et al. (1969) reported that induction of tumors in (C57BL/6 × C3H/Anf)f$_1$ (strain X) and C57BL/y × AKR)f$_1$ (strain Y) of mice treated with the fungicide ethyl selenac (selenium diethyldithiocarbamate) at a daily dosage of 26 ppm in the diet given *ad libitum* (after 28 days of age) following the administration of ethyl selenac at 10 mg/kg in 0.5% gelatin by stomach tubing to weanlings (days 7-28). Twelve of 18 male mice and three female mice of 17 strain X compared with three of 17 male and zero of 17 female strain Y had hepatomas at the end of approximately 21 months.

Schroeder et al. (1970) reported a 38% incidence of neoplasia in older male rats fed 3 ppm selenite and a 75% incidence in older female rats. Malignancy was present in 14 and 52% of the male and female rats, respectively, combined with a high rate of metastases. Although both subcutaneous and mammary tumors occurred, no location or histology of the lesion was noted.

In regard to the previous studies on carcinogenicity of selenium, Shapiro (1972) has observed that the presence of localized, but clearly aneoplastic, changes must be explained in part due to the age, sex, strain, diet, or medications used in animals consuming selenium; the specific influence of selenium is as yet undetermined; and the time interval before onset of neoplasia in each of these studies merits greater investigation and precludes conclusions regarding the carcinogenicity of selenium in animals that have not been under study for at least 18 months.

In an earlier section of this chapter, reference was made to the study of Schroeder and Mitchener (1971a,b) reporting the toxic effects of selenate, selenite, and tellurite when administered in drinking water to 313 rats from the time of weaning until natural death. It was ultimately found that selenate in older animals was both tumorigenic and carcinogenic (e.g., 16.9% malignant tumors in the controls vs. 41.7% in the selenate-treated group). This study has not been amenable to critical analysis, since the selenium-treated animals lived longer than the control animals; hence, the tumor incidence may thus have been due to the increased life span. A later study by Schroeder and Mitchener (1972) conducted on mice failed to show any increases in tumor incidence caused by selenium administration.

Jaffe et al. (1972) described studies in which rats were kept for up to 5 wk on diets containing 4.5 or 10 ppm organic selenium. The animals showed decreases in hemoglobin, hematocrit, fibrinogen, and prothrombin activity. At a higher level, growth inhibition, hepatic lesions, and spleen hyperplasia were also observed. When the 10 ppm Se diet was fed for 7 wk, a moderate reticulocytosis and a decrease in the time of erythrocyte survival were also noted. The incidence of pathological symptoms was higher in rats maintained on a normal (18%) protein diet than in animals receiving a high-protein (26%) diet or in those that had been bred on a 4.5 ppm Se diet.

Teratogenic effects. Franke and Tulley (1935) originally reported the production of deformed chick embryos in eggs from hens fed a diet containing a toxic principle. After it was demonstrated (Franke et al., 1936) that this toxic principle was selenium, similar defects were produced by injecting selenite or selenate into the air cells of fertile eggs. The teratogenicity of inorganic selenium salts to chick embryos was also demonstrated by Halverson et al. (1965).

Palmer et al. (1973) recently elaborated the toxicity of various selenium derivatives to 4-day-old chick embryos. The most toxic compound studied was methyl selenic acid. Embryonic monstrosities were produced especially by the injection of sodium selenite or selenomethionine. The most common deformities were underdevelopment of beak and abnormal development of feet and legs. This study also demonstrated the low toxicity of trimethyl selenium ion, $(CH_3)_3 Se^+$ (TMSe) to developing chick embryos, supporting the hypothesis that it is a detoxification product of selenium as has been suggested for rats (Byard, 1969).

Selenium has been shown to cross the placenta in rats and cats (Westfall et al., 1938), and its passage can be inferred in other species from the symptoms of selenium poisoning observed in the offspring of animals grazing on seleniferous rangeland (Rosenfeld and Beath, 1964). For example, foals and calves in affected areas are sometimes born with deformed hooves. The offspring of pigs (Wahlstrom and Olson, 1959), sheep (Rosenfeld and Beath, 1947), and rats (Rosenfeld and Beath, 1954) have all exhibited abnormalities following administration of selenium during pregnancy. However, Holmberg and Ferm (1969) reported that 2 mg sodium selenite/kg (an almost lethal dose) administered iv had no teratogenic or embryotoxic effects in the hamster. Toxic effects of six trace metals on the reproduction of mice and rats were elaborated by Schroeder and Mitchener (1971b). Breeding mice of the Charles Rivers CD strain and rats of the Long-Evans (BLU:LE) strain were exposed to low doses of selenium, lead, cadmium, molybdenum, titanium, and nickel in drinking water in an environment controlled as to contaminating trace metals. Each group was carried through three generations. Compared with control mice given only doubly deionized water, selenate (3 ppm selenium) resulted in excess deaths before weanling, runts, and failures to breed. Feeding the six trace elements resulted in relative toxicities in the following order: lead > cadmium > selenium > nickel, arsenic, titanium, and molybdenum. Exposure of breeding mice and rats resulted in the following relative toxicities: lead > cadmium > selenium > nickel > titanium > molybdenum > arsenic.

While developing males *in utero* are believed to be more vulnerable to toxic substances than are developing females, Schroeder and Mitchener (1971b) found, however, the ratio of males to females born increased in mice exposed to selenium and arsenic, compared with controls. It was concluded that certain trace elements fed to rats and mice in doses that do not interfere with growth or survival are clearly intolerable for normal reproduction.

The interrelationships of selenium, cadmium, and arsenic in mammalian teratogenesis were reported by Holmberg and Ferm (1969). Cadmium sulfate and sodium arsenate injected into pregnant golden hamsters (*Cricetus auratus*) are independently teratogenic. However, sodium selenite, which is not teratogenic under similar conditions (2 mg/kg administered iv during day 8 of pregnancy), does provide significant protection against the malformations induced by cadmium or arsenic when injected simultaneously with either of these teratogens. The effect of a time interval between the injection of cadmium and selenium revealed that the protective effect of selenium was still marked at a 30-min interval, but decreased markedly at 2- to 4-hr intervals. At the 30-min interval, the protective effect was more marked when the selenium was injected first rather than the reciprocal method.

The significant decrease in the teratogenic toxicity of both cadmium and arsenic, which occurred when either was injected simultaneously with selenium into pregnant hamsters, suggests a metabolic antagonism between selenium and both cadmium and arsenic. Evidence of such metabolic antagonism between selenium and both cadmium and arsenic is scarce. Sodium selenite has been reported to reverse the degenerative vascular changes that cadmium chloride induces on the rat gonad (Kar et al., 1960; Mason et al., 1964). Further evidence suggests that selenium protects against cadmium damage to the rat testis by chelating with the cadmium bound to sulfur on sulfhydryl thiol groups (Gunn and Gould, 1967). Both selenium and cadmium follow very closely the biochemical pathway of sulfur (Ganther, 1965; Robertson, 1970), and it has been suggested (Holmberg and Ferm, 1969) that the mechanism of selenium protection from the teratogenic effect of cadmium on the hamster embryo may very well depend on macromolecules containing sulfur groups.

It was noted by Holmberg and Ferm (1969) that while the levels of the three metals used in their studies exceed the usual levels in environmental contamination, a possibility exists of accidental exposure to high levels of one of these metals during early human gestation.

Robertson (1970) suggested that selenium may be a possible teratogen in humans. Out of one possible and four certain pregnancies among women exposed to selenite-containing powder, an ingredient in a medium for salmonella culture, only one pregnancy went to term; the infant showed bilateral clubfoot. Of the other pregnancies, two could have terminated because of other clinical factors. Clegg (1971) reported that widespread inquiries in other laboratories where exposure to selenite could occur have not revealed any similar pattern, except for one miscarriage in the only pregnancy in 5 yr in one small laboratory. It is apparent that evidence is equivocal and difficult to relate to the possible effects of selenium on human embryonic development. Clegg (1971) suggests that it would appear unlikely that food contamination with selenium would be a major problem in the cities because of the wide variations in the source of foodstuffs. However, he cautions that a close watch should be kept on rural populations, especially in areas where the soil has a high selenium content.

The placental transmission of selenium in humans has been measured by Hadjimarkos et al. (1959). Shamberger (1971) in a study of neonatal death rates in humans from high and low selenium areas concluded that there was no detectable teratogenic threat to humans from selenium in the natural environment.

Cytogenic and mutagenic effects. Selenium has been shown to affect the genetic process in barley (Walker and Ting, 1967) and in *Drosophila melanogaster* (Ting and Walker, 1969; Walker and Bradley, 1969). Treatment with sodium selenite before meiosis decreased the genetic recombination in barley and caused structural alterations in the meiotic chromatin (Walker and Ting, 1967). Selenoamino acids also reduced the genetic crossing-over in *D. melanogaster.* For example, selenocystine at 2 μM had a significant effect on crossing-over in the X chromosome of *D. melanogaster* (Ting and Walker, 1969; Walker and Bradley, 1969). It was also found that interacting effects of urethane and selenocystine were at certain levels antagonistic and at other levels synergistic.

Sentein (1967) described the action of selenates and selenites on the segmentation mitoses, which was similar to that of SeO_2:polar dissociation with conserved dominance of the principal pole, stickiness, and clumping of chromosomes. Fukina and Kudryavtseva (1969) reported that sodium selenite solution added to kidney tissue cultures of rabbits and pigs in 10^{-1} to 10^{-4} dilutions caused cell degenerative changes and decreased the mitotic activity. However, application of sodium selenite in considerations of 10^{-5} did not affect the tissue culture.

The effects of a variety of metal compound (including sodium selenate, sodium selenite, sodium tellurite, ammonium tellurate, sodium arsenate, antimony sodium tartrate, cadmium chloride, cobalt nitrate, nickel powder, nickel oxide, beryllium sulfate, sodium metavanadate, sodium orthovanadate, and mercuric chloride) on chromosomes in cultures of human leukocytes and diploid fibroblasts were elaborated by Paton and Allison (1972). Subtoxic doses of metal salts were added to leukocyte cultures and to fibroblast cells at various times between 2 and 24 hr before fixation. An increased incidence of chromosome breakage was found in leukocytes treated for 48 hr with arsenic, antimony, and tellurium salts. Chromosome aberrations were not seen after addition of the salts of selenium, cadmium, cobalt, nickel, iron, vanadium, and mercury to leukocyte cultures. (Cells were exposed to at least two concentrations of each metal salt for 24, 48, or 72 hr.)

Selenomethionine was reported by Craddock (1972) to act as a methyl donor in the methylation of DNA, transfer RNA, and ribosomal RNA in the intact rat. The relative amounts of the different methylated bases formed in each nucleic acid were similar to those found after injection of [^{14}C] methyl-methionine. It was thus considered probable that selenoadenosylseleno-methionine (Se-A SeM), which is known to be formed from selenomethionine *in vivo*, is a methyl donor in no way different from *S*-adenosylmethionine in the reactions catalyzed by the nucleic acid methylating enzymes.

Anticarcinogenic Effects of Selenium

A number of studies have demonstrated the inhibitory effect of selenium on carcinogenesis. For example, Shamberger and Rudolf (1966) reported that sodium selenide, one of the most powerful antioxidants known (Tappel, 1962), greatly reduced the number of tumors in ICR Swiss female mice ($p < 0.05$) and the total number of tumors produced when administered concomitantly with croton oil to 7,12-dimethyl[a] anthracene-treated mouse skin. Shamberger (1969) also showed that both sodium selenide (0.0005%) and oral sodium selenite (1.0 ppm) were effective in decreasing the incidence of skin papillomas produced in mice by the inducer–promoter system di-methylbenzanthracene croton oil, or by benzopyrene. In later studies, Shamberger (1970) reported that 0.0005% sodium selenide significantly reduced the numbers of tumors in female albino ICR Swiss mice in five of six nondietary promotion experiments. In two of these six experiments, vitamin E also significantly reduced the number of animals with tumors, and in all six experiments, both selenium and vitamin E decreased the total number of papillomas, while other lysosomal stabilizers and antioxidants such as hydro-cortisone and chloroquin were ineffective. After 19 wk of administration, sodium selenide did not significantly lower the number of mice with 3-methylcholanthrene-induced papillomas, and after 30 wk of administration it did not reduce the number of cancers. Torula yeast diets containing 1.0 ppm sodium selenite administered to mice markedly decreased the number of skin tumors induced by 7,12-dimethylbenz[a] anthracene plus croton oil ($p < 0.05$) and benzo[a] pyrene ($p < 0.001$). However, Torula diets containing 0.1 ppm sodium selenite and also a commercial diet (Rockland) failed to decrease tumor incidence.

A 50% decrease in dimethylaminoazobenzene-induced liver tumors in groups of male rats fed diets with 5 ppm sodium selenite has been reported by Clayton and Bauman (1949). Von Euler et al. (1956) described the regression of rat ascites tumors by feeding selenite together with 4-methylbenzeneselenic acid. Gusberg et al. (1941), however, reported both selenide and selenate to be ineffective in preventing growth of sarcoma in rats.

Riley (1969) studied the relationship between subcutaneous mast cells and papilloma and found that topical selenium not only decreased the incidence of papillomas but also prevented the accumulation of mast cells at the base of the lesions in carcinogen-treated animals.

The effect of dietary selenium N-2-fluorencylacetamide (FAA)-induced cancer in vitamin E-supplemented selenium OSU-Brown rats was studied by Harr et al. (1972). Mammary adenocarcinomas and/or hepatic carcinomas developed more slowly in groups of animals treated with 150 ppm FAA and 0.5–2.5 ppm selenite than in animals treated with 0.0–0.1 ppm added selenium and 150 ppm FAA. The reduction was 85–95% after 200 days of exposure to 150 ppm FAA and 20–30% after 320 days of exposure.

Harr et al. (1973) in a corollary to the above study also reported that the concentration of selenium in the liver of rats fed selenite-supplemented rations was inversely correlated with the incidence of hepatic and mammary cancer. Chronic exposure to various concentrations of selenite in the feed did not increase the concentration of selenium in skeletal muscle of rats. The concentration of selenium in the liver of rats fed nutritionally adequate amounts of selenite increased slowly, even though a portion of the exposure regimen included 40% of the LD50 dose. In these groups of animals, the addition of selenite to the basal selenium-deficient ration increased longevity and decreased cancer induction.

It is interesting to note that selenium has been proposed as an antitumor agent for more than 60 yr. For example, Dalbert (1913), Walker and Klein (1915), Lockhart-Mummery (1935), Todd et al. (1935), and Hernaman-Johnson (1935) proposed the combined use of selenium as lead selenide or sulfur–selenium colloid but recognized that the utility of these compounds would be severely limited by their systemic toxicity.

Weisberger and Suhrlaud (1956a,b) demonstrated the growth-inhibiting effects of selenocystine on Murphy lymphosarcoma cells in the rat. In four patients with acute and chronic leukemia treated with selenocystine, prompt reduction in leukocyte count occurred, but marked systemic toxicity precluded further treatment.

The mechanism by which selenium prevented tumor formation in the above studies is not known. Despite extensive study, the biologically active form of selenium is not definitively known, nor are its apparent metabolic relationships with vitamin E or coenzyme Q.

Selenium analogs of sulfur compounds of biological interest have been reported to occur in animal tissues, for example, selenomethionine, selenocystine, selenocoenzyme A, and selenotaurine (Shrift, 1961; Olson, 1965).

Inorganic selenium is believed to be associated with the sulfur of proteins through loose ionic bonds that can be dissociated rather than be incorporated into proteins in animals through protein synthesis of its amino acid analogs—selenomethionine and selenocystine (Cummins and Martin, 1967; Jenkins, 1951).

Shamberger (1970) suggested that inorganic selenium may seek out the protein of rapidly growing tumors, where it prevents or retards further growth, thus causing the tumor inhibition observed. The attachment of inorganic selenium to the —SH— groups of the protein may prevent or diminish the ability of carcinogens per se to attach themselves to these sites.

Toxic Effects in Humans

Industrial exposure. Although selenium has been used commercially for many years with no reported long-term systemic effects to date on industrial workers, the element has long been regarded as a toxic substance. It is most essential to designate the form of the element in regard to the toxicity of selenium.

Elemental Selenium. The increasing use of selenium in a variety of manufacturing processes represents an important potential industrial health hazard involving essentially two major groups of industries: those that extract, mine, treat, or process selenium-bearing minerals and those that utilize selenium in manufacturing. Table 7 lists the potential industrial hazards in terms of the industry, source of hazard, and type of hazard. The numbers of workers affected, especially in the manufacturing category, are considered relatively low (Dudley, 1938a,b; Hamilton and Hardy, 1949) with the extent of the hazard depending primarily on the types of processes being carried out with a specific physical or chemical form of selenium and/or its derivatives.

TABLE 7 Potential Industrial Hazards

Industry	Source of hazard	Type of hazard
Primary industries		
Copper	Ore concentrate and flue dusts, sludges	Se, SeO_2 and mixed dusts
Lead and zinc	Ore concentrate and flue dusts, sludges	Se, SeO_2 and mixed dusts
Pyrites roasting	Roasting towers, sludges	Mixed dusts, Se
Lime and cement (certain areas)	Dust, kiln gases	Mixed dusts, SeO_2
Secondary industries		
Glass, ceramics	Melting pots, furnaces	Fumes of Se, SeO_2
Rubber	Vulcanizing and curing processes	Organic vapors, H_2Se
Steel and brass	Alloy furnaces	Dusts, Se, SeO_2 fumes
Paint and ink pigments	Pigment compounding and mixing[a]	H_2Se, SeO_2 soluble dusts
Plastics	Mixers, presses	Organic vapors
Photoelectric	Melting and casting operations	Vapors, Se, SeO_2
Chemicals	Mixing, melting, synthesis	Se, SeO_2, H_2Se, organic vapors

[a]Imprinting with selenium-bearing ink may be added as a source of hazard in the paint and ink pigment industry.

Thus, for example, selenium-containing dusts, fumes, vapors, or liquids may present definite hazards, the degree of which depend on the specific processes involved or dissipation of the noxious materials. Soluble dusts, for example, such as selenium dioxide, selenium trioxide, and certain halogen compounds, may prove toxic due to the ease with which they are absorbed by the tissues of the skin, the lung, and alimentary canal. The most toxic vapors include hydrogen selenide, methyl selenide, ethyl selenide, and various aromatic selenides. Conversely, certain dusts, such as that of elemental selenium, may be of such composition that on inhalation no soluble selenium compound is liberated.

Exposure to elemental selenium in manufacturing processes has produced intense and immediate irritation of eyes, nose, and throat, severe burning sensations of the nostrils, immediate sneezing, coughing, nasal congestion, dizziness, and redness of the eye. While the most heavily exposed have had slight difficulty in breathing, frontal headaches, severe dyspnea, and edema of the uvula (Clinton, 1947), it is considered doubtful that all the observed symptoms are due to elemental selenium per se; the high temperature used in many processes may have converted elemental selenium to selenium dioxide. It has been reported by Nagai (1959) that long-continued exposure of workers engaged in the manufacture of selenium rectifiers in Japan has resulted in hypochronic anemia and drastic leukopenia, with increasing numbers of female workers having irregular menses or menostasis.

Symptoms of chronic selenium poisoning occurring in industry (copper refining), first recognized by Hamilton in 1917 (Hamilton, 1927, 1934), were characterized by bronchial irritation, gastrointestinal disturbances, nasopharyngeal irritation, and a persistent garlic odor on the breath.

Acute selenium poisoning in industry has not been widespread, despite the widespread and increasing use of this metal. Selenium fume exposure has been reported by Clinton (1947) to have occurred in a plant engaged in smelting scrap aluminum, where a furnace was charged with 20,000 lb of aluminous crop containing about 200–300 lb of selenium rectifier plates. Exposure to selenium fumes resulted in severe local irritation rather similar to that produced by sulfur dioxide, but did not produce selenium poisoning. Thus far, the studies would indicate that elemental selenium is relatively nontoxic to humans. The only forms of elemental selenium where irritation has been observed are fine dust and selenium fumes. On rare occasions a skin reaction has been observed in industrial workers, and a few cases have been recorded of dermatitis on the backs of hands of workers who had handled selenium (Amor and Pringle, 1945).

Selenium Dioxide. Selenium dioxide is the main selenium problem in industry since it is formed whenever selenium is boiled in the presence of air. It is a white powder, which forms selenous acid in contact with water or sweat, and this is about as irritating as sulfur dioxide and sulfurous acid. In regard to acute local effects, the sudden inhalation of large quantities of SeO_2

may produce pulmonary edema, due to the local irritant effect on the alveoli in the lungs. The other acute effects are those where SeO_2 comes into contact with skin, nails, and eyes, resulting in intense local irritation and inflammation.

The toxic effects of SeO_2 dust may be due to its rapid conversion to selenous acid when the dust comes into contact with moist surfaces of the tissues. Cutaneous disorders in workers handling SeO_2 have indicated that the selenite ion was responsible for its toxic action (Hoger and Bohm, 1944). Skin lesions produced by SeO_2 resembling hematomas in a state of resorption were also described (Frant, 1949).

In regard to generalized systemic effects, the first and most characteristic sign of selenium absorption is a garlic odor of the breath. (It should be noted that the garlic odor of the breath is not a reliable guide to selenium absorption as it can be absent while an individual is excreting as much as 0.1 mg/liter selenium in the urine) (Glover, 1970). The odor is generally believed to be due to dimethyl selenium. The second most prominent systemic effect is a metallic taste of the tongue, which is a more subtle and earlier sign than the garlic odor on the breath. There have been suggestions that the garlic odor may be due to the contamination of selenium with tellurium (Watkins et al., 1942). The metallic taste is less dramatic and often overlooked by those exposed and, hence, is an unreliable sign of selenium exposure. The third group of generalized effects are classified as indefinite sociopsychological effects that include symptoms of pallor, lassitude, and irritability. Glover (1970) noted that all these acute local and generalized effects due to selenium clear dramatically when the person is removed from or protected against contact with selenium dioxide.

The possible long-term effects, however, would especially require scrutiny in regard to liver or renal damage. As in animals both the liver and kidneys appear to be especially sensitive to selenium absorption. Porphinuria has been said to be associated with selenosis, especially when skin lesions are present and may be increased four times the normal (Holstein, 1951).

Glover (1970) additionally analyzed the observed and expected 17 deaths, which were the only ones that were traceable among the 300 workers during the 26-yr history of a particular selenium rectifier process in the United Kingdom. One-third of the deaths were cancerous, which was in line with the national average, and the cancers were from differing organs of origin, two from the lungs, and one each from the stomach, colon, ovary, and testes.

Hydrogen Selenide. Hydrogen selenide, a very offensive smelling gas, is probably one of the most toxic and most irritating selenium compounds known. It is formed by the action of acids or, in some cases, water on inorganic substances. It is also formed by the reaction of selenium with organic matter and the direct reaction of the elements. Buchan (1947) described five cases of industrial selenosis (subacute intoxication) due to less than 0.2 ppm of hydrogen selenite (which was believed generated from the use of selenous acid solution in an etching and imprinting operation). Predominant symptoms were nausea, vomiting, metallic taste in the mouth, dizziness, and extreme lassitude and fatigability. The average excretion of selenium in the exposed workers was 20–150 μg/liter urine [within the normal limits given by Sterner and Lidfeldt (1941)]. No significant level in the urine was observed. Acute irritation of the mucous membrane of the respiratory tract, pulmonary edema, severe bronchitis, and bronchial pneumonia have been described in acute hydrogen selenide intoxication (Symanski, 1950). Long-term systemic effects might possibly result if the exposure is repetitive. Rohmer et al. (1950) reported on a new aspect of selenium poisoning in a chemist exposed to a high concentration of hydrogen selenide who developed a severe hyperglycemia, which could only be controlled by increasingly large doses of insulin.

Selenium Oxychloride. Selenium oxychloride ($SeOCl_2$) is an excellent solvent for many substances, including metals. It is a corrosive liquid with severe vesicant properties capable of producing third-degree burns (Dudley, 1938b), which heal very slowly. The vapors of selenium oxychloride are toxic and decompose readily in air; hence, their irritant and corrosive action on the respiratory tract is not as severe as on the skin. The low vapor pressure limits the concentration to the order of 60–70 ppm when in contact with dry recirculated air (Patty, 1949).

Nonindustrial exposure. Few cases of acute nonindustrial poisoning from selenium have been recorded. Rosenheim and Tunnicliffe (1901) described one epidemic due to beer containing selenic acid derived from impure sulfuric acid used in its manufacture. However, it was admitted that the influence of selenium was secondary to that of sulfuric acid present in samples of beer at levels three to four times that of selenium.

Chronic selenium poisoning from ingestion of foods grown on seleniferous soils has been suggested chiefly on the grounds of absorption as indicated by the urinary excretion of selenium (Dudley, 1938b). Investigations of the sources of selenium to which humans are exposed in seleniferous regions have shown wide occurrence of the element in foods of animal origin, such as milk, eggs, and meat, as well as in vegetables and cereal grains. It is believed that a concentration of 5 ppm in common foods or one-tenth this concentration in milk or water is potentially dangerous (Rosenfeld and Beath, 1964b).

Smith and co-workers (1936) attempted to ascertain the effects of selenium on the rural populations of seleniferous areas in selected localities in Wyoming, South Dakota, and Nebraska. Examination of members of 111 families revealed no symptoms or groups of symptoms that could be considered decisively characteristic of selenium poisoning in man. The more pronounced manifestations of disease seen in the 111 families (exclusive of vague symptoms of anorexia, indigestion, general pallor, malnutrition) were the following: bad teeth seen in one or more members of 48 families, yellowish discoloration of the skin (in many cases a very definite icterus) seen in 46 subjects, skin eruptions (in 20 subjects), chronic arthritis (15 subjects), diseased nails of the fingers and toes (in eight subjects), and subcutaneous edema (in five cases). An attempt was made to correlate the clinical findings with the urinary selenium concentrations, which, however, did not reveal a constant causal association of health disturbances with concentration of selenium in the urine. Little difference was found in the percentage of symptomatic cases through a wide selenium range in the urine from a trace to 1.33 ppm. It can be assumed, however, that a higher concentration of selenium in the urine probably represents a higher level of intake and a correspondingly higher concentration in the tissues. It should be noted that the high incidence of symptoms in groups of individuals excreting small quantities of selenium might be a manifestation of irreparable damage due to the intake of selenium in the higher concentrations at some time in the past.

The studies of Smith et al. (1936, 1937a,b) indicate that the incidence of primary jaundice, recurrent jaundice, and sallow color of younger individuals is higher than the probable occurrence in people not living on seleniferous land. In experimental animals with mild or moderate chronic selenium poisoning from inorganic, as well as from organic, selenium marked bilirubinemia has been observed (Fimiani, 1949). In chronic poisoning in experimental animals with doses of inorganic selenium of similar magnitude to those probably absorbed by the human subjects, Smith et al. (1973b) noted the frequency of microscopic lesions of the liver. Experimental studies of the chronic poisoning of animals have shown that the excretion levels of selenium in the urine bear a rather definitive relationship to the daily dose ingested. It can be inferred that the human subjects investigated in the studies of Smith et al. (1936, 1937a) were absorbing from 0.01 to 0.1 or possibly as much as 0.2 mg selenium/kg body weight per day. The daily intake of 1.0 mg selenium/kg may produce chronic toxicity in humans. Smith and Lillie (1940) caution that the continued ingestion of food selenium as low as 0.2 mg/kg may be harmful. It should also be noted that more recent surveys have suggested that a high incidence of bad teeth may be correlated with selenium intake. For example, a survey on white native born and reared boys and girls from 10 to 18 yr of age living in seleniferous areas in Wyoming showed that a higher percentage of children had gingivitis in the seleniferous than in the non-seleniferous area. The urinary excretion ranged from 0.02 to 1.12 ppm selenium (Tang and Storvick, 1960). Additional studies of Hadjimarkos et al. (1952, 1958) also indicate that selenium may be a factor contributing toward the increased susceptibility to dental caries.

Aside from unusual industrial exposure and exposure of individuals residing in highly seleniferous areas, humans consume a modest amount of selenium in their daily diet. Grains, skimmed milk, brown sugar, pork, kidney, seafoods, egg yolk, beef, chicken, mushrooms, and garlic contain considerable amounts of selenium (Morris and Levander, 1970; Underwood, 1971). Woolrich (1973) estimates that the percent of total human intake of selenium from air is 0.07 $\mu g/m^3$, which represents a 0.1% intake from air. In comparison, the daily human intake of selenium from food and water is 68 μg.

Miscellaneous selenium compounds that have been classified as toxic in humans include sodium selenate (toxic dose estimated to be 0.2 g) (Thienes and Haley, 1972).

EPIDEMIOLOGY STUDIES

Despite the apparent utility of epidemiological methods and techniques for studying chronic selenium toxicity in human beings, it is somewhat surprising that scant few such studies have been reported. For example, two studies were reported in the United States (Smith et al., 1936; Smith and Westfall, 1937) almost 40 yr ago among farm people living in high seleniferous regions of South Dakota, Wyoming, and Nebraska, where selenium toxicity was prevalent in livestock. Routine physical examinations among the participants revealed various symptoms of ill health indicative of selenium toxicity, such as dermatitis, gastrointestinal disturbances, jaundice, arthritis, and sloughing of finger- and toenails. It is also of interest to note that one of the most frequent symptoms of disease observed was a high prevalence of dental caries.

Hadjimarkos and co-workers (Hadjimarkos, 1956, 1970; Hadjimarkos and Bonhorst, 1958, 1961) initially reported epidemiological studies on the association between selenium intake and dental caries among children who were life-long residents in the state of Oregon. The consumption of small amounts of dietary selenium during the period of tooth formation increases the prevalence of dental caries. A direct relationship was found between prevalence of dental caries in children and levels of selenium in urine, eggs, milk, and water in three counties in Oregon. There was a tenfold difference in the selenium content of milk and eggs between the counties of low and high caries that were statistically significant ($p < 0.01$). It is also important to note that water is not an important source of selenium intake. The findings in Oregon [described by Hadjimarkos and co-workers (1956, 1958, 1961, 1970)] on the direct association between levels of dietary selenium and prevalence of dental caries were corroborated later by two additional studies conducted by independent investigators among children who were life-long residents in certain localities in the states of Wyoming, Montana, Oregon, and South Dakota (Tang and Storvick, 1960; Ludwig and Bibby, 1969).

The various studies on the association between selenium intake and caries, both among children and with laboratory animals, have been extensively reviewed by Hadjimarkos (1965, 1968, 1970). The mode of action of selenium in increasing the susceptibility to dental caries has not been resolved.

Selenium may alter primarily the the chemical makeup of the protein fraction of enamel or enhance acid decalcification of the inorganic components of the enamel.

Because of the striking inhibitory effects of selenium in animals against a variety of carcinogens and the possibility that selenium at higher levels might be enhancing carcinogenesis, it is intriguing to determine whether selenium occurrence might be correlated with human cancer mortality. Shamberger and Frost (1969) recently described a very high negative correlation ($R = -0.96$; $p < 0.001$) between the selenium level of the blood of adult males in the United States and human cancer death rates. This correlation was based on the values for 10 cities or counties with populations of 40,000–70,000. The selenium levels in the blood were those secured by Allaway et al. (1968) with normal blood selenium values ranging from 10 to 34 μg/100 ml, depending on the geographic residence of the individual. Age-adjusted cancer death rates were used when available; the cancer death rate up to age 40 may be only 10–150/100,000, while from age 40 to 100 the cancer death rate increases dramatically from 150 to 1,000/100,000, depending on the age group.

Selenium is known to be concentrated in the liver and kidney as reported by Hirooka and Galambos (1966). Shamberger and Willis (1971) suggest that if selenium has an effect on cancer and the protection mechanism depends on its concentration, it should be anticipated that the incidence of human renal and hepatic cancer would be sharply reduced in areas with adequate selenium. To eliminate certain variables such as the degree of urbanization, the amount of air pollution, and racial makeup, 17 large cities with a population of over 80,000 in the high-selenium areas were matched with 17 large cities in the low-selenium area within about 15% at random on the basis of their white, age-adjusted, 1959–1961 expected deaths (Shamberger and Frost, 1969).

The salient features of the above epidemiology study of Shamberger and Willis (1971) are the following: (1) when low-selenium areas are compared with those having medium and high selenium, striking differences were observed in the total cancer death rates for both males and females (Table 8). Several of the death rates that decreased for both males and females included lymphomas and cancers of the digestive organs and peritoneum, lungs, and breasts. [These results with lymphomas are in essential agreement with study of Mautner et al. (1963) that indicated that selenoguanine has a higher therapeutic index against mouse lymphomas than does thioguanine.] Decreases were observed in the male, but only in the case of urinary organ cancer and cancer of the buccal cavity and pharynx. Female death rates from cancer of genital organs were greater in the low-selenium areas than in the medium- and the high-selenium areas. (2) There is a decrease in the death rate of five high-selenium counties compared with those rates in the five low-selenium counties in the same state. A similar pattern is observed in California where

TABLE 8 Selenium Distribution in Forage Crops and the 1965 Age-Adjusted Human Cancer Death Rates for Selected Cancer Sites for Each Sex

| | Cancer deaths/100,000 ± SE | | | | | |
| | Male | | | Female | | |
Types of cancer	Low Se	Medium	High	Low Se	Medium	High
All	201.78 ± 0.66	161.12 ± 0.83	166.28 ± 0.92	169.65 ± 0.61	135.96 ± 0.76	136.47 ± 0.82
Digestive organs and peritoneum	67.38 ± 0.38	47.89 ± 0.45	48.59 ± 0.50	54.71 ± 0.34	39.71 ± 0.41	41.32 ± 0.45
Respiratory	54.82 ± 0.35	43.49 ± 0.43	44.39 ± 0.48	11.08 ± 0.15	8.79 ± 0.19	7.98 ± 0.20
Genital organs	19.25 ± 0.20	18.40 ± 0.28	19.28 ± 0.31	27.78 ± 0.24	24.37 ± 0.32	23.40 ± 0.34
Breast	0.33 ± 0.03	0.25 ± 0.03	0.16 ± 0.03	33.55 ± 0.27	25.93 ± 0.33	24.39 ± 0.35
Lymphomas	10.29 ± 0.15	9.32 ± 0.20	8.78 ± 0.21	7.59 ± 0.13	6.54 ± 0.17	6.38 ± 0.18
Urinary organs	12.00 ± 0.16	8.78 ± 0.19	8.72 ± 0.21	5.75 ± 0.11	4.82 ± 0.14	5.63 ± 0.17
Leukemia and aleukemia	0.09 ± 0.14	8.46 ± 0.19	9.10 ± 0.22	6.49 ± 0.12	6.30 ± 0.16	6.32 ± 0.18
Buccal cavity and pharynx	6.04 ± 0.11	5.08 ± 0.15	4.52 ± 0.15	1.75 ± 0.06	1.86 ± 0.09	1.61 ± 0.09

selenium exposure diminishes from south to north. The low-selenium cities have a much higher death rate than the high- or medium-selenium cities. (3) When the blood selenium content is correlated with the human cancer mortality in 19 cities, a correlation coefficient of -0.434 ($p < 0.05$) is obtained. Eliminating two cities (Lubbock and El Paso) located in a high-selenium region responsible for most of the variance and where the selenium blood content of the individuals tested was relatively low, the remaining 17 cities then would have a correlation coefficient of -0.81 ($p < 0.001$). (4) The rank correlation between the selenium content of milk in 1965 and the 1965 human cancer mortality is -0.60 ($p < 0.05$), which is barely significant. (5) Seventeen states low in selenium occurrence had a renal cancer death rate of 2.98 ± 0.07 and the 26 states from the high- and medium-selenium areas have a death rate of 2.58 ± 0.098 ($p < 0.01$). The renal cancer death ratio is reduced for both sexes, but the male liver death ratio is not significantly reduced, while the female liver death ratio reduction is significant. It is not considered likely that the mechanism of selenium protection against human cancer mortality is dependent on concentration of selenium in the liver and kidney. Shamberger and Willis (1971) consider the protection mechanism in both sexes to be primarily a contact phenomenon rather than a concentration phenomenon. (6) In general, the terminology "high," "medium," and "low" selenium concentration refers to average forage crop selenium concentrations in various states and do not denote amounts necessary for prevention of selenium-deficient diseases of animals or amounts necessary to possibly prevent human cancer. (7) The results of the human epidemiological studies of selenium effects appear to be consistent with the topical application (Riley, 1969; Shamberger, 1970) and the laboratory animal dietary studies with no evidence seen of any enhancement of human cancer mortality (Shamberger and Willis, 1971).

Shamberger et al. (1973a,b) recently measured blood selenium in normal males and females in different age groups and in patients with cancer and other diseases. Normal values for 48 males and females were 22.9 ± 3.52 (SD) $\mu g/100$ ml. Values were almost the same in the 18–49 and 50–80 age ranges for both males and females. Patients with gastrointestinal cancer or metastases to gastrointestinal organs had significantly lower ($p < 0.001$) values. Patients with Hodgkins disease, cirrhosis, and hepatitis also had low blood selenium. In carcinoma of the breast, most blood cancers, Crohn's disease, other gastrointestinal disorders, certain sarcomas, lipomas, most patients with rectal cancer, and patients with several other diseases, selenium values were normal.

It was suggested by the authors (Shamberger et al., 1973a) that measurements of blood selenium may be consistent with epidemiology studies cited earlier (Shamberger and Willis, 1971).

TOXICITY OF TELLURIUM

Salient features of the chemical properties (Bagnall, 1966; Elkin and Margrave, 1968), analysis (Koirtyohann and Feldman, 1964; Sindeeva, 1964; Bagnall, 1966; Chakrabarti, 1967; Severne and Brooks, 1972), metabolic and biochemical effects (DeMeio, 1946; DeMeio and Henriques, 1947; DeMeio and Jetter, 1948; Cerwenka and Cooper, 1961; Vignoli and Defretin, 1964; Schroeder et al., 1967; Hollins, 1969; Lenchenko and Plotko, 1969; Cravioto et al., 1970; Slouka, 1970; Slouka and Hradil, 1970; Agnew and Cheng, 1971), natural occurrence (Hampel, 1971; Mason, 1952; Lansche, 1967), industrial production and utilization patterns (Lansche, 1967; Elkin and Margrave, 1968; Davidson and Lakin, 1972), environmental exposure levels (Schroeder et al., 1967; McBride and Wolfe, 1971; Selyankina and Alekseeva, 1970; Davidson and Lakin, 1972), levels in foods (Schroeder et al., 1967; Underwood, 1971), and levels in human tissues (Nason and Schroeder, 1967; Schroeder et al., 1967) of tellurium have all been reported.

The gastrointestinal absorption of both metallic tellurium and tellurium dioxide is poor compared to that of ingested tellurites and tellurates, which are absorbed to the extent of approximately 25%. Intravenously and subcutaneously administered tellurites and tellurates are readily absorbed and distributed to the tissues. In the blood, tellurium is complexed to plasma proteins.

The excretion patterns of tellurium are similar to those of selenium and depend on chemical nature, dose, and aqueous solubility of the tellurium compound and route of administration. Orally ingested tellurium compounds are excreted mainly in the feces, while iv administered tellurium compounds are excreted in the urine.

Toxic Effects in Laboratory Animals

Acute toxicity. The tellurites have been found to be more toxic than the tellurates, regardless of the route of administration. Muehlberger and Schrenk (1928) reported the minimum lethal dose (MLD) was 1.4 mg/kg for sodium tellurite and 30.5 for the tellurate given intravenously to the rat. The MLDs for oral administration to the rabbit were 31 mg/kg and 56 mg/kg for the tellurite and tellurate, respectively. The range of dosage killing 75% of rats within 48 hr after intraperitoneal injection was 2.25-2.50 mg/kg for sodium tellurite and 20.0-30.0 mg/kg for sodium tellurate (Franke and Moxon, 1937). Franke and Moxon (1937) found that dietary tellurium as sodium tellurite and sodium tellurate was toxic at levels of 25 and 13 ppm, respectively, growth was reduced, and there was no effect on hematopoiesis. The minimal lethal doses (LD75) of several elements in their highest and next to highest states of oxidation (by ip injection of rats) were tabulated with the following ratios of toxicity: selenite:selenate, 0.61; arsenite:arsenate, 0.26; tellurite:tellurate, 0.08. Elemental tellurium was found to be much less toxic (DeMeio,

1946) than either tellurite or tellurate. The toxicity of Na_2TeO_3 is ten times the toxicity of Na_2TeO_4 when given by intraperitoneal injection, but only twice as toxic when fed in chronic studies. Oral toxicity is somewhat lower than parenteral, presumably because of reduction to metallic tellurium in the gastrointestinal tract.

Acute poisoning in animals produces tremors, convulsions, respiratory arrest, and death. Amdur (1958) found that guinea pigs that died 48 hr after intramuscular injection of 75 mg telluric oxide showed hemorrhage and necrosis of the kidneys. In those that survived no residual evidence of toxicity and no abnormality of liver or kidneys was noted.

Pathomorphological changes of the internal organs of rats and rabbits were studied by El'Nichnykh and Lenchenko (1969) after peroral introduction of sodium tellurite (0.5 and 10 mg/kg), sodium tellurate (1, 10, 25, and 50 mg/kg), and chronic intoxication of sodium tellurite (0.0005, 0.005, 0.05, and 0.5 mg/kg). The introduction of sodium tellurite 0.5 and 10 mg/kg and that of sodium tellurate (1, 10, 25, and 50 mg/kg) caused degenerative changes in kidneys of the investigated animals. In all cases sodium tellurite was more toxic than sodium tellurate.

When fed to Peking ducks, 500–1,000 ppm of tellurium tetrachloride in food was acutely toxic (Carlton and Kelly, 1967) with markedly reduced growth and heavy mortality during the second week. At levels of 50 and 250 ppm, growth was not affected but mortality reached 50 and 70% by the fourth week with hydropericardium, patchy necrosis of the heart muscle, liver, and cerebrum, and dark-appearing kidneys. Tellurium was found in brain cells and duodenal mucosa.

Chronic effects. Apart from garlic breath, the main effects of chronic tellurium intoxication in animals are digestive disturbance, lack of growth, emaciation, somnolence, and loss of hair. Loss of hair appears in animals given tellurium orally or parenterally; elementary tellurium, especially administered orally, is much less toxic in this respect than its compounds and had no unfavorable effect on growth and no ill-effects on internal organs except in concentrations of 1,500 ppm in food. Elemental tellurium, when admixed in the diet, is absorbed after transformation to tellurium compounds (DeMeio, 1946).

Lampert et al. (1970) found that preweanling, Long-Evans rats fed 1 and 1.25% diets containing elemental tellurium became paralyzed within a few days. The paralysis was caused by segmented demyelination in spinal roots secondary to the degeneration of Schwann cells.

Schroeder and Mitchener (1971a, 1972) studied the effects on growth, survival, and tumors in Swiss mice of the Charles River CD strain given tellurite and tellurate (as well as selenite and selenate) in drinking water for life at 2 ppm Te and 3 ppm Se. The diet contained about 0.16 μg Te/g wet weight (analyzed by atomic absorption spectrophotometry) and 0.06 μg Se (analyzed by photofluorometry). The doses of tellurium and selenium given were at the subtoxic or barely toxic levels. Tetravalent tellurium appeared to exert little effect on body weight. In terms of life-spans, no toxic effects were evident. The hexavalent tellurium did not affect body weight or growth rates appreciably in either sex. Fasting serum glucose levels were higher than controls in older tellurium-fed males and in younger tellurium-fed females, without excess glycosurial. Fasting serum cholesterol levels were higher than the controls in tellurium-fed rats of both sexes and aortic lipids were considerably increased. In older animals, selenate was both tumorigenic and carcinogenic, while tellurite was not. There was 16.9% malignant tumors in the controls, 41.7% in the selenate, and 18.2% in the tellurite groups.

Carcinogenic, Teratogenic, and Mutagenic Effects

Reference has been made to the studies of Schroeder and Mitchener (1971a,b, 1972) involving the administration of tellurite and tellurate in drinking water at 2 ppm tellurium for life. In 90 mice fed tellurium, there were 18 tumors (20%) of which 12 were malignant (13%). [Of 119 controls, there were 23 spontaneous tumors (19%) of which 10 were malignant (43%).] The predominant tumors in the treated animals were lympholeukemia, of which there were seven; other tumors included five carcinomas and an adenoma of lung, several fibromas, and a benign adrenocortical adenoma. Both hexavalent and tetravalent forms of Te had little tumorigenic or carcinogenic activity.

While acute toxic effects of tellurium compounds in adult animals have been described, of particular potential importance in the context of environmental exposure and contaminations is the induction of hydrocephalus in newborn rats of mothers ingesting small amounts of tellurium. The reasons for the unique susceptibility of the fetal brain tissues apparently are still obscure as the molecular basis for the toxic effects exhibited by tellurium in the adult. Table 9 presents some of the results obtained in the elemental tellurium-induced hydrocephalus studies described. The amount of tellurium needed in the diet of the mother to produce hydrocephalus varied from 500 to 3,500 ppm of the diet (Garro and Pentschew, 1964; Agnew et al., 1968; Duckett, 1968; Agnew, 1972). Fifty to 100% of the gestating rats that ingested tellurium gave birth to hydrocephalic animals and 10–100% of the animals in the litter were hydrocephalic. The incidence of hydrocephalus was dose related. No apparent anomalies other than hydrocephalus have been observed. The period for tellurium-induced hydrocephalus in the rat is between days 10 and 15 of gestation. When given for periods of less than 6 days (during the sensitive period), no hydrocephalus was detected in the offspring (Duckett et al., 1971).

Agnew (1972) described the transplacental uptake of [127]Te by whole-body autoradiography in rat fetuses at various stages of gestation following ip injection of the dams with a single dose of [127m]Te (0.25–0.35 mg/kg, 600 μCi) on day 7, 8, 9, 10, 11, 14, 15, 17, or 20. No grossly toxic effects were observed in the 24 hr following injection. The levels of radioactivity in all the

TABLE 9 Comparative Information Concerning the Methods, Materials, and Results of the Previous Reported Experiments, Where Hydrocephalus is Induced in Offspring of Gestating Rats Fed Tellurium during Pregnancy

Investigator	Rat strain	Number of pregnant animals	Amount of tellurium in diet (ppm)	% of litters with hydrocephalus	% of hydrocephalus in affected litters	Survival time
Garro and Pentschew, 1964	Long-Evans	Not quoted	500	60	100	99% lethality; longest survival period was 1 month
			1,250	60–90	100	
			2,500	100	100	
Agnew et al., 1968	Wistar	Not quoted	1,250	No hydrocephalus		Not quoted
			2,500	No hydrocephalus		
			3,300	80	25–100	
Duckett 1971	Wistar	100	3,000	52	100	Up to 2 yr; about two-thirds of the offspring died within a few days
		30	3,500	Multiple abortions; average litter numbered 13	None	

225

tissues studied, including maternal blood, were less than 1% of the injected dose. Tellurium was shown to have reached the fetus by days 9 and 10 of gestation (the period of sensitivity for hydrocephalus induction on tellurium transport was exercised by the placenta, the mean offspring/maternal blood ratio of radioactivity decreased from 0.3 on days 8–11 to 0.16 on days 12–21. Autoradiographs of late fetuses showed a marked accumulation of tellurium in the choroid plexus of the lateral and fourth ventricles and the development of a blood–brain barrier to tellurium was apparent by day 18; relatively high tellurium levels in liver were also demonstrated from day 12. The demonstration by Agnew (1972) of the presence of tellurium in the early embryo in significant quantities suggests that tellurium may act directly on embryonic, rather than on maternal processes. The preferential uptake of tellurium by the developing choroid plexus further suggested the possibility that tellurium may interfere directly with choroid-plexus function. Agnew and Curry (1972) recently confirmed the period of teratogenic vulnerability of rat embryo to induction of hydrocephalus by tellurium to be day 9 and 10 of gestation following im injection of 13 mg/kg doses of tellurium to pregnant Long-Evans rats.

Chromosome damage in human cell cultures induced by metal salts has been reported by Paton and Allison (1972). An increased incidence of chromosome breakage was found in leukocytes treated 48 hr with sodium tellurite $(1.2 \times 10^{-8} M)$ and ammonium tellurate $(2.4 \times 10^{-7} M)$; the percentage of cells with chromatid breaks found was 13 and 8%, respectively, for the ammonium tellurate and sodium tellurite exposures. However, the mechanism by which Te affects chromosomes remains obscure.

Toxic Effects in Humans

Acute effects. Considerably less is known about the toxic effects of tellurium than about selenium. Generally, tellurium appears to be less toxic than selenium, and tellurium has not been found to be a very hazardous element in the industrial sense (Amdur, 1947; Blackadder and Manderson, 1975). Elemental tellurium is the least poisonous form of tellurium; although reports of cases of occupational tellurium poisoning are rare (Shie and Deeds, 1920; Steinberg et al., 1942; Amdur, 1947), tellurium is recognized as a potential industrial hazard. The threshold limit value in the United States and United Kingdom is 0.1 mg/m^3, while the maximum allowable concentration in the Soviet Union is 0.01 mg/m^3.

Most industrial exposures have resulted from exposure of both tellurium and tellurium dioxide fumes that are readily absorbed through the lungs. The body decomposes absorbed tellurium compounds into elemental tellurium and the garlic-odored compound dimethyl telluride. Tellurium compounds taken into the body orally are reduced to elemental tellurium by the naturally occurring bacteria in the intestines and eliminated via the feces. Hydrogen telluride, the tellurides, and telluric acid and its anhydrides and salts are considered the most toxic and tellurates, the least harmful. Only tellurium dioxide (TeO_2), hydrogen telluride (H_2Te), and potassium tellurite (K_2TeO_3) are of industrial health significance. One reason for the decreased hazard of tellurium in comparison to that of selenium is that tellurium forms its oxide only over 450°C and the dioxide formed is almost insoluble in water and body fluids.

The symptoms of acute tellurium poisoning in humans include a strong and lasting garlic odor about the body and mouth, a metallic taste and dryness in the mouth, nausea, restlessness, giddiness, sleepiness, headaches, diminished reflexes, tremors, convulsion, unconsciousness, and finally death. The outstanding symptom that led to the recognition that tellurium might rank as an industrial poison was the strong garlic odor of the breath of workers in contact with it. The term "mild tellurism" has been coined for a syndrome occurring principally where fumes of tellurium (probably in the form of hydrogen telluride) are emitted.

In three cases of exposure to the tellurium fume that occurred accidentally during the pouring of a tellurium–copper alloy, the only significant symptom was garlic odor (Amdur, 1947). Tellurium was present in the urine of the exposed workers in amounts from 0.008 to 0.016 mg/liter. Similar symptoms were described by Steinberg et al. (1942) of workers exposed to dense white fumes during the addition of metallic tellurium to molten iron. The concentration of tellurium in the atmosphere varied from 0.1 to 1.0 mg/10 m^3 and the urinary concentration ranged from 0 to 0.06 mg/liter. Blood cell counts, hemoglobin, and routine urine analysis were all normal. The only symptoms regarded as significant of tellurium poisoning were the garlic odor of the breath and somnolence. The garlic odor appeared more rapidly after consumption of alcohol, whereas the somnolence only appeared when the urinary level of tellurium was above 0.01 mg/liter.

Blackadder and Manderson (1975) described two cases involving the occupational accidental exposure to 50 g tellurium hexafluoride gas. Both individuals were handling volatile liquid esters that are readily absorbed through intact skin, possibly resulting in the decomposition of elemental tellurium in the dermis and subcutaneous tissues.

Both showed typical signs and symptoms of intoxication, and the stench of sour garlic was noted on breath and from excreta. An unusual feature was the bluish-black discoloration of the webs of the fingers and streaks on the face and neck. No permanent damage was noted and each patient made a spontaneous recovery without treatment.

Three cases of accidental poisoning, resulting from the mistaken administration of sodium tellurite for sodium iodide during retrograde pyelography, have been reported (Keall et al., 1946). In two of the cases, death occurred after approximately 6 hr. The symptoms in order of appearance were cyanosis, vomiting, loss of consciousness, and death. Autopsy revealed pulmonary congestion and distinct fatty change of the liver.

Chronic effects. It has been reported (Steinberg et al., 1942) that workers exposed 2 yr to amounts ranging from 0.1 to 7.4 mg of tellurium and tellurium oxide/10 m^3 air exhibited, in decreasing frequency, garlic odor of the breath, dryness of the mouth, metallic taste, somnolence, garlic odor of the sweat, loss of appetite, and nausea. There was no suppression of sweat nor evidence of intoxication. Approximately 90% of the air samples ranged between 0.1 and 1.0 mg Te/100 m^3 air. It should be noted that the symptoms recorded over the 2-yr period could also occur from brief exposure to higher concentrations. Thus, it is not possible to conclude that concentrations below 1 mg/10 m^3 would produce all of the above symptoms. It was suggested that where concentrations do not rise above 1 mg/m^3, serious illness would not be expected.

Ingestion of tellurium compounds causes loss of appetite, nausea, and vomiting, a metallic taste in the mouth, and malaise in addition to blackening of the teeth and tongue. Severe poisoning results in respiratory depression and circulatory collapse.

SUMMARY

In many considerations of the potential hazards of selenium and tellurium, it is essential to designate the form of the element or the compound in question. This may be said to be particularly relevant for selenium. Although elemental selenium per se is relatively nontoxic, certain selenium compounds (e.g., hydrogen selenide) are comparatively toxic to humans.

The use applications of selenium and its derivatives are increasing and present a potential health hazard. The two major groups of industries are those that extract, mine, treat, or process selenium-bearing minerals and those that utilize selenium in manufacturing (e.g., pigments, rectifiers, photoelectric cells, vulcanizing agents).

Soluble dusts (e.g., selenium dioxide, selenium trioxide) and certain halogen compounds may be toxic due to the relative ease with which they can be absorbed via the lungs, alimentary canal, and the skin.

Levels of selenium required by different species suggest that the critical level for dietary selenium, below which deficiency symptoms are observed, is apparently 0.02 ppm for ruminants and about 0.03–0.05 ppm for poultry.

In mammals, excess selenium has been shown to be teratogenic, hepatotoxic, and neurotoxic, retarding growth and causing muscular weakness. Selenium deficiency, on the other hand, is associated with hepatic necrosis, retarded growth, muscular degeneration, and infertility.

Aspects of both the carcinogenic and anticarcinogenic effects of compounds of selenium remain equivocal and are still to be resolved unambiguously, hopefully with more rigorous definitive experimentation. It should be noted that neoplasia is not observed among the specific lesions attributed to selenium deficiency in animals, nor has any significant increase or decrease in the incidence of neoplasia been observed in industrial workers exposed to selenium.

Evidence is controversial and difficult to relate to the possible effects of selenium on human embryonic development. Causative relationships between selenium intake and susceptibility to dental caries in both humans and experimental animals have not been definitively and unambiguously established.

There are no apparent reports of untoward effects of selenium on human health when it has been used at nutritional levels in producing food animals. There have been relatively few cases of selenium toxicity among humans even in seleniferous areas. This may be due more to diversity of the human diet than to inherent differences in tolerance to selenium. Episodes of human toxicity are comparatively rare from industrial exposure to selenium and its compounds.

The use of tellurium and its derivatives at present (and perhaps the immediate future) is considered minor and not hazardous per se in terms of direct industrial use exposure. Although tellurium and its compounds possess low orders of acute toxicity, relatively little is known regarding their chronic effects in low doses. Of perhaps greater potential concern are considerations of the embryonic and cytogenetic effects of tellurium and its compounds. In addition, no biological role of tellurium for humans has been ascertained. This question has been inadequately explored and needs more definitive elaboration.

REFERENCES

Agnew, W. F. 1972. Transplacental uptake of ^{127m}tellurium studied by whole-body autoradiography. *Teratology* 6:331.

Agnew, W. T. and Cheng, J. T. 1971. Protein binding of tellurium-127m by maternal and fetal tissues of the rat. *Toxicol. Appl. Pharmacol.* 20: 346–356.

Agnew, W. F. and Curry, E. 1972. Period of teratogenic vulnerability of rat embryo to induction of hydrocephalus by tellurium. *Experientia* 28:1444–1445.

Agnew, W. F., Fauvre, F. M. and Pudenz, P. H. 1968. Tellurium hydrocephalus: Distribution of tellurium-127m between maternal, fetal and neonatal tissues of the rat. *Exp. Neurol.* 21:120–132.

Allaway, W. H. 1969. Control of the environmental levels of selenium. In *Trace substances in environmental health*, ed. D. V. Hemphill, vol. 2, pp. 181–206. Columbia: University of Missouri.

Allaway, W. H., Kubota, J., Losee, F. and Roth, M. 1968. Selenium, molybdenum and vanadium in human blood. *Arch. Environ. Health* 16:342–350.

Amdur, M. L. 1947. Tellurium. *J. Occup. Med.* 3:386–391.

Amdur, M. L. 1958. Tellurium oxide, an animal study in acute toxicity. *Arch. Ind. Health* 17:665.

Amor, A. J. and Pringle, P. 1945. A review of selenium as an industrial hazard. *Bull. Hyg.* 20:239–241.

Anspaug, L. R. and Robison, W. L. 1971. *Prog. Atomic Med.* 3:63–138.

Arthur, D. 1972. The selenium content of Canadian foods. *Can. Inst. Food Sci. Technol. J.* 5:165–169.

Bagnall, K. W. 1966. *The chemistry of selenium, tellurium, and polonium.* Amsterdam: Elsevier.

Blackadder, E. S. and Manderson, W. G. 1975. Occupational absorption of tellurium: A report of two cases. *Br. J. Ind. Med.* 32:59–61.

Blaxter, K. 1962. *Vitam. Horm.* 20:633–643.

Bowen, H. J. M. and Cawse, P. A. 1963. The determination of selenium in biological material by radioactivation. *Analyst* 88:721–725.

Brune, D., Samsahl, K. and Westov, P. O., 1966. A comparison between the amounts of As, Au, Br, Cu, Fe, Mo, Se and Zn in normal and uraemic human whole blood by means of neutron activation analysis. *Clin. Chim. Acta* 13:285–289.

Buchan, R. F. 1947. Industrial selenosis. *Occup. Med.* 3:439–456.

Buck, W. B. and Ewan, R. C. 1973. Toxicology and adverse effects of mineral imbalance. *Clin. Toxicol.* 6:459–485.

Burk, R. F., Jr., Pearson, W. N., Wood, R. F. and Viteri, F. 1967. *Am. J. Clin. Nutr.* 20:723–733.

Byard, J. L. 1969. Trimethyl selenide, a urinary metabolite of selenite. *Arch. Biochem. Biophys.* 130:556–560.

Cameron, G. R. 1947. Liver atrophy produced by chronic selenium intoxication. *J. Pathol. Bacteriol.* 59:539–545.

Caravaggi, C., Clark, F. L. and Jackson, A. R. B. 1970. Acute selenium toxicity in lambs following intramuscular injection of sodium selenite. *Res. Vet. Sci.* 2:146–149.

Carlton, W. W. and Kelly, W. A. 1967. Tellurium toxicosis in Peking ducks. *Toxicol. Appl. Pharmacol.* 11:203–214.

Cerwenka, E. A. and Cooper, W. C. 1961. Toxicology of selenium and tellurium and their compounds. *Arch. Environ. Health* 3:189–199.

Chakrabarti, C. L. 1967. The atomic absorption spectroscopy of tellurium. *Anal. Chim. Acta* 39:293.

Clayton, C. C. and Bauman, C. A. 1949. Diet and azo dye tumors, effect of diet during a period when the dye is not fed. *Cancer Res.* 9:575–582.

Clegg, D. J. 1971. Embryotoxicity of chemical contaminants of foods. *Food Cosmet. Toxicol.* 9:195–205.

Clinton, M., Jr. 1947. Selenium fume exposure. *J. Ind. Hyg. Toxicol.* 29:225–226.

Craddock, V. M. 1972. Reactivity of seleno methionine in nucleic acid methylene reactions in the rat. *Chem. Biol. Interreactions* 5:207–211.

Cravioto, H., Agnew, W. F., Carregal, E. J. A. and Pudenz, R. H. 1970. Distribution of tellurium in the nervous system of the rat—An ultrastructural study. *J. Neuropathol. Exp. Neurol.* 29:158.

Cummins, C. M. and Kimura, E. T. 1971. Safety evaluation of selenium sulfide antidandruff shampoos. *Toxicol. Appl. Pharmacol.* 20:89–96.

Cummins, L. M. and Martin, J. L. 1967. Are selenocystine and selenomethionine synthesized *in vivo* from sodium selenite in mammals? *Biochemistry* 6:3162–3168.

Dalbert, P. 1913. *Bull. Assoc. Fr. Etude Cancer* 5:121.

Davidson, D. F. and Lakin, H. W. 1972. U.S. Geol. Surv., Prof. Paper 820, 627–630.

Davis, W. E. 1972. U.S. Natl. Tech. Inform. Serv. P.B. rep. no. 210679, pp. 1–57.

DeMeio, R. H. 1946. Tellurium, toxicity of ingested elemental tellurium for rats and rat tissues. *J. Ind. Hyg. Toxicol.* 29:393–395.

DeMeio, R. H. and Henriques, F. C., Jr. 1947. Tellurium excretion and distribution in tissues. *J. Biol. Chem.* 169:609.

DeMeio, R. H. and Jetter, W. W. 1948. Tellurium, III. Toxicity of ingested tellurium dioxide for rats. *J. Ind. Hyg. Toxicol.* 30:53.

Dennis, D. G. 1971. *Trans. 35th Wildlife Conf. (Toronto)*, pp. 14–26.

Devasconcellos, H. M. and Hampe, O. G. 1963. *Arq. Inst. Biol. (Saõ Paulo)*, 35:45–59.

Dickson, R. C. and Tomlinson, R. H. 1967. Selenium in blood and human tissues. *Clin. Chim. Acta* 16:311–321.

Dinkel, C. A., Minyard, J. A., Whitehead, E. I. and Olson, O. E. 1957. *S.D. Agric. Exp. Stn. Circ.* 135:4–7.

Diplock, A. T. 1970. In *Trace element metabolism in animals*, ed. C. F. Mills, pp. 190–204. Edinburgh: Livingstone.

Diplock, A. T., Baum, H. and Lucy, J. A. 1971. The effect of vitamin E on the oxidation state of selenium in rat liver. *Biochem. J.* 123:721–729.

Duckett, S. 1968. The germinal layer of the growing human brain during early fetal life. *Anat. Rec.* 161:231–240.

Duckett, S. 1970a. Fetal encephalopathy following ingestion of tellurium. *Experientia* 26:1239–1241.

Duckett, S. 1970b. Teratogenesis caused by tellurium. *Ann. N.Y. Acad. Sci.* 192:220–226.

Duckett, S., Sandler, A. and Scott, T. 1971. The target period during fetal life for the production of tellurium hydrocephalis. *Experientia* 27: 1064–1065.

Dudley, H. C. 1938a. Toxicology of selenium. *U.S. Public Health Rep.* 53:281–292.

Dudley, H. C. 1938b. Toxicology of selenium, toxic and vesicant properties of Se oxychloride. *U.S. Public Health Rep.* 53:94–98.

Dudley, H. C. and Miller, J. W. 1937. Toxicology of selenium. IV. Effects of exposure to hydrogen selenide. *U.S. Public Health Rep.* 52:1217–1231.

Dudley, H. C. and Miller, J. W. 1941. Toxicology of selenium, effects of sub-acute exposure to hydrogen selenide. *J. Ind. Hyg. Toxicol.* 23:470–477.

Elkin, E. M. 1969. Tellurium and tellurium compounds. In *Kirk-Othmer encyclopedia of chemical technology*, 2d ed., vol. 19, pp. 756–774. New York: Interscience.

Elkin, E. M. and Margrave, J. L. 1968. Selenium and selenium compounds. In *Kirk-Othmer encyclopedia of chemical technology*, 2d ed., vol. 17, pp. 809–833. New York: Interscience.

El'Nichnykh, L. N. and Lenchenko, V. G. 1969. Histomorphological changes in animal organs during poisoning with tellurium compounds. *Klin. Patog. Profil. Prof. Zabol. Khim. Etiol. Predpr. Tsvet. Chern. Med.* 2:155–160. [*Chem. Abstr.* 74:97299T (1969)]

Ewan, R. C., Baumann, C. A. and Pope, A. L. 1968. Determination of selenium in biological materials. *J. Agric. Food Chem.* 16:212–215.

Filatova, U. S. 1951. The toxicity of selenium anhydride. *Gig. Sanit.* 5:18–23.

Fimiani, R. 1949. Ascorbic acid in blood and urine in chronic experimental selenium poisoning. *Folia Med.* 32:459–460.

Fitzhugh, O. G., Nelson, A. A. and Bliss, C. I. 1944. The chronic toxicity of selenium. *J. Pharmacol. Exp. Ther.* 80:289–299.

Food and Drug Administration. 1974. Selenium in animal feed. *Fed. Reg.* 39:1355-1358.

Franke, K. W. and Moxon, A. L. 1936. Comparison of the minimal fatal doses of Se, Te, As, and V. *J. Pharmacol. Exp. Ther.* 58:454-459.

Franke, K. W. and Moxon, A. L. 1937. Toxicity of orally ingested As, Sr, Te, V and Mo. *J. Pharmacol. Exp. Ther.* 61:89-102.

Franke, K. W. and Tully, W. C. 1935. A new toxicant occurring naturally in certain samples of plant foodstuffs. V. Low hatchability due to deformities in chicks. *Poult. Sci.* 14:273-279.

Franke, K. W., Moxon, A. L., Poley, W. E. and Tully, W. C. 1936. A new toxicant occurring naturally in certain samples of plant foodstuffs. XII. Monstrosities produced by the injection of selenium salts into hen's eggs. *Anat. Rec.* 65:15-22.

Frant, R. 1949. Schadelijke werking van seleendioxide op de menselitke huid. *Ned. Tijdschr. Geneesk.* 93:874-876.

Fukina, A. M., and Kudryavtseva, T. P. 1969. *Mikroelem med. Zhivotnovod*, ed. M. Kruming. USSR: Elm Baku. [*Chem. Abstr.* 73:129114B (1970)]

Fuller, J. M., Beckman, E. D., Goldman, M. and Bustad, L. K. 1967. In *Selenium in biomedicine*, ed. O. H. Muth, pp. 119-123. Westport, Conn.: Avi.

Frost, D. V. 1971. The case for selenite as a feed additive. *Feedstuffs* 43:12.

Ganther, H. E. 1965. The fate of selenium in animals. *World Rev. Nutr. Diet.* 5:338-366.

Ganther, H. E. 1970. In *Trace element metabolism in animals*, ed. C. F. Mills, pp. 212-215. Edinburgh: Livingstone.

Ganther, H. E. 1973. In *Selenium*. New York: Reinhold.

Ganther, H. E. and Corcoran, C. 1969. Seleno trisulfides. II. Cross linking of reduced pancreatis ribonuclease with selenium. *Biochemistry* 8: 2557-2563.

Ganther, H. E., Levander, O. A. and Baumann, C. A. 1966. Dietary control of selenium volatilization in the rat. *J. Nutr.* 88:55-61.

Ganther, H. E., Wagner, P. A., Sunde, M. C. and Hoekstra, W. G. 1971. *Trace substances in environmental health*, ed. D. D. Hempitill, vol. 6, pp. 247-252. Columbia: University of Missouri.

Gardiner, M. R., Armstrong, J., Fels, H. and Glencross, R. N. 1962. A preliminary report on selenium and animal health in western Australia. *Aust. J. Exp. Agric. Anim. Husb.* 2:261-269.

Garro, F. and Pentschew, A. 1964. Neonatal hydrocephalus in the offspring of rats fed during pregnancy non-toxic amounts of tellurium. *Arch. Psych. Z. Ges. Neurol.* 206:272-280.

Glenn, M. W., Ensen, R. J. and Griner, L. A. 1964. Sodium selenate toxicosis. The effects of extended oral administration of sodium selenate on mortality, clinical signs, fertility and early embryonic development in sheep. *Am. J. Vet. Res.* 25:1486-1494.

Glover, J. R. 1970. Selenium and its industrial toxicology. *Ind. Med.* 39:50.

Green, T. E. and Turley, M. 1961. In *Treatise on analytical chemistry*, ed. I. M. Kolthoff and P. J. Elving, vol. 7, part 2, pp. 137-204. New York: Interscience.

Gunn, S. A. and Gould, J. C. 1967. In *Selenium in Biomedicine*, ed. O. H. Muth, pp. 395-413. Westport, Conn.: Avi.

Gusberg, S. B., Zamecnik, P. and Aub, J. C. 1941. The distribution of injected organic diselenides in tissues of tumor bearing animals. *J. Pharmacol. Exp. Ther.* 71:239-244.

Hadjimarkos, D. M. 1956. Geographic variations of dental caries in Oregon. VII. Caries prevalence among children in the Blue Mountain region. *J. Pediatr.* 48:195–201.

Hadjimarkos, D. M. 1965. Effect of selenium on dental caries. *Arch. Environ. Health* 10:893.

Hadjimarkos, D. M. 1968. In *Advances in oral biology*, ed. P. H. Staple, vol. 3, pp. 253–292. New York: Academic Press.

Hadjimarkos, D. M. 1970. *Trace substances in environmental health*, ed. D. D. Hemphill, vol. 4, pp. 301–306. Columbia: University of Missouri.

Hadjimarkos, D. M. and Bonhorst, C. W. 1958. The trace element selenium and its influence on dental caries susceptibility. *J. Pediatr.* 52:274–278.

Hadjimarkos, D. M. and Bonhorst, C. W. 1961. The selenium content of eggs, milk and water in relation to dental caries in children. *J. Pediatr.* 59:256.

Hadjimarkos, D. M., Storvick, C. A. and Remmert, L. F. 1952. Selenium and dental caries. *J. Pediatr.* 40:451–455.

Hadjimarkos, D. M., Bonhorst, C. W. and Mattice, J. J. 1959. The selenium concentration in placental tissue and fetal cord blood. *J. Pediatr.* 54:296–298.

Hall, R. H., Laskin, S., Frank, P., Maynard, E. A. and Hodge, H. C. 1951. Toxicity of elemental selenium. *Arch. Ind. Hyg. Occup. Med.* 4: 458–464.

Halverson, A. W., Jerde, L. G. and Hills, C. L. 1965. Toxicity of inorganic selenium salts to chick embryos. *Toxicol. Appl. Pharmacol.* 7:675–679.

Halverson, A. W., Palmer, I. S. and Guss, P. C. 1966. Toxicity of selenium to post-weanling rats. *Toxicol. Appl. Pharmacol.* 9:477–484.

Halverson, A. W., Tsay, D. T., Triebwasser, K. C. and Whitehead, E. I. 1970. Development of hemolytic anemia in rats fed selenite. *Toxicol. Appl. Pharmacol.* 17:158–159.

Hamilton, A. 1927. *Industrial poisons in the United States*, p. 111. New York: Macmillan.

Hamilton, A. 1934. *Industrial toxicology*, p. 303. New York: Harper.

Hamilton, A. and Hardy, H. C. 1949. *Industrial toxicology*, 2d ed. New York: Paul B. Hoeber.

Hampel, C. A. 1961. *Rare metals handbook*, 2d ed. New York: Reinhold.

Harr, J. R., Bone, J. F. and Tinsley, I. J. 1967a. Selenium toxicity in rats. In *Selenium biomedicine*, ed. O. H. Muth, pp. 179–184. Westport, Conn.: Avi.

Harr, J. R., Bone, J. F., Tinsley, J. J., Weswig, P. H. and Yamamoto, R. S. 1967b. In *Selenium in biomedicine*, ed. O. H. Muth, pp. 141–152. Westport, Conn.: Avi.

Harr, J. R., Exon, J. H., Whanger, P. D. and Weswig, P. H. 1972. Effect of dietary selenium on *N*-2-fluorenyl acetamide (FAA)-induced cancer in vitamin E supplemented selenium depleted rats. *Clin. Toxicol.* 5: 187–194.

Harr, J. R., Exon, J. H., Weswig, P. H. and Whanger, P. D. 1973. Relationship of dietary selenium concentration, chemical cancer induction, and tissue concentration of selenium in rats. *Clin. Toxicol.* 6:487–495.

Hartley, W. J. and Grant, A. B. 1961. A review of selenium responsive diseases of New Zealand livestock. *Fed. Proc.* 20:679–688.

Hashimoto, Y., Hwang, J. Y. and Yanagijawa, S. 1970. Possible source of atmospheric pollution of selenium. *Environ. Sci. Technol.* 4:157–158.

Heinrich, M. and MacCanon, D. M. 1957. Toxicity of intravenous selenite in dogs. *Proc. S.D. Acad. Sci.* 36:173–177.

Henschler, D. and Kirschner, V. 1969. Zur Resorption und Toxicitat von Selensulfid. *Arch. Toxicol.* 24:341–344.

Hernaman-Johnson, F. 1935. *Br. Med. J.* 1:1052.

Hirooka, T. and Galambos, J. T. 1966. Selenium metabolism. II. Effects of injected selenium compounds and of liver injury. *Proc. Soc. Exp. Biol. Med.* 121:723.

Hoger, D. and Bohm, C. 1944. Über Hautschadigungen durch Selenite. *Dermatologica* 90:217–223.

Hogue, D. E., Procter, J. F., Warner, R. G. and Loosli, J. K. 1962. Relation of selenium, vitamin E and an unidentified factor to muscular dystrophy in the lamb. *J. Anim. Sci.* 21:25–30.

Hollins, J. G. 1969. The metabolism of tellurium in rats. *Health Phys.* 17:497–505.

Holmberg, R. E., Jr., and Ferm, V. H. 1969. Interrelationships of selenium, cadmium and arsenic in mammalian teratogenesis. *Arch. Environ. Health* 18:873–877.

Holstein, E. 1951. Berufliche Selene in Wirkungen. *Zb. Arbeitsmed.* 1:102.

Innes, J. R. M., Ulland, B. M., Balerio, M. J., Petrucelli, L., Fishbein, L., Hart, E. R., Pallota, A. J., Bates, R. R., Falk, H. C., Gart, J. J., Klein, M., Mitchell, I. and Peters, J. 1969. Bioassay of pesticides and industrial chemicals for tumorigenicity in mice. A preliminary note. *J. Natl. Cancer Inst.* 42:1101–1114.

Jaffe, W. G. and Mondragon, M. C. 1969. Adaptation of rats to selenium intake. *J. Nutr.* 97:431–436.

Jaffe, W. G., Mondragon, M. C., Layrisse, M. and Ojeda, A. A. 1972. Toxicity symptoms in rats fed organic selenium. *Arch. Latin Am. Nutr.* 22:467–474.

Jahnel, F., Page, I. H. and Muller, B. 1932. Uber die Beziehungen des Tellurs zum Nerven System. *Z. Ges. Neurol. Psychiat.* 142:214–222.

Jenkins, K. R. 1951. Evidence for the absence of seleno-cystine and seleno-methionine in the serum proteins of chicks administered selenite. *Can. J. Biochem.* 46:1417–1425.

Jenkins, K. J., Hidiroglou, M. and Ryan, J. F. 1969. Intravascular transport of selenium by chick serum proteins. *Can. J. Physiol. Pharmacol.* 47: 459–467.

Johnson, H. 1970. Determination of selenium in solid wastes. *Environ. Sci. Technol.* 4:850–853.

Kar, A. B. and Das, R. P. 1963. The nature of the protective action of selenium on cadmium induced degeneration of the rat testes. *Proc. Natl. Inst. Sci. India B* 29:297–302.

Kar, A. B., Das, R. P. and Muker, F. N. I., VI. 1960. Prevention of cadmium induced changes in the gonads of rat by zinc and selenium. A study in antagonism between metals in the biological system. *Proc. Natl. Inst. Sci. India* 26:40–50.

Keall, J. H. H., Martin, N. H. and Tunbridge, R. E. 1946. Three cases of accidental poisoning by sodium tellurite. *Br. J. Ind. Med.* 3:175.

Kimmerle, G. 1960. Comparative investigation into the inhalation toxicity of the hexafluorides of sulfur, selenium and tellurium. *Arch. Toxikol.* 18:140–146.

Klug, H. L., Moxon, A. L., Petersen, D. F. and Painter, E. P. 1953. Inhibition of rat liver succinic dehydrogenase by selenium compounds. *J. Pharmacol. Exp. Ther.* 108:437–441.

Koeman, J. H., Peeters, W. H. M., Koudstaal-Hol, C. H. M. C., Thoe, P. S.

and DeGoeij, J. J. M. 1973. Mercury–selenium correlations in marine mammals. *Nature* 245:385–386.

Koirtyohann, S. R. and Feldman, C. 1964. In *Developments in applied spectroscopy*, ed. J. E. Forrette and E. Banterman, vol. 3. New York: Plenum Press.

Kosta, L., Byrne, A. R. and Zelenko, V. 1975. Correlation between selenium and mercury in man following exposure to inorganic mercury. *Nature* 254:238–239.

Kubota, J. and Allaway, W. H. 1971. Personal communication, cited in Shamberger, R. J. and Willis, C. E. (1971), *Crit. Rev. Clin. Lab. Sci.* 2:211–222.

Lakin, H. W. and Davidson, D. F. 1967. The relation of the geochemistry of selenium to its occurrence in soils. In *Selenium in medicine*, ed. O. H. Muth, J. E. Oldfield, and P. H. Weswig, pp. 27–57. Westport, Conn.: Avi.

Lampert, P., Garro, F. and Pentschew, A. 1970. Tellurium neuropathy. *Acta Neuropathol.* 15:308–317.

Lansche, A. M. 1967. Selenium and tellurium—A materials survey. U.S. Bur. Mines Inform. Circ. No. 8340:1–46.

Lenchenko, V. G. and Plotko, E. G. 1969. Toxicological characteristics of tellurium compounds on oral administration. *Profazbol. Khim. Etiol. Predpr. Tsvet. Chern. Met.* 2:137–154. [*Chem. Abstr.* 73:118655C (1969)]

Levander, O. A. 1972. Metabolic interrelationships and adaptations in selenium toxicity. *Ann. N.Y. Acad. Sci.* 192:181–192.

Levander, O. A. and Argrett, C. C. 1969. Effects of arsenic, mercury, thallium, and lead on selenium metabolism in rats. *Toxicol. Appl. Pharmacol.* 14:308–314.

Levander, O. A. and Morris, V. C. 1970. Interreactions of methionine, vitamin E, and antioxidants in selenium toxicity in the rat. *J. Nutr.* 100:1111–1118.

Lillie, R. D. and Smith, M. I. 1940. Histogenesis of hepatic cirrhosis in chronic food selenosis. *Am. J. Pathol.* 16:223–230.

Lindberg, P. and Siren, M. 1965. Fluorometric selenium determinations in the liver of pigs and pigs affected with nutritional muscular dystrophy and in liver dystrophy. *Acta Vet. Scand.* 6:59–64.

Lockhart-Mummery, J. 1935. *Br. Med. J.* 1:867.

Ludwig, T. G. and Bibby, B. G. 1969. *Caries Res.* 3:32.

Maag, D. D. and Glenn, M. W. 1964. Toxicity of selenium: Farm animals. In *Selenium in biomedicine*, ed. O. H. Muth, pp. 127–140. Westport, Conn.: Avi.

Martin, J. L. and Gerlach, M. L. 1972. Selenium metabolism in animals. *Ann. N.Y. Acad. Sci.* 192:193–199.

Mason, B. H. 1952. *Principles of geochemistry.* New York: Wiley.

Mason, K. E., Young, J. O. and Brown, J. E. 1964. Effectiveness of selenium and zinc in protecting against cadmium induced injury of the rat testis. *Anat. Rec.* 148:309.

Mathias, M. M., Allaway, W. H., Hogue, D. E., Marion, M. U. and Gardner, R. W. 1965. Value of selenium in alfalfa for the prevention of selenium deficiencies in chicks and rats. *J. Nutr.* 86:213–218.

Mattii, M., Badiello, R. and Cecchetti, E. 1967. Toxicity of selenium containing analogs of cystamine and taurine in rats. *Radiobio. Radioter. Fis. Med.* 22:56–63.

Mautner, H. G., Shih-Hsi, C., Jaffee, J. J. and Jartorelli, A. C. 1963. The

236   *L. Fishbein*

synthesis and anti-neoplastic properties of selenoguanine, selenocystine and related compounds. *J. Med. Chem.* 6:36–43.

McBride, B. C. and Wolfe, B. C. 1971. Biosynthesis of dimethyl arsine by methanobacterium. *Biochemistry* 10:4312–4317.

McConnell, K. P. and Carpenter, D. M. 1971. Interrelationship between selenium and specific trace elements. *Proc. Soc. Exp. Biol. Med.* 137:996–1001.

McConnell, K. P. and Portman, W. 1952. Toxicity of dimethyl selenide in the rat and mouse. *Proc. Soc. Exp. Biol. Med.* 79:230–231.

Miller, W. T. and Williams, K. T. 1940. Minimum lethal doses of selenium as sodium selenite to horses, mules, cattle, and swine. *J. Agric. Res.* 60:163–167.

Morris, V. C. and Levander, O. A. 1970. Selenium content of foods. *J. Nutr.* 100:1383–1388.

Morrow, D. A. 1968. Acute selenite toxicosis in lambs. *J. Am. Vet. Med. Assoc.* 152:1625–1629.

Moxon, A. C. 1940. Toxicity of selenium-cystine and some other organic selenium compounds. *J. Am. Pharm. Assoc. Sci. Ed.* 29:249–251.

Moxon, A. L. and Rhian, M. A. 1943. Selenium poisoning. *Physiol. Rev.* 23:305–337.

Moxon, A. L., Anderson, H. D. and Painter, E. P. 1938. The toxicity of some organic selenium compounds. *J. Pharmacol. Exp. Ther.* 63:357–368.

Muehlberger, C. W. and Schrenk, H. H. 1928. Effect of the state of oxidation on toxicity of certain elements. *J. Pharmacol. Exp. Ther.* 33:270–271.

Muth, O. H. and Binns, W. 1964. Selenium toxicity in domestic animals. *Ann. N.Y. Acad. Sci.* 111:583–590.

Muth, O. H., Oldfield, J. E., Remmert, L. F. and Schubert, J. R. 1968. Effects of selenium and vitamin E in white muscle disease. *Science* 128:1090.

Nagai, I. 1959. An experimental study of selenium poisoning. *Tagaku Kenku (Acta Med.)* 29:1505–1532.

Nason, A. P. and Schroeder, H. A. 1967. Erratum: Abnormal trace elements in man: Tellurium. *J. Chron. Dis.* 20:671.

Nelson, A. A., Fitzhugh, O. G. and Calvery, H. O. 1943. Liver tumors following cirrhosis caused by selenium in rats. *Cancer Res.* 3:230–236.

Oelschlager, V. W. and Menke, K. H. 1969. Uber Selengehalte Pflanzlicher Tierischer und Anderer Stoffe. *Z. Ernahrwiss.* 9:216.

Olson, R. E. 1965. Interrelationships among vitamin E, coenzyme Q and selenium. *Fed. Prod.* 24:85–92.

Olson, O. E. and Frost, D. V. 1970. Selenium in papers and tobacco. *Environ. Sci. Technol.* 4:686–687.

Painter, E. P. 1941. The chemistry and toxicity of selenium compounds with special reference to the selenium problem. *Chem. Rev.* 28:179–213.

Palmer, I. S., Fischer, D. D., Halverson, A. W. and Olson, O. E. 1969. Identification of a major selenium excretor product in rat urine. *Biochim. Biophys. Acta* 177:336–341.

Palmer, I. S., Gunsalus, R. P., Halverson, A. W. and Olson, O. E. 1970. Trimethyl selenonium ion as a general excretory product from selenium metabolism in the rat. *Biochim. Biophys. Acta* 208:260–266.

Palmer, I. S., Arnold, R. C. and Carlson, C. W. 1973. *Poult. Sci.* 52: 1841–1846.

Parizek, J. 1971. In *Newer trace elements in nutrition*, ed. W. Mertz and W. E. Cornatzer, pp. 85–119. New York: Marcel Dekker.

Parizek, J. and Ostadalova, R. 1967. The protective effect of small amounts of selenite in sublimate intoxication. *Experientia* 23:142.

Patrias, G. 1969. Selenium, a missing link in animal nutrition. *Feedstuffs* 41:24–25.

Paton, G. R. and Allison, A. C. 1972. Chromosome damage in human cell cultures induced by metal salts. *Mutat. Res.* 16:332–336.

Patty, F. H. 1949. *Industrial hygiene and toxicology*, vol. II. New York: Wiley.

Paulson, G. D., Baumann, C. A. and Pope, A. L. 1968. Metabolism of ^{75}Se-selenite, ^{75}Se-selenate, ^{75}Se-selenomethionine and ^{35}S-sulfate by rumen microorganisms *in vitro. J. Anim. Sci.* 27:497–504.

Pillay, K. K. S., Thomas, C. C., Jr., and Kaminski, J. W. 1969. Neutron activation analysis of the selenium content of fossil fuels. *Nucl. Appl. Technol.* 7:478–483.

Pletnikova, I. P. 1970. Biological action and tolerable level of selenium entering the organism with drinking water. *Gig. Sanit.* 35:14–19.

Potter, V. R. and Elvehjem, C. A. 1937. The effect of inhibitors of succinoxidase. *J. Biol. Chem.* 117:341–349.

Proctor, J. F., Hogue, D. E. and Warner, R. G. 1958. Selenium, vitamin E, and linseed oil meal as preventatives of muscular dystrophy in lambs. *J. Anim. Sci.* 17:1183–1187.

Radeleff, R. D. 1964. *Veterinary toxicology*, pp. 158–161. Philadelphia: Lea and Febiger.

Riley, J. F. 1969. Mast cells, cocarcinogenesis and anticarcinogenesis in the skin of mice. *Experientia* 24:1237–1241.

Robertson, D. S. F. 1970. Selenium—A possible teratogen? *Lancet* 1:518.

Rohmer, R., Carrot, E. and Gouffault, J. 1950. Poisoning by selenium compounds. *Bull. Soc. Chim. Fr.* pp. 275–278.

Rosenfeld, I. and Beath, O. A. 1945. The elimination and distribution of selenium in the tissues in experimental selenium poisoning. *J. Nutr.* 30:443–449.

Rosenfeld, I. and Beath, O. A. 1947. Congenital malformations of eyes of sheep. *J. Agric. Res.* 75:93–103.

Rosenfeld, I. and Beath, O. A. 1954. Effect of selenium on reproduction in rats. *Proc. Soc. Exp. Biol. Med.* 87:295–297.

Rosenfeld, I. and Beath, O. A. 1964a. Metabolic effects of selenium in animals. *Wyo. Agric. Exp. Stn. Bull.* 414:1–64.

Rosenfeld, I., and Beath, O. A. 1964b. *Selenium—Geobotany, biochemistry, toxicity and nutrition*. New York: Academic Press.

Rosenheim, O. and Tunnicliffe, F. W. 1901. Selenium compounds as factors in the recent beer poisoning epidemic. *Lancet* 1:318, 434, 927.

Rotruck, J. J., Pope, A. L., Ganther, H. E. and Hoekstra, W. G. 1972. Prevention of oxidative damage to rat erythrocytes by dietary selenium. *J. Nutr.* 102:689–696.

Schroeder, H. A. 1967. Effects of selenate, selenite and tellurite on the growth and early survival of mice and rats. *J. Nutr.* 92:334–338.

Schroeder, H. A. and Mitchener, M. 1971a. Selenium and tellurium in rats. *J. Nutr.* 101:1531–1540.

Schroeder, H. A. and Mitchener, M. 1971b. Toxic effects of trace elements on the reproduction of mice and rats. *Arch. Environ. Health*, 23:102–106.

Schroeder, H. A. and Mitchener, M. 1972. Selenium and tellurium in mice. *Arch. Environ. Health* 24:66–71.

Schroeder, H. A., Buckman, J. and Balassa, J. J. 1967. Abnormal trace elements in man: Tellurium. *J. Chron. Dis.* 20:147–161.

Schroeder, H. A., Frost, D. V. and Balassa, J. J. 1970. Essential trace metals in man: Selenium. *J. Chron. Dis.* 23:227–243.

Schwarz, K. and Foltz, C. M. 1957. Selenium as an integral part of factor 3 against necrotic liver degeneration. *J. Am. Chem. Soc.* 79:3292–3293.

Seifter, J., Ehrlich, P., Hudgma, G. and Mueller, G. 1946. Thyroid adenomas in rats receiving selenium. *Science* 103:762.

Selyankina, K. P. and Alekseeva, L. S. 1970. Selenium and tellurium in the atmosphere in the vicinity of copper electrolytic refining plants. *Gig. Sanit.* 35:95–96 [*Chem. Abstr.* 73:69096 (1970)].

Sentein, P. 1967. *Chromosoma* 23:95–136.

Severne, B. C. and Brooks, R. R. 1972. Rapid determination of selenium and tellurium by atomic-absorption spectrophotometry. *Talanta* 19: 1467–1470.

Shamberger, R. J. 1969. *Proc. Am. Assoc. Cancer Res.* 10:311.

Shamberger, R. J. 1970. Relationship of selenium to cancer. I. Inhibitory effect of selenium in carcinogenesis. *J. Natl. Cancer Inst.* 44:931–936.

Shamberger, R. J. 1971. Is selenium a teratogen? *Lancet* 2:1316.

Shamberger, R. J. and Frost, D. V. 1969. Possible protective effect of selenium against human cancer. *Can. Med. Assoc. J.* 100:682.

Shamberger, R. J. and Rudolf, J. 1966. Protection against cocarcinogenesis by antioxidants. *Experientia* 22:116–122.

Shamberger, R. P. and Willis, C. E. 1971. Selenium distribution and human cancer mortality. *Crit. Rev. Clin. Lab. Sci.* 2:211–221.

Shamberger, R. J., Tytko, S., and Willis, C. E. 1973a. Selenium in the blood of normals, cancer patients, and patients with other diseases. *Clin. Chem.* 19:672–675.

Shamberger, R. J., Rukovena, E., Longfield, A. K., Tytko, S. A., Deodhar, S. and Willis, C. E. 1973b. Antioxidants and cancer. I. Selenium in the blood of normals and cancer patients. *J. Natl. Cancer Inst.* 50:863–870.

Shapiro, J. R. 1972. Selenium and carcinogenesis. *Ann. N.Y. Acad. Sci.* 192:215–219.

Shendrikar, A. D. 1974. A critical evaluation of analytical methods for the determination of selenium in air, water, and biological materials. *Sci. Total Environ.* 3:155–168.

Shendrikar, A. D. and West, P. W. 1973. Determination of selenium in the smoke from trash burning. *Environ. Lett.* 5:29–35.

Shie, M. D. and Deeds, F. E. 1920. The importance of tellurium as a health hazard in industry. *Public Health Rep.* 35:939–942.

Shrift, A. 1961. Biochemical interrelations between selenium and sulfur in plants and microorganisms. *Fed. Proc.* 20:695–702.

Sindeeva, N. D. 1964. *Mineralogy and types of deposits of selenium and tellurium.* New York: Interscience.

Slouka, V. 1970. Distribution and excretion of intravenous radiotellurium in the rat. *Sb. Ved. Pr. Vlvdu Hradci. Kralove* 47:3–19 [*Chem. Abstr.* 75:59454C (1970)].

Slouka, V. and Hradil, J. 1970. Kinetics of orally administered radio tellurium in the rat. *Sb. Ved. Pr. Vlvdu Hradci. Kralove* 47:21–43 [*Chem. Abstr.* 75:59472G (1970)].

Smith, M. I. 1939. The influence of diet on the chronic toxicity of selenium. *U.S. Public Health Rept.* 54:1441.

Smith, M. I. and Lillie, R. D. 1940. The chronic toxicity of naturally occurring food selenium. *U.S. Public Health Serv. Natl. Inst. Health Bull.* 174:1–13.

Smith, M. I. and Westfall, B. B. 1937. Further field studies on the selenium problem in relation to public health. *U.S. Public Health Rep.* 52: 1375–1384.

Smith, M. I., Franke, K. W. and Westfall, B. B. 1936. The selenium problem in relation to public health. *U.S. Public Health Rep.* 51:1496–1505.

Smith, M. I., Stohlman, E. F. and Lillie, R. D. 1937a. The toxicity and pathology of selenium. *J. Pharmacol. Exp. Ther.* 60:449–471.

Smith, M. I., Westfall, B. B. and Stohlman, E. F. 1937b. Elimination of Se and its distribution in tissues. *U.S. Public Health Rep.* 52:1171.

Smith, M. I., Westfall, B. B. and Stohlman, E. F. 1938. Studies on the fate of Se in the organism. *U.S. Public Health Rep.* 53:1199–1216.

Stahl, A. R. 1969. Preliminary air pollution survey of selenium and its compounds. Washington, D.C.: U.S. Dep. of Health, Education, and Welfare, NAPCA.

Steinberg, H. H., Massari, S. C., Miner, A. C. and Rink, R. 1942. Industrial exposure to tellurium. *J. Ind. Hyg. Toxicol.* 24:183–192.

Sterner, J. H. and Lidfeldt, V. 1941. The selenium content of normal urine. *J. Pharmacol.* 73:205–210.

Symanski, H. 1950. Ein Fallvon Selenwaser Stoffvergiftung. *Deutsch. Med. Wochschr.* 75:1730–1731.

Tang, G. and Storvick, C. A. 1960. Effect of naturally occurring selenium and vanadium on dental caries. *J. Dental Res.* 39:473–488.

Tappel, A. L. 1962. Vitamin E as the biological lipid antioxidant. *Vitam. Horm.* 20:493.

Thienes, C. H. and Haley, T. J. 1972. *Clinical toxicology*, 5th ed., p. 208. Philadelphia: Lea and Febiger.

Thompson, J. N. and Scott, M. C. 1969. Role of selenium in the nutrition of the chick. *J. Nutr.* 97:335–342.

Ting, K. P. and Walker, G. W. R. 1969. The distributive effect of selenoamino acid treatment on crossing-over in *Drosophila melanogaster*. *Genetics* 61:141–155.

Todd, A. T., Scott, S. G. and Coke, H. 1935. Discussion on prevention and treatment of metastases in carcinoma mammae. *Proc. R. Soc. Med.* 28:681–694.

Tscherkes, L. A., Aptekar, S. G. and Volgarev, M. N. 1961. Hepatic tumors induced by selenium. *Byull. Edsper. Biol. Med.* 53:78–82.

Tsuzuki, H., Okawa, K. and Hosoya, T. 1960. Experimental selenium poisoning. *Yakohama Med. Bull.* 11:368.

Underwood, E. J. 1971. *Trace elements in human and animal nutrition*, 3d ed., pp. 323–368. New York: Academic Press.

Vesce, C. A. 1947. Intossicazione sperimentale da selenio. *Intossicazione Sperimentale da Selenio, Folia Med. (NAPOLI)* 33:209–212.

Vignoli, L. and Defretin, J. P. 1964. La toxicologie du tellure. *Ann. Biol. Clin.* 22:399–417.

Volgarev, M. N. and Tscherkes, C. A. 1967. Further studies in tissue changes associated with sodium selenate. In *Selenium in biomedicine*, ed. O. H. Muth, pp. 179–184. Westport, Conn.: Avi.

Von Euler, V. H., Hasselquist, H. and Von Euler, B. 1956. Biologically active trace elements in organic combination. *Ark. Kemi.* 9:583–591.

Von Lehmden, D. J., Junger, R. H. and Lee, R. E., Jr. 1974. Determination of trace elements in coal, flyash, fuel oil and gasoline. A preliminary comparison of selected analytical techniques. *Anal. Chem.* 46:239–245.

Wahlstrom, R. C. and Olson, O. E. 1959. The effect of selenium on reproduction in swine. *J. Anim. Sci.* 18:141–145.

Walker, C. H. and Klein, F. 1915. *Am. Med.* 21:628–633.

Walker, G. W. R. and Bradley, A. M. 1969. Interacting effects of sodium monohydrogen arsenate and selenocystine on crossing over in *Drosophila melanogaster. Can. J. Genet. Cytol.* 11:677–688.

Walker, G. W. R. and Ting, K. P. 1967. Effects of selenium on recombination in barley. *Can. J. Genet. Cytol.* 9:314–320.

Watkins, G. R., Bearse, A. E. and Shutt, R. 1942. Industrial utilization of selenium and tellurium. *Ind. Eng. Chem.* 34:899–903.

Watkinson, J. H. 1967. Analytical methods for selenium in biological material. In *Selenium in biomedicine*, ed. O. H. Muth, J. E. Oldfield, and P. H. Weswig. Westport, Conn.: Avi.

Weisberger, A. S. and Suhrlaud, L. G. 1956a. Studies on analogs of L-cysteine. The effect of selenium cystine on Murphy lymphosarcoma tumor cells in the rat. *Blood* 11:11–18.

Weisberger, A. S. and Suhrlaud, L. G. 1956b. Studies on analogs of L-cysteine and L'-cystine. *Blood* 11:19–24.

West, P. W. 1967. Selenium-containing inorganics in paper may play cancer role. *Chem. Eng. News* 45:12–13.

Westfall, B. B., Stohlman, E. F. and Smith, M. I., 1938. Placental transmission of selenium. *J. Pharmacol.* 64:55–57.

Woolrich, P. F. 1973. *Am. Ind. Hyg. Assoc. J.* 34:216–217.

Chapter 8

NUTRIENT INTERACTIONS WITH TOXIC ELEMENTS

H. H. Sandstead
United States Department of Agriculture
Agricultural Research Service
Human Nutrition Laboratory
Grand Forks, North Dakota

INTRODUCTION

In nonindustrial situations, the major exposure of humans to toxic elements occurs principally through their normal food supply. Exceptions almost always result from environmental contamination or the inappropriate use of seeds that have been treated with heavy metal fungicides for animal or human food. The estimated usual adult intakes of lead, cadmium, and mercury from food and water are 200, 50, and 17 μg, respectively (Shibko et al., 1976). These levels are substantially below those generally considered toxic.

The significance of low concentrations of toxic metals in foods is, at present, unknown. While such concentrations do not appear hazardous in the usual sense, observations in experimental models have suggested interference with the metabolism of certain essential elements and other nutrients. Therefore, the presence of toxic metals at low levels in human food might have biological implications if the essential nutrients with which they interfere are present in marginal amounts. Some of the observations discussed subsequently tend to support this hypothesis.

Toxic elements of major concern in the environment are lead, cadmium, and mercury, and this review is limited to these elements. Most of the studies pertinent to the subject of interactions have been done on experimental animals fed levels of the toxic elements that substantially exceed human exposure. Therefore, the findings must be interpreted with caution if an attempt is made to apply them to humans. Even so, the findings show that toxic elements interact with essential nutrients and disrupt metabolic processes. They also show that specific essential nutrients have protective properties against specific elements.

241

Many of the nutritional factors that appear to influence the susceptibility of humans and animals to lead have been reviewed by Levander (1976), Sandstead (1976), Pond (1975), Goyer and Rhyne (1973), and the National Research Council (1972). Reviews pertinent to cadmium include those of Levander (1976), Sandstead (1976), Parizek (1976), Pond (1975), Fox (1974, 1976), and Friberg et al. (1971). The most comprehensive review on mercury is that of Friberg and Vostal (1972). Parizek (1976) has reviewed some of the more recent findings on interactions of mercury with other trace elements. The reader is referred to these reviews for additional information on the many interactions that have been observed, with this chapter providing an introduction to the subject.

INTERACTIONS OF LEAD WITH NUTRIENTS

The average daily U.S. dietary exposure to lead has been estimated to be 100 μg for infants and children and 200 μg for adults. Thus, children ingest substantially more lead in terms of body weight than do adults. Presumably, therefore, interactions between lead and essential nutrients are biologically of more significance for the young. A second factor of toxicological importance is the apparently greater intestinal absorption of lead by infants and children (Kolbye et al., 1974). Among the nutrients with which lead interacts, either directly or indirectly, are calcium, vitamin D, iron, zinc, copper, iodine, sodium and potassium, chromium, protein and amino acids, ascorbic acid, vitamin E and selenium, and niacin. Nonnutrients in the diet that can bind lead and presumably decrease its availability include phytate and vegetable gums (Shibko et al., 1976).

Calcium

Interactions of calcium with lead have been reviewed by Mahaffey (1974). Low intakes of calcium increase the susceptibility of rats to lead. For example, a lead intake of 12 ppm in drinking water by rats fed a low calcium diet resulted in findings similar to those produced by 200 ppm in animals fed adequate calcium. The lead content of kidneys of the former group was 10 times the level present in the latter. In addition, animals fed the low calcium diet and no lead in drinking water had nearly four times as much lead in their kidneys as did the controls. These findings suggest that the uptake of background lead was increased by calcium deficiency. Calcium deficiency also appears to cause weanling rats to increase their intake of lead-containing solutions (Snowdon and Sanderson, 1974).

In contrast to a low calcium dietary intake, a high calcium dietary intake inhibits lead absorption by rats (Kostial et al., 1971). At the cellular level calcium also appears to be protective. For example, 1 mM calcium has been found to inhibit the effect of 8×10^{-5} M lead on the myoneural function of the phrenic nerve-stimulated rat hemidiaphragm (Silbergeld et al., 1974).

Vitamin D

Some of the interactions of lead with calcium may be influenced by vitamin D or its metabolites. It has been reported that vitamin D-treated rats had higher tissue concentrations of lead than did vitamin D-deprived rats (Sobel et al., 1940). Although not established, it appears that lead might have adverse effects on the formation of 1,25-dihydroxycholecalciferol by renal tubular cells and thus cause a decreased intestinal absorption of calcium. This suggestion is supported by the known injurious effect of lead on the proximal renal tubular cells (Goyer and Rhyne, 1973) and the observed impairment of 1,25-dihydroxycholecalciferol synthesis by another toxic metal, cadmium (Suda et al., 1974).

Iron

The adverse effect of lead on iron metabolism is perhaps most evident in the bone marrow. For example, lead inhibits heme synthesis in reticulocytes (Waxman and Rabinoritz, 1966) and causes an anemia that is morphologically similar to the anemia of iron deficiency. Ferrous iron appears to have a protective effect against the adverse effect of lead, while iron deficiency seems to increase the susceptibility of the reticulocyte to lead. The several steps in the hemoglobin synthesis pathway sensitive to lead have been reviewed by Goyer and Rhyne (1973).

Increased bone and soft tissue lead has been found in iron-deficient rats given lead in drinking water. However, the uptake of lead by the iron-deficient animal as reflected by soft tissue (kidney) accumulation appears to be substantially less than in the calcium-deficient animal (Goyer and Rhyne, 1973).

Lead can apparently displace iron from tissues. For example, rabbits given tetraethyl lead have been observed to lose 2 mol iron from their brains for each mole of lead retained (Niklowitz and Yeager, 1973). Presumably this apparent displacement occurs as part of the normal turnover of iron and an inhibition of uptake of new iron by lead similar to that which occurs in lead-poisoned reticulocytes.

Young children not living in the central area of a city were found to have iron deficiency associated with increased concentrations of blood lead. Following treatment with iron the concentrations of lead in their blood decreased (Smith and Goldsmith, 1976). The relationship of dietary intake of iron, calcium, and phosphorus of children living in the central area of a city to level of blood lead was also studied. Increased blood lead (>40 $\mu g/100$ ml) was not associated with a low intake of iron in these children, but was associated with low intakes of dietary calcium and phosphorus (Mahaffey et al., 1976). Because more than 65% of low income children aged 24–36 months examined in the U.S. Ten State Nutrition Survey (1968–1970) were found to have an intake of iron nearly one-half the recommended allowance for iron and because 30% of the children had a calcium intake one-half or less the recommended allowance, it seems possible, in view of the findings of Smith and Goldsmith (1976) and of Mahaffey et al. (1976), that marginal and deficient intakes of iron and/or calcium might contribute to the occurrence of lead poisoning in children exposed to increased environmental lead. Children particularly at risk of lead poisoning are those who ingest dust and dirt of the central city as well as paint chips. According to Lepow et al. (1974) the lead intake of such children is sometimes above 500 μg/day, far exceeding published estimates of maximal permissible daily intakes for 2-yr-old children, which range from 70 to 500 μg. The studies of Smith and Goldsmith (1976) and of Mahaffey et al. (1976) tend to support these speculations.

Zinc

Interactions of lead with zinc are less well described than those with calcium and iron. According to Finnelli et al. (1974) aminolevulinic acid dehydrase of the heme synthesis pathway is a zinc-dependent enzyme. The inhibition of the enzyme by lead is apparently alleviated by zinc (Finnelli et al., 1975).

Zinc has also been reported to provide some protection to horses grazed on pasture contaminated with lead and zinc from refinery effluent. Although their tissue content of lead was nearly doubled, they showed fewer signs of intoxication (Willoughby et al., 1972) than animals exposed to lead alone. This report is in apparent contrast to findings in swine exposed to increased levels of lead and zinc in the presence of high or low calcium intakes. In these experiments high zinc enhanced the toxic effects of lead and increased lead retention (Hsu et al., 1975). Findings for rats contrast with findings for large animals in both of the above studies. Increased dietary zinc impaired intestinal absorption of lead and thus protected from dietary lead. Injected zinc was not protective (Cerklewski and Forbes, 1976).

As was noted previously for iron, lead displaces zinc or prevents its uptake by the brain. Rabbits exposed to toxic levels of tetraethyl lead lost 0.5 mol zinc from the brain for each mole of lead they retained (Niklowitz and Yeager, 1973).

Copper

An adverse effect of lead on copper metabolism is implied by the decrease in serum ceruloplasmin that occurred in rats exposed to lead and fed adequate levels of copper. Copper deficiency accentuated the toxic effects of lead and of lead accumulated in the livers and kidneys. Anemia was also present and growth was retarded (Petering, 1973).

Iodine

The adverse effect of lead on the uptake of iodine by the thyroid and the conversion of iodine to protein-bound iodine in rats suggest that lead interferes with iodine metabolism (Sandstead, 1967). Supportive observations have been reported on lead-intoxicated sheep (Goldatovic et al., 1966). Humans intoxicated with lead from industrial exposures or from ingestion of lead-contaminated illicitly distilled whisky (Sandstead et al., 1969) have also been found to display impaired uptakes of iodine by the thyroid gland.

Sodium and Potassium

In humans the renal tubular mechanism for sodium retention is susceptible to lead (Sandstead et al., 1970). Patients with chronic plumbism due to ingestion of illicitly distilled whisky have been observed to lose inappropriate amounts of sodium in their urine when deprived of dietary sodium. Possible mechanisms include renal tubular injury (Goyer and Rhyne, 1973) and/or an impaired endocrine response. The latter phenomenon is characterized by a failure of plasma renin and aldosterone to increase as expected and an apparent lack of aldosterone stimulation of the distal convoluted tubule to conserve sodium (Sandstead et al., 1970; McAllister et al., 1971).

In addition to its interference with sodium reabsorption by the kidney, lead appears to impair the energy-dependent mechanism of red blood cells for controlling sodium and potassium exchange (Angle and McIntire, 1974). These *in vitro* studies presumably reflect similar *in vivo* phenomena.

Chromium

An interaction between chromium and lead is implied by studies in rats that have revealed an apparent protective effect of chromium against chronic lead exposure. The life-shortening effect of 25 ppm lead in drinking water of male rats was decreased by the addition of 1 ppm chromium (Schroeder et al., 1970). The mechanism of the effect was not defined.

Protein and Amino Acids

A possible relationship between susceptibility to lead toxicity and the dietary content of protein and certain amino acids has been observed. According to Baernstein and Grand (1942) a 20% casein diet protected rats from the adverse effects of dietary lead to a greater extent than did a 6 or 13% casein diet. The addition of methionine or cysteine to the 6% diet improved the weight gain and mortality of both the lead-poisoned and the control rats. More contemporary work supports these early findings (Gontzeer et al., 1964). Evidence of a protective effect of cysteine against lead has also been provided by *in vitro* studies of rabbit liver aminolevulinic acid dehydrase exposed to lead in the presence of cysteine (DeBarreiro, 1969).

The influence of dietary protein on the intestinal absorption of lead is incompletely defined. Miler et al. (1970) reported that rats fed a protein-free diet retain twice as much lead as rats fed a 20% casein diet. Others have reported higher tissue lead in rats fed a 4% protein diet and injected with 100 μg of lead acetate daily than in controls. The protein-deprived, lead-exposed animals showed atrophy of the sexual organs and failure of spermatogenesis, and appeared to have an increased susceptibility to infection (Der et al., 1974). They also displayed a marked inhibition of hepatic demethylation of p-chloro-N-methylalanine indicating injury to hepatic microsomes.

Vitamin E and Selenium

Vitamin E (α-tocopherol) has been found to protect rats against the hemolytic effects of 250 ppm lead in drinking water (Levander et al., 1974). The erythrocytes of the vitamin E-deficient rats showed increased mechanical fragility and were more susceptible to lipid peroxidation in the presence of lead. Levander speculated that the findings might have relevance for children of low income families living in the inner city because 2.5–6.1% of such children show biochemical evidence consistent with an inadequate vitamin E nutriture. Levander also tested selenium in his experimental system and found no protective effect against lead-induced hemolysis (O. Levander, 1976, personal communication).

Ascorbic Acid

Pharmacological doses (200–800 mg daily) of ascorbic acid (vitamin C) have been found to protect lambs from lead in mine tailings (Clegg and Rylands, 1966). The vitamin has also been shown to protect rats from the effects of 30 ppm dietary lead on heme synthesis (Kao and Forbes, 1973). According to Graber and Wei (1974) these findings do not imply that citrus fruits are protective, as citric acid apparently increases the intestinal absorption of lead by rats.

Niacin

Niacin has been reported to improve heme synthesis in lead-intoxicated rats (Kao and Forbes, 1973). This beneficial effect apparently does not decrease coproporphyrin excretion (Acocella, 1966).

Tenconi and Acocella (1966) believe that lead alters the metabolism of the niacin precursor, tryptophan, without significantly impairing synthesis of niacin. Others, however, are at variance with this interpretation (Pecora et al., 1966).

Effect on Food Efficiency

The influence of lead on nutrient utilization has not been defined. Because mitochondrial oxidation is adversely affected by lead (Goyer and Rhyne, 1973), impaired utilization of nutrients might be expected. Other apparent effects of lead that might impair efficiency of food utilization include inhibition of pituitary-adrenal homeostasis and suppression of thyroid function (Sandstead, 1973). Men with chronic plumbism from drinking illicitly distilled lead-contaminated whisky were found to have a decreased adrenal and pituitary responsiveness to hypoglycemia and decreased ACTH release by the pituitary gland in response to metyrapone. The abnormalities in iodine metabolism and thyroid function have been noted previously (Sandstead et al., 1969). Although efficiency of utilization of nutrients was not tested in these men, it seems possible it was decreased.

INTERACTIONS OF CADMIUM WITH NUTRIENTS

Studies of cadmium toxicity in experimental animals have almost always been done with levels of cadmium that greatly exceed usual levels of human exposure on a mg/kg dry, fiber-free diet basis (Fox, 1974). With few exceptions the studies have been of short duration. Therefore their applicability to humans might be limited. Even so, some of the interactions among nutrients that have been revealed by these experiments may have relevance. Nutrients that interact with cadmium and influence its toxicity include zinc, copper, iron, selenium, manganese, calcium, vitamin D, protein, ascorbic acid, and pyridoxine.

Zinc

Parizek (1957) was one of the first to show that zinc can antagonize toxic effects of cadmium when he injected 3.0 mol zinc/kg body weight into rats simultaneously with 0.03 mol cadmium/kg and prevented testicular necrosis. Subsequently it was found that prior treatment with zinc was also protective, apparently as a consequence of the induction of metallothionein synthesis by zinc (Nordberg et al., 1972; Webb, 1972a; Bremner and Davis, 1975; Bremner, 1976) and the subsequent binding of cadmium by the metallothionein (Kagi and Vallee, 1961; Webb, 1972a).

Studies in rats have shown that the dietary level of zinc can influence susceptibility to cadmium. When the molar ratio of zinc to cadmium in drinking water was 1:1, signs of cadmium toxicity were much greater than when the molar ratio of zinc to cadmium was 4:1 (Petering et al., 1971). These findings suggest that the ratio of zinc to cadmium in critical organs might be a factor in cadmium toxicity. In support of this concept is the finding that hypertension in rats chronically exposed to cadmium was related to an increase in the molar ratio of cadmium relative to zinc in renal cortex (Schroeder et al., 1966). More recent studies of cadmium-induced hypertension support this interpretation (H. M. Perry, personal communication, 1974, cited by Sandstead, 1976). When rats were fed a diet containing 22.3 ppm zinc and water containing 100 ppm zinc, the induction of increased blood pressure by 2.5 ppm cadmium in drinking water was prevented. Limited autopsy studies in humans also support this concept. Renal tissues from hypertensive humans were found to have an increased molar ratio of cadmium to zinc (Schroeder, 1967; Lener and Bibr, 1971) and the severity of atherosclerosis apparently correlated with an increased renal cortical cadmium-to-zinc ratio (Voors et al., 1973). All studies of humans, however, have not been supportive (Morgan, 1969).

The interaction of cadmium with zinc appears to be competitive. Thus, when the amount of zinc is marginal, toxic effects of cadmium increase and signs of zinc deficiency appear (Petering et al., 1971). One of the effects of cadmium that might be related to its inhibition of zinc-dependent processes was its apparent inhibitory effect on oral glucose tolerance in rats given 17.2 ppm cadmium, 10.0 ppm zinc, and 0.25 ppm copper in drinking water and a zinc-deficient, copper-deficient diet based on sprayed egg white (Petering, 1973). An impairment in insulin release was not observed except in the rats with a deficient intake of zinc. Perhaps the selenium present in the egg white was in part protective in this experiment.

In vitro cadmium interferes with the function of certain zinc-dependent enzymes (Vallee and Ulmer, 1972). Similar phenomena presumably occur *in vivo*. Thus, at levels substantially below those that cause testicular necrosis, cadmium blocked the incorporation of thymidine into testicular DNA (Lee and Dixon, 1973) and inhibited certain zinc-dependent enzymes (Webb, 1972b).

It seems likely that the adverse effects of cadmium on fetuses are due to an inhibition of zinc-dependent processes (Ferm and Carpenter, 1968; Gale, 1973), since the teratogenic effects of cadmium are similar to those of severe zinc deficiency (Hurley and Swenerton, 1966). At present it does not appear that the effects of cadmium on rat fetuses in the above experiments are applicable to human fetuses exposed to ambient levels of cadmium. Analysis of cadmium in Japanese fetuses has shown the levels to be extremely low (Chaube et al., 1973).

Copper

An early demonstration of the inhibitory effect of cadmium on copper-mediated processes was the finding that 100 ppm cadmium fed to chicks produced histological abnormalities in aorta similar to those caused by copper deficiency (Hill et al., 1963). Cadmium presumably impaired the copper-dependent cross-linking of elastin and collagen (O'Dell et al., 1961; Rucker et al., 1971).

Much smaller amounts of cadmium have been shown to impair copper metabolism in the liver. Rats fed a diet containing 1.5 ppm cadmium, 30 ppm zinc, and 2.6 ppm copper displayed a significant decrease in plasma ceruloplasmin, while 18.0 ppm cadmium caused a marked reduction in liver copper (Campbell and Mills, 1973). In contrast to the reduction of copper in liver, copper was increased in kidney by cadmium poisoning (Freeland and Cousins, 1973; Schroeder and Nason, 1974). The increase was presumably due to a release of copper from other tissues and its subsequent binding to metallothionein in the renal parenchyma. A report of increased copper in urine of hypertensive humans (McKenzie and Kay, 1973) is of interest in view of the hypothesis that cadmium has a causative role in human hypertension and because of the effects of cadmium on levels of copper in renal tissue noted above.

The protective effects of copper against cadmium poisoning are not as impressive as those of zinc. It has been reported that copper with appropriate amounts of zinc and manganese significantly decreased the accumulation of cadmium by quail fed 0.02 ppm to 1.02 ppm cadmium (Fox, 1976). These findings have been interpreted as showing that copper and other trace elements might act together to protect against the adverse effects of cadmium.

Selenium

Selenium has been known since 1960 to protect against testicular necrosis from studies of its simultaneous injection with cadmium into rats (Kar et al., 1960; Mason et al., 1964). Subsequently, selenium was found to protect from abortion 21-day pregnant rats injected with cadmium and other toxic sequelae (Parizek et al., 1968).

Selenium has also been found to prevent increased blood pressure in rats given 10 ppm cadmium in drinking water for 6 months. The level of selenium required was 3.5 ppm in the drinking water and 0.5 ppm in the diet. In the absence of cadmium this intake of selenium would be expected to cause selenium toxicity (Perry and Erlanger, 1974).

Another example of the protective effect of selenium is the prevention of cadmium injury to pancreatic beta cells, which would result in impaired insulin release, as well as prevention of cadmium induction of hepatic gluconeogenic enzymes (Merali and Singhal, 1975).

Iron

Cadmium-induced anemia in quail can be prevented by dietary ferrous iron. In contrast, ferric iron is much less effective (Fox et al., 1971). These findings are presumably related to the greater absorbability of ferrous iron. Similar observations have been reported in rats (Pond and Walker, 1972) and swine (Pond et al., 1973).

At the cellular level toxic effects of cadmium on iron metabolism have been shown in rats fed 10 ppm cadmium. Iron content of liver microsomes and soluble fraction were decreased (Whanger, 1973).

Manganese

Manganese appears to have a protective effect against low levels of cadmium when it is fed with appropriate amounts of zinc and copper (Fox, 1976). Hepatic and renal manganese are apparently increased by cadmium (Schroeder and Nason, 1974). The significance of this finding is unknown.

Calcium

Calcium deficiency has been found to increase the intestinal absorption of cadmium and its subsequent deposition in bone and soft tissues (Larson and Piscator, 1971; Itokawa et al., 1974; Pond and Walker, 1975). When protein deficiency is also present, bone mineralization is apparently decreased by cadmium. It has been suggested that the most distressing symptom of Itai-Itai disease, bone pain, is due to a combined deficiency of calcium and protein along with cadmium poisoning (Itokawa et al., 1973).

Studies of swine (Hennig and Anke, 1964) and chicks (Stancer and Dardzonov, 1967) have shown that cadmium has a potent inhibitory effect on calcium incorporation into bone even when dietary calcium is adequate. The effect appears at least in part to be at the level of the intestinal epithelium and the absorption mechanism. A factor that seems partly responsible is the inhibition of 1,25-dihydroxycholecalciferol synthesis by renal tubules (Suda et al., 1974). This hormone facilitates the intestinal absorption of calcium.

Vitamin D

Interference with synthesis of 1,25-dihydroxycholecalciferol is an important effect of cadmium on vitamin D metabolism. The consequences of this phenomenon for humans are incompletely defined, although it may in part contribute to the morbidity of Itai-Itai. Vitamin D has marginal effects on cadmium absorption in rachitic chicks (Cousins and Feldman, 1973). Therefore, it seems unlikely that usual levels of vitamin D in human diets have an influence on the absorption of dietary cadmium.

Vitamin C

Large amounts of ascorbic acid have been found to prevent signs of cadmium poisoning in quail (Fox and Fry, 1970). Beneficial effects included suppression of anemia, improved growth, and improved bone mineralization and mineral element composition (Fox et al., 1971). Combined ascorbic acid and iron supplementation have similar effects (Maji and Yoshida, 1974).

Protein

The type of protein in experimental diets appears to influence the susceptibility of animals to cadmium. When quail were fed casein, gelatin, or soybean protein, they were more severely affected by cadmium than were quail fed dried egg white (Fox et al., 1973). These findings might have resulted from the relatively high levels of selenium present in some egg white. Rats fed a diet containing 3.5 ppm selenium primarily from egg white for 2.5 yr with 2.5 ppm cadmium in their drinking water displayed no apparent injury (H. H. Sandstead, unpublished data, 1976).

The combined interaction of low dietary protein and calcium with cadmium causing osteopenia in rats was noted above (Itokawa et al., 1974).

Pyridoxine

In contrast to the nutrients discussed above, pyridoxine appears to increase the toxic effects of cadmium, presumably because it is essential for the absorption of cadmium by the intestine. Pyridoxine deficiency decreases the intestinal absorption of cadmium (Stowe et al., 1974). These findings illustrate how cadmium absorption is dependent on normal intestinal cell function and suggest that cadmium might compete with an essential trace element such as copper for absorption (Evans and Grace, 1975; Hahn and Evans, 1975).

Lead

Although lead is not considered a nutrient, effects of simultaneous exposure to lead and cadmium must be noted. Teratogenic effects are more severe in hamsters when both metals are given (Ferm, 1969). This finding might have application in livestock and humans exposed to effluent from lead-zinc refineries or mines.

INTERACTIONS OF MERCURY WITH NUTRIENTS

Mercury occurs in human diets primarily as methyl and inorganic mercury. Outbreaks of human poisoning due to methyl mercury in fish have been reported from Minamata and Niigata, Japan (Irukayama et al., 1962; Takeuchi et al., 1962; Uchida and Inoue, 1962). These reports have been reviewed by Friberg and Vostal (1972) and by Kitamura et al. (1976). Consumption of wheat treated with a methyl mercury fungicide has resulted in poisoning of farmers and of livestock. The largest of these accidents occurred in Iraq in 1972 (Bakir et al., 1973; Clarkson and Marsh, 1976).

Selenium

A protective effect of selenium against methyl mercury intoxication (Ganther et al., 1972; Newbern et al., 1972; Ganther and Sunde, 1974) and other forms of mercury (Johnson and Pond, 1974; Potter and Matrone, 1974) has been shown in experimental animals. It appears that selenium is the substance in tuna that protects against the relatively high levels of methyl mercury present (Ganther et al., 1972; Ganther and Sunde, 1974).

On the other hand, mercury apparently protects against toxic levels of selenium (Hill, 1974; Potter and Matrone, 1974). The formation of a mercury-selenium complex is presumably the explanation for detoxification. This concept is supported by observations on renal tubular epithelium and reticuloendothelial cells of animals exposed simultaneously to mercury and selenium. Inclusion bodies containing mercury and selenium were found in nuclei and cytoplasm of the cells (Groth et al., 1976).

The protective effect of selenium might account for the lack of clinically evident mercury poisoning in persons who frequently consume methyl mercury-containing tuna, freshwater fish, or fish protein concentrate (FPC). Rats fed FPC that provides a weekly intake of 12 μg mercury/kg body weight did not show evidence of mercury poisoning. This amount of FPC was equivalent to 1.85 lb fresh fish, containing 1 μg mercury/kg (Newbern et al., 1972).

Eighty-eight Korean fishermen and 45 Samoan islanders, who consumed large amounts of tuna and other sea fish, displayed no evidence of mercury poisoning (Clarkson and Marsh, 1976). Eleven of them had blood levels of mercury greater than 120 μg/ml; all but one, however, had levels below 200 μg/ml. These levels were below those found in mercury-intoxicated persons in Iraq.

Zinc

Zinc appears to have a protective effect against mercury in certain experimental situations. In sufficient concentrations mercury is a potent teratogen. When 2 mg zinc/kg body weight was injected into pregnant golden hamsters on the eighth day of gestation with either 2 or 4 mg mercury/kg, fewer fetal anomalies were observed than when mercury was given alone (Gale, 1973). In addition, pretreatment with zinc 1 hr before dosing with 2 mg mercury/kg increased the protection. This effect was not observed when the 4 mg mercury/kg was given. It seems likely the mercury was complexed by metallothionein, which had been induced by the zinc (Bremner, 1976).

Copper

Mercury appears to inhibit copper metabolism. When rats given mercury were tested for copper absorption from an isolated *in vivo* gut loop, the uptake of copper-64 was modestly, though not statistically, depressed. In addition, the proportion of copper-64 retained by blood and liver was decreased, and the copper-64 levels were increased in the kidneys of the rats (Van Campen, 1966).

SUMMARY

Interactions of lead, cadmium, and mercury with essential nutrients have been reviewed. The interactions that appear to have the greatest health implications for humans are those of lead with calcium and iron; cadmium with zinc, copper, and selenium; and mercury with selenium. Future studies in which the levels of the toxic elements more closely approximate human exposure and the composition of the experimental diets are better defined are necessary to test the hypothesis that interactions between the toxic elements and essential nutrients are indeed of importance for human health.

REFERENCES

Acocella, G. 1966. *Acta Vitaminol.* 20:195–202.
Angle, C. and McIntire, M. C. 1974. *Environ. Health Perspect.* 7:133–137.
Baernstein, H. D. and Grand, J. A. 1942. *J. Pharmacol. Exp. Ther.* 74:18–24.
Bakir, F., Damluji, S. F., Amiu-Zaki, L., Murtadha, M., Khalidi, A., Al-Raivi, N. Y., Tikriti, S., Dhakir, H. I., Clarkson, T. W., Smith, J. C. and Doherty, R. A. 1973. *Science* 181:230–241.

Bremner, I. 1976. *Br. J. Nutr.* 35:245–252.

Bremner, I. and Davis, N. T. 1975. *Biochem. J.* 149:733–738.

Campbell, J. K. and Mills, C. F. 1973. *Proc. Nutr. Soc.* 33:15A.

Cerklewski, F. L. and Forbes, R. M. 1976. *J. Nutr.* 106:689–696.

Chaube, J., Nishimura, H. and Swinyard, C. A. 1973. *Arch. Environ. Health* 26:237–240.

Clarkson, T. W. and Marsh, D. 1976. In *Effects and dose response relationships of toxic metals*, ed. G. F. Nordberg, pp. 246–261. Amsterdam: Elsevier.

Clegg, F. G. and Rylands, J. M. 1966. *J. Comp. Pathol.* 76:15–22.

Cousins, R. J. and Feldman, J. L. 1973. *Nutr. Rep. Int.* 8:363–369.

DeBarreiro, O. C. 1969. *Biochem. Pharmacol.* 18:2267–2271.

Der, R., Hilderbrand, D., Fahim, Z., Griffin, W. T. and Fahim, M. S. 1974. In *Trace substances in environmental health–VIII*, ed. D. D. Hemphill, pp. 417–431. Columbia: University of Missouri Press.

Evans, G. W. and Grace, C. I. 1975. *Proc. Soc. Exp. Biol. Med.* 147:687–689.

Ferm, V. H. 1969. *Experientia* 25:56–57.

Ferm, V. H. and Carpenter, S. J. 1968. *Lab. Invest.* 18:429–432.

Finnelli, V. N., Murthy, L. and Peirano, W. B. 1974. *Biochem. Biophys. Res. Commun.* 60:1418–1424.

Finnelli, V. N., Klauder, D. S., Karaffa, M. A. and Petering, H. G. 1975. *Biochem. Biophys. Res. Commun.* 65:303–311.

Fox, M. R. S. 1974. *J. Food Sci.* 39:321–324.

Fox, M. R. S. 1976. In *Trace elements in human health and disease*, vol. II, ed. A. S. Prasad, pp. 401–416 New York: Academic Press.

Fox, M. R. S. and Fry, B. E., Jr. 1970. *Science* 196:989–991.

Fox, M. R. S., Fry, B. E., Jr., Harland, B. F., Schertel, M. E. and Weeks, C. E. 1971. *J. Nutr.* 101:1295–1305.

Fox, M. R. S., Jacobs, R. M., Fry, B. E., Jr. and Harland, B. F. 1973. *Fed. Proc.* 32:924 (Abstr.).

Freeland, J. H. and Cousins, R. J. 1973. *Nutr. Rep. Int.* 8:337–347.

Friberg, L. and Vostal, J. 1972. *Mercury and the environment.* Cleveland: CRC Press.

Friberg, L., Piscator, M. and Nordberg, G. 1971. *Cadmium in the environment.* Cleveland: CRC Press.

Gale, T. F. 1973. *Environ. Res.* 6:95–105.

Ganther, H. E. and Sunde, M. L. 1974. *J. Food Sci.* 39:1–5.

Ganther, H. E., Goudie, C., Sunde, M. L., Kopecky, M. S., Wagner, P., Oh, S. H. and Hoekstra, W. G. 1972. *Science* 175:1122–1124.

Goldatovic, D., Djurovic, M. and Petrovic, C. 1966. *Arch. Farm.* 16:151–154.

Gontzeer, I., Gutzesco, P., Corona, D. and Lungu, D. 1964. *Arch. Sci. Physiol. (Paris)* 18:211–224.

Goyer, R. A. and Rhyne, B. 1973. *Int. Rev. Exp. Pathol.* 12:1–77.

Graber, B. T. and Wei, E. 1974. *Toxicol. Appl. Pharmacol.* 27:685–691.

Groth, D. H., Stettler, L. and MacKay, G. 1976. In *Effects and dose response relationships of toxic metals*, ed. G. Nordberg, pp. 525–543. Amsterdam: Elsevier.

Hahn, C. and Evans, G. W. 1975. *Am. J. Physiol.* 228:1020–1023.

Hennig, A. and Anke, M. 1964. *Arch. Tierernaehr.* 14:55–57.

Hill, C. H. 1974. *J. Nutr.* 104:593–598.

Hill, C. H., Matrone, G., Payne, W. L. and Barber, C. W. 1963. *J. Nutr.* 80:227–235.

Hsu, F. S., Krook, L., Pond, W. G. and Duncan, J. R. 1975. *J. Nutr.* 105:112–118.

Hurley, L. S. and Swenerton, H. 1966. *Proc. Soc. Exp. Biol. Med.* 123: 692–697.

Irukayama, K., Fujiki, M., Kai, F. and Kondo, T. 1962. *Kumamoto Med. J.* 15:1–12, 57–68.

Itokawa, Y., Abe, T. and Tanaka, S. 1974. *Arch. Environ. Health* 28:149–154.

Johnson, S. L. and Pond, W. G. 1974. *Nutr. Rep. Int.* 9:135–147.

Kagi, J. H. R. and Vallee, B. L. 1961. *J. Biol. Chem.* 236:2435–2442.

Kao, R. L. C. and Forbes, R. M. 1973. *Arch. Environ. Health* 27:31–35.

Kar, A. B., Das, R. P. and Mukery'i, F. N. I. 1960. *Proc. Natl. Inst. Sci. Ind., B, Biol. Sci.*, Suppl. 26:40–50.

Kitamura, S., Sumino, K., Hayakawa, K. and Shibata, T. 1976. In *Effects and dose response relationships of toxic metals*, ed. G. F. Nordberg, pp. 262–272. Amsterdam: Elsevier.

Kolbye, A. C., Mahaffey, K. R., Fioino, J. A., Corneliussan, P. C. and Jelinek, C. F. 1974. *Environ. Health Perspect.* No. 7:65–74.

Kostial, K., Simonovic, I. and Pisonic, M. 1971. *Environ. Res.* 4:360–366.

Larson, S. E. and Piscator, S. 1971. *Israel J. Med. Sci.* 7:495–497.

Lee, I. P. and Dixon, R. L. 1973. *J. Pharmacol. Exp. Ther.* 87:641–652.

Lener, J. and Bibr, B. 1971. *Lancet* 1:970.

Lepow, M. L., Brockman, L., Rubino, R. A., Markowitz, S., Gillette, M. and Kapish, J. 1974. *Environ. Health Perspect.* No. 7:99–102.

Levander, O. 1976. *Fed. Proc.* In press.

Levander, O., Morris, V. C., Higgs, D. J. and Fernetti, R. J. 1974. *J. Nutr.* 105:1481–1485.

Mahaffey, K. R. 1974. *Environ. Health Perspect.* No. 7:107–112.

Mahaffey, K. R., Treloar, S. and Banks, T. 1976. *J. Nutr.* 106(7): Abstr. 53.

Maji, T. and Yoshida, H. 1974. *Nutr. Rep. Int.* 10:139–149.

Mason, K. E., Young, J. O. and Brown, J. A. 1964. *Anat. Rec.* 148:309.

McAllister, R. M., Michelakis, A. M. and Sandstead, H. H. 1971. *Am. Med. Assoc. Arch. Int. Med.* 127:919–923.

McKenzie, J. M. and Kay, D. L. 1973. *N.Z. Med. J.* 78:68–70.

Merali, Z. and Singhal, R. L. 1975. *J. Pharmacol. Exp. Ther.* 195:58–66.

Miler, N., Sattler, E. L. and Menden, E. 1970. *Med. Ernaehr.* 11:29–32.

Morgan, J. M. 1969. *Arch. Intern. Med.* 123:405–408.

National Research Council. 1972. *Lead, airborne lead in perspective*, pp. 1–330. Washington, D.C.: National Academy of Sciences.

Newbern, P. M., Glaser, O., Friedman, L. and Stillings, B. R. 1972. *Nature* 237:40–41.

Niklowitz, W. J. and Yeager, D. W. 1973. *Life Sci.* 13:897–905.

Nordberg, G. F., Nordberg, M., Piscator, M. and Vesterberg, O. 1972. *Biochem. J.* 126:491–498.

O'Dell, B. L., Harwick, B. C., Reynolds, G. and Savage, J. E. 1961. *Proc. Soc. Exp. Biol. Med.* 108:402–405.

Parizek, J. 1957. *J. Endocrinol.* 15:56–63.

Parizek, J. 1976. In *Effects and dose response relationship of toxic metals*, ed. G. F. Nordberg, pp. 498–510. Amsterdam: Elsevier.

Parizek, J., Ostadalora, I., Benes, I. and Babicky, A. 1968. *J. Reprod. Fertil.* 16:507–509.

Pecora, L., Silveestrone, A. and Branchccio, A. 1966. *Panminerva Med.* 8:284 (Kettering Abstr. N1 429, 192).

Perry, H. M., Jr. and Erlanger, M. W. 1974. *J. Lab. Clin. Med.* 83:510–515.

Petering, H. G. 1973. In *Trace element metabolism in animals—2*, ed. W. G. Hoekstra, J. W. Suttie, H. E. Ganther and W. Mertz, pp. 311–325. Baltimore: University Park Press.

Petering, H. G., Johnson, M. A. and Stemmer, K. L. 1971. *Arch. Environ. Health* 23:93–101.

Pond, W. G. 1975. *Cornell Vet.* 65:441–456.

Pond, W. G. and Walker, E. F., Jr. 1972. *Nutr. Rep. Int.* 5:365–370.

Pond, W. G. and Walker, E. F., Jr. 1975. *Proc. Soc. Exp. Biol. Med.* 148:665–668.

Pond, W. G., Walker, E. F., Jr. and Kirtland, D. 1973. *J. Anim. Sci.* 36:1122–1124.

Potter, S. and Matrone, G. 1974. *J. Nutr.* 104:638–647.

Rucker, R. B., O'Dell, B. L., Parker, H. E. and Rogler, J. D. 1971. In *Trace substances in environmental health—IV*, ed. D. D. Hemphill, pp. 255–259. Columbia: University of Missouri Press.

Sandstead, H. H. 1967. *Proc. Soc. Exp. Biol. Med.* 124:18–20.

Sandstead, H. H. 1973. In *Trace substances in environmental health—VI*, ed. D. D. Hemphill, pp. 223–236. Columbia: University of Missouri Press.

Sandstead, H. H. 1976. In *Effects and dose-response relationships of toxic metals*, ed. G. F. Nordberg, pp. 511–526. Amsterdam: Elsevier.

Sandstead, H. H., Stant, E. G., Brill, A. B., Aries, L. and Terry, R. T. 1969. *Am. Med. Assoc. Arch. Int. Med.* 123:632–635.

Sandstead, H. H., Michelakis, A. M. and Temple, T. E. 1970. *Am. Med. Assoc. Arch. Environ. Health* 20:356–363.

Schroeder, H. A. 1967. *Circulation* 35:57082.

Schroeder, H. A. and Nason, A. P. 1974. *J. Nutr.* 104:167–178.

Schroeder, H. A., Knoll, S. S., Little, J. W., Livingston, P. O. and Myers, M. A. G. 1966. *Arch. Environ. Health* 13:788–789.

Schroeder, H. H., Mitchner, M. and Nason, A. P. 1970. *J. Nutr.* 100:5968.

Shibko, S. I., Shapieo, R. E. and Kolbye, A. C., Jr. 1976. In *Effects and dose-response relationships of toxic metals*, ed. G. F. Nordberg, pp. 199–206. Amsterdam: Elsevier.

Silbergeld, E. K., Fales, J. R. and Goldberg, A. M. 1974. *Nature* 247:49–50.

Smith, J. 1976. In *Trace elements in human health and disease*, vol. II, ed. A. S. Prasad, pp. 443–452 New York: Academic Press.

Snowdon, C. T. and Sanderson, B. A. 1974. *Science* 183:92–94.

Sobel, A. E., Yuska, H., Peters, D. D. and Kramer, B. 1940. *J. Biol. Chem.* 132:239–265.

Stancer, H. and Dardzonov, T. 1967. *Zhivotnovud. Nauki* 4(6):97–103 (Abstr. 4102 in annotated bibliography 21 of the CAB, Bucksburn, Aberdeen AB2 9SB).

Stowe, H. D., Goyer, R. A., Medhey, P. and Cates, M. 1974. *Arch. Environ. Health* 28:209–216.

Suda, T., Horiuchi, H., Ogata, E., Ezawa, I., Otaki, N. and Kimura, M. 1974. *FEBS Lett.* 42:23–26.

Takeuchi, T., Morikowa, N., Matsumoto, H. and Shiraishi, Y. A. 1962. *Acta Neuropathol.* 2:40–57.

Tenconi, L. T. and Acocella, G. 1966. *Acta Vitaminol.* 20:189–194.

Ten State Nutrition Survey. 1968–1970. U.S. Department of Health, Education, and Welfare Publ. No. (HSM)72-8134.

Uchida, M. and Inoue, T. 1962. *Kumamoto Med. J.* 15:149–153.

Vallee, B. L. and Ulmer, D. D. 1972. *Ann. Rev. Biochem.* 41:71–128.

Van Campen, D. R. 1966. *J. Nutr.* 88:125–130.

Voors, A. W., Shuman, M. S. and Gallagher, N. P. 1973. In *Trace substances in environmental health—IV*, ed. D. D. Hemphill, pp. 215–222. Columbia: University of Missouri Press.

Waxman, H. S. and Rabinoritz, M. 1966. *Biochem. Biophys. Acta* 129:369.

Webb, M. 1972a. *Biochem. Pharmacol.* 21:2767–2771.

Webb, M. 1972b. *J. Reprod. Fertil.* 30:83–98.

Whanger, P. D. 1973. *Res. Commun. Pathol. Pharmacol.* 5:733–740.

Willoughby, R. A., McDonald, E. and McSherry, B. J. 1972. *Vet. Rec.* Oct. 14, p. 382.

Chapter 9

METAL CARCINOGENESIS

F. William Sunderman, Jr.
Department of Laboratory Medicine
University of Connecticut School of Medicine
Farmington, Connecticut

INTRODUCTION

In view of the prevalent public and scientific concern about environmental and occupational factors in the etiology of human cancer, it is timely to attempt to summarize current knowledge regarding the carcinogenic properties of metallic compounds. This review is intended to bring up to date the résumés of metal carcinogenesis which have been published by Furst (1971), Furst and Haro (1969a, 1969b), Hueper (1966), Kanisawa (1971), Roe and Lancaster (1963), Sunderman (1971), Williams (1972) and by the International Agency for Research on Cancer (1973). The aims of this review are (1) to summarize concisely the epidemiological evidence that certain metals may be carcinogenic for man; (2) to present comprehensive compilations of investigations which have demonstrated the carcinogenicity of metallic compounds in experimental animals; and (3) to discuss the experimental studies which pertain to possible mechanisms of metal carcinogenesis. This review does not consider related topics, such as asbestos carcinogenesis, solid-state or smooth-surface carcinogenesis, or the carcinogenicity of radioactive metals. Particular emphasis is given to nickel carcinogenesis, since the carcinogenicity of nickel has been more thoroughly investigated than that of any other metal.

ARSENIC

Epidemiological studies. Human subjects who have been chronically exposed to arsenic compounds by oral or respiratory routes have a

This study was supported by U.S. Atomic Energy Commission Grant No. AT(11-1)-3140 and by U.S. Public Health Service Contract No. HSM 99-72-24 from the National Institute of Occupational Safety and Health.

The author is grateful to Nancy M. Harvey for assistance in the preparation of this manuscript.

significantly increased incidence of epidermoid carcinomas of the skin and lungs, and of precancerous dermal keratoses (Bowen's disease) (Dobson and Pinto, 1966; Ehlers, 1974; Friedrich, 1972; Goldman, 1973; Lee and Fraumeni, 1969; Minkowitz, 1964; National Institute for Occupational Safety and Health, 1973; Ott et al., 1974; Yeh, 1973; Zachariae, 1972). In addition, there has been a case report of leukemia which arose as a complication of aplastic anemia induced by arsenic toxicity (Rosen, 1971). Human cancers have been observed following exposure to arsenic in drinking water (Yeh, 1973) and medications (Ehlers, 1974; Goldman, 1973; Minkowitz, 1964) as well as in metallurgical (Lee and Fraumeni, 1969), chemical (Ott et al., 1974), and agricultural (Friedrich, 1972; Zachariae, 1972) workers.

Experimental carcinogenesis. There is little evidence that arsenic compounds are carcinogenic in experimental animals (Baroni et al., 1963; Frost, 1967; Hueper and Payne, 1962; Milner, 1969). Osswald and Goerttler (1971) observed that the incidence of lymphocytic leukemias and malignant lymphomas in female Swiss mice and their offspring was considerably increased, as compared with nontreated animals, by subcutaneous injection of a sodium arsenic compound (valence not specified) during pregnancy. The incidence of these cancers was further increased by additional postnatal injections of the arsenic compound.

Possible mechanisms of arsenic carcinogenesis. Jung and Trachsel (1970) found that sodium arsenate inhibited methylthymidine uptake into human dermal cells *in vitro*, consistent with suppression of DNA synthesis. Paton and Allison (1972) found chromosomal aberrations in human leukocyte and dermal fibrocyte cultures exposed to sodium arsenite. Petres et al. (1970) and Rosen (1971) have suggested that arsenic may substitute for phosphorous in DNA, causing a weak bond in the DNA chain.

BERYLLIUM

Epidemiological studies. Mancuso (1970) observed that the incidence of respiratory cancer was significantly increased among beryllium workers who were employed for less than 15 months, especially in those with prior respiratory illnesses (bronchitis and pneumonitis). Hasan and Kazemi (1974) reported an equivocal increase in the incidence of respiratory cancers in patients with chronic pulmonary berylliosis. On the other hand, Bayliss (1972) investigated the causes of deaths among 3900 men who had worked in beryllium factories, and he found no increase in the incidence of respiratory cancers.

Experimental carcinogenesis. As indicated in Table 1, intravenous injections of suspensions of beryllium salts to rabbits and mice have been shown to induce osteogenic sarcomas, and inhalation of aerosols of beryllium oxide, phosphate, and sulfate has been found to cause pulmonary carcinomas in rats and monkeys (Gardner and Heslington, 1946; Cloudman et al., 1949; Nash, 1950; Dutra and Largent, 1950; Dutra et al., 1951; Sissons, 1950;

TABLE 1 Beryllium Carcinogenesis in Experimental Animals

Authors	Animals	Compounds and routes	Tumors
Gardner and Heslington, 1946	Rabbits	$ZnBeSiO_3$, BeO (iv)	Osteosarcomas
Cloudman et al., 1949	Rabbits, mice	$ZnBeSiO_3$, BeO (iv)	Osteosarcomas
Nash, 1950	Rabbits	$ZnBeSiO_3$, BeO (iv)	Osteosarcomas
Dutra and Largent, 1950; Dutra et al., 1951	Rabbits	$ZnBeSiO_3$, BeO (iv, inhalation)	Osteosarcomas
Sissons, 1950; Barnes et al., 1950	Rabbits	$ZnBeSiO_3$, BeO, $BeSiO_3$ (iv)	Osteosarcomas
Hoagland et al., 1950	Rabbits	$BeHPO_4$, $ZnBeSiO_3$, BeO (iv)	Osteosarcomas
Janes et al., 1954, 1956; Kelly et al., 1961	Rabbits	$ZnBeSiO_3$ (iv)	Osteosarcomas
Araki et al., 1954	Rabbits	BeO (iv)	Osteosarcomas
Vorwald and Reeves, 1959; Vorwald et al., 1966	Monkeys, rats	BeO, $BeSO_4$ (inhalation)	Pulmonary carcinomas
Schepers, 1961, 1964	Monkeys, rats	$BeSO_4$, $BeHPO_4$ (inhalation)	Pulmonary carcinomas
Yamaguchi, 1963	Rabbits	BeO (iv)	Osteosarcomas
Reeves, 1967; Reeves and Vorwald, 1967; Reeves et al., 1967	Rats	$BeSO_4$ (inhalation)	Pulmonary carcinomas
Komitowski, 1968	Rabbits	BeO (iv)	Osteosarcomas

259

Barnes et al., 1950; Hoagland et al., 1950; Janes et al., 1954, 1956; Kelly et al., 1961; Araki et al., 1954; Vorwald and Reeves, 1959; Vorwald et al., 1966; Schepers, 1961, 1964; Yamaguchi, 1963; Reeves, 1967; Reeves et al., 1967; Reeves and Vorwald, 1967; Komitowski, 1968).

Possible mechanisms of beryllium carcinogenesis. Beryllium becomes localized in the nucleus of regenerating rat hepatocytes (Witschi and Aldridge, 1968) and inhibits thymidine incorporation into hepatic DNA (Witschi, 1968). In regenerating rat liver, beryllium blocks the induction of certain enzymes needed for DNA synthesis (thymidine kinase, thymidylate kinase, thymidylate synthetase, deoxythymidylate deaminase and DNA polymerase), without affecting the activity of various other enzymes (Witschi, 1970). Beryllium also inhibits the induction of several hepatic enzymes following administration of phenobarbital, methylcholanthrene, cortisol, and tryptophan (Witschi and Marchand, 1971). Inhibition of hepatic enzyme induction apparently is not mediated by direct inhibition of RNA or protein synthesis, since beryllium does not impair incorporation of orotic acid into hepatic RNA or incorporation of leucine into hepatic proteins (Marcotte and Witschi, 1972). Marcotte and Witschi (1972) have proposed that beryllium binding to nuclear proteins or chromatin may block the expression of repressed portions of the genome, without influencing the transcription of portions of the genome which have already been derepressed.

CADMIUM

Epidemiological studies. Malcolm (1972) and Winkelstein and Kantor (1969) have reviewed the tenuous evidence that occupational exposure to cadmium might be associated with increased incidence of prostatic cancer. Potts (1965) reported 3 cases of prostatic cancer and 2 cases of other forms of cancer in a group of 74 cadmium workers. Kipling and Waterhouse (1967) studied 248 cadmium workers (including the subjects reported by Potts) and found 4 cases of prostatic cancer and 8 cases of other forms of cancer. Holden (1969) observed 1 case of prostatic cancer and 1 case of pulmonary cancer among 42 cadmium workers. On the basis of these reports, it appears dubious that there is any significant association between occupational exposure to cadmium and prostatic cancer, although there has not been any large-scale epidemiological survey to rule out this possibility.

Experimental carcinogenesis. As indicated in Table 2, parenteral injections of cadmium powder, cadmium sulfide, oxide, sulfate, and chloride in rodents have been shown to induce sarcomas at the sites of injection (Heath et al., 1962; Heath and Daniel, 1964a; Kazantzis, 1963; Kazantzis and Hanbury, 1966; Haddow et al., 1964a; Roe et al., 1964a; Guthrie, 1964b; Gunn et al., 1963, 1964, 1965, 1967; Nazari et al., 1967; Favino et al., 1968; Knorre, 1970, 1971; Lucius et al., 1972). Moreover, subcutaneous injection of $CdCl_2$ causes acute testicular necrosis, which is followed by Leydig cell regeneration and hyperplasia, and ultimately induces tumors of the Leydig

TABLE 2 Cadmium Carcinogenesis in Experimental Animals

Authors	Animals	Compounds and routes	Tumors
Heath et al., 1962; Heath and Daniel, 1964a	Rats	Cd powder in fowl serum (im)	Sarcomas
Kazantzis, 1963; Kazantzis and Hanbury, 1966	Rats	CdS, CdO (sc)	Sarcomas
Haddow et al., 1964a; Roe et al., 1964a	Rats	$CdSO_4$, $CdCl_2$ (sc)	Sarcomas and Leydigiomas
Guthrie, 1964b	Chickens	$CdCl_2$ (intratesticular)	Teratoma
Gunn et al., 1963, 1964, 1965, 1967	Rats, mice	$CdCl_2$ (im)	Sarcomas and Leydigiomas
Nazari et al., 1967; Favino et al., 1968	Rats	$CdCl_2$ (sc)	Sarcomas and Leydigiomas
Knorre, 1970, 1971	Rats	$CdCl_2$ (sc)	Sarcomas and Leydigiomas
Lucis et al., 1972	Rats	$CdCl_2$ (sc, intrahepatic)	Sarcomas and Leydigiomas
Reddy et al., 1973	Rats	$CdCl_2$ (sc)	Leydigiomas

cells (interstitial cells of the testis) (Haddow et al., 1964a; Roe et al., 1964a; Guthrie, 1964b; Gunn et al., 1963, 1964, 1965, 1967; Nazari et al., 1967; Favino et al., 1968; Knorre, 1970, 1971; Lucis et al., 1972; Reddy et al., 1973). For example, in a study by Reddy et al. (1973), 80% of rats which were given a single injection of $CdCl_2$ developed Leydigiomas within 1 year, compared to no tumors in saline-treated control rats. The Leydig cell tumors induced by $CdCl_2$ are endocrinologically functional, based upon histochemical studies and measurements of androgen excretion (Gunn et al., 1965; Favino et al., 1968; Lucis et al., 1972; Reddy et al., 1973). Experimental carcinogenesis by cadmium salts has recently been reviewed by Lacassagne (1971) and Flick et al. (1971).

Possible mechanisms of cadmium carcinogenesis. Gunn et al. (1963, 1964, 1965, 1967) showed that the acute testicular injury produced by cadmium chloride is mediated by vascular damage to the internal spermatic artery and pampiniform venous plexus, and that simultaneous administration of zinc acetate protects these vessels from acute damage and thereby prevents the ultimate development of Leydigiomas. Gunn et al. (1964) reported that zinc acetate also prevents the induction of sarcomas at the site of injection; however, this finding has not been confirmed by Furst and Casetta (1972). Pulido et al. (1966), Kägi et al. (1973, 1974) and Bühler and Kägi (1974) have isolated and characterized metallothionein, a cysteine-rich protein (mol. wt. = 6,600), which binds zinc, cadmium, mercury, and copper and is present in many tissues of the body. Repeated administration of cadmium, mercury, or zinc induces the biosynthesis of hepatic metallothionein (Winge and Rajagopolan, 1972; Nordberg et al., 1972; Chen et al., 1974; Piotrowski et al., 1974). Webb (1972) has reported that induced biosynthesis of hepatic metallothionein is apparently controlled at the translational level, since it is inhibited by cyclohexamide but not by actinomycin D. Webb (1972) has also found that metallothionein does not bind cobalt or nickel *in vivo.* Nordberg (1971) has shown that pretreatment of mice with repeated small doses of cadmium chloride gives protection against acute testicular necrosis produced by a single, larger dose of cadmium chloride, and he has attributed this protective effect to the induction of metallothionein. It may be speculated that the inhibitory effect of zinc upon cadmium induction of Leydigiomas is also related to induced biosynthesis of metallothionein.

CHROMIUM

Epidemiological studies. The predilection of workers in chromium refineries for the development of respiratory cancers (lung, nose, pharynx, and sinuses) has been thoroughly documented in several studies (Machle and Gregorius, 1948; Baetjer, 1950; Mancuso and Hueper, 1951; Bidstrup, 1951; Bidstrup and Case, 1956; Taylor, 1966) and has recently been reviewed by Enterline (1974) and by Baetjer et al. (1974). During the period from 1930 to 1947, workers in chromate plants in the United States had a relative risk of dying from respiratory cancer which was twenty times the rate for a control population (Machle and Gregorius, 1948). The epidemiological studies have suggested that hexavalent chromium compounds may be responsible for respiratory neoplasia (Baetjer et al., 1974).

Experimental carcinogenesis. Intraosseous, intramuscular, subcutaneous, intrapleural, and intraperitoneal injections of chromium powder and of hexavalent chromium compounds have produced local sarcomas in rats, mice, and rabbits (Table 3). Laskin et al. (1970) and Kuschner and Laskin (1971) have succeeded in inducing squamous cell carcinomas and adenocarcinomas in lungs of rats following intrabronchial implantation of pellets of calcium chromate. These workers compared the carcinogenicities of $CaCrO_4$, $Cr_2(CrO_4)_3$, Cr_2O_3, and CrO_3; they found that calcium chromate was by far the most potent respiratory carcinogen (Laskin et al., 1970; Kuschner and Laskin, 1971). Nettesheim et al. (1971) observed alveologenic adenomas and adenocarcinomas in mice exposed to chronic inhalation of calcium chromate, but they did not find any bronchogenic tumors.

Possible mechanisms of chromate carcinogenesis. Venitt and Levy (1974) attempted to elucidate possible mechanisms of chromate carcinogenesis by investigating the mutagenicity of chromates in bacteria. They found that hexavalent (but *not* trivalent) chromium compounds were mutagenic in certain strains of *Escherichia coli*. The mutagenicity of chromates was not modified by the genetic absence of pathways for repair of DNA. They concluded that chromates are among the mutagens which exert their effects by directly modifying DNA bases so that base-pair errors arise at subsequent cell divisions. Although their evidence is indirect, Venitt and Levy (1974) speculated that their observations may point to the mechanism of chromate carcinogenesis.

COBALT

Epidemiological studies. There is no evidence that occupational exposures to cobalt are associated with increased risk of neoplasia.

Experimental carcinogenesis. As summarized in Table 4, cobalt powder, cobalt oxide, and cobalt sulfide are capable of inducing local sarcomas following parenteral administration to rabbits and rats. Although not included in Table 4, Heath et al. (1971) and Swanson et al. (1973) have found that wear particles from prostheses made from a cobalt-chromium alloy are carcinogenic for rat muscle.

Possible mechanisms of cobalt carcinogoenesis. Heath et al. (1969) have found that, during the course of induction of rhabdomyosarcomas by implanted cobalt powder, the metal slowly dissolves and disappears from the injection site. They have shown by *in vitro* experiments that metallic cobalt slowly reacts with serum proteins to form soluble nondialyzable complexes which are less toxic for rat myoblasts in culture than the equivalent amount of ionic Co(II). Heath et al. (1969) have suggested that cobalt-protein complexes, possibly adsorbed on the surface of myoblasts, may enter the cell

TABLE 3 Chromium Carcinogenesis in Experimental Animals

Authors	Animals	Compounds and routes	Tumors
Vollmann, 1940; Schinz and Uehlinger, 1942	Rabbits	Metallic Cr (intraosseous)	Sarcomas
Hueper, 1958; Hueper and Payne, 1959, 1962; Payne, 1960	Rats, mice	Roasted chromite ore, $CaCrO_4$, CrO_3, $Na_2Cr_2O_7$ (im, sc, and intrapleural)	Sarcomas and squamous cell carcinomas
Dvizhkov and Fedorova, 1967	Rats	Cr_2O_3 (ip and intrapleural)	Sarcomas
Roe and Carter, 1969	Rats	$CaCrO_4$ (sc)	Sarcomas
Laskin et al., 1970; Kuschner and Laskin, 1971	Rats	$CaCrO_4$ (intrabronchial)	Squamous cell and adenocarcinomas
Nettesheim et al., 1971	Mice	$CaCrO_4$ (inhalation)	Adenomas and adenocarcinomas

TABLE 4 Cobalt Carcinogenesis in Experimental Animals

Authors	Animals	Compounds and routes	Tumors
Vollmann, 1940; Schinz and Uehlinger, 1942	Rabbits	Metallic Co (intraosseous)	Sarcomas
Thomas and Thiery, 1953	Rabbits	Co powder (sc)	Sarcomas
Heath, 1956, 1960; Daniel et al., 1967; Heath et al., 1969	Rats	Co powder in fowl serum (im)	Sarcomas
Gilman, 1962; Gilman and Ruckerbauer, 1962	Rats	CoO, CoS (im)	Sarcomas

by endocytosis, and that subsequent digestion of the carrier proteins by lysosomal proteinases leads to intracellular liberation and redistribution of Co(II).

IRON

Epidemiological studies. Several investigators (Turner and Grace, 1938; Faulds and Stewart, 1956; McLaughlin and Harding, 1956; Braun et al., 1960; Monlibert and Roubille, 1960; Roussel et al., 1964; Boyd et al., 1970) have suggested that hematite miners have slightly increased risk of developing pulmonary cancers. For example, Boyd et al. (1970) observed 36 deaths from lung cancer among underground iron miners in West Cumberland, Great Britain, which was significantly greater than the 21 deaths expected from local experience among nonminers. They concluded that the iron miners suffer from lung cancer mortality which is about 70% higher than normal, and they speculated that this might be due to a carcinogenic effect of iron oxide or to other factors, including radioactivity and cigarette smoking. Robinson et al. (1960) described a patient who developed a soft-tissue sarcoma in the deltoid muscle at the site of a previous injection of iron-dextran. The authors demonstrated the persistence of excess iron at the site of the tumor. MacKinnon and Bancewicz (1973) reported two patients who developed sarcomas in the buttocks at the sites of previous injections of iron-dextran. The possible hazard of sarcoma induction in man by parenteral administration of iron-dextran complexes has been reviewed by Roe (1967) and Myhre (1973).

Experimental carcinogenesis. Iron-polysaccharide complexes such as iron-dextran have been shown by many investigators to induce local sarcomas in mice, rats, and rabbits following parenteral administration in high dosages (Table 5). Iron is essential for the carcinogenicity, since the carbohydrate moieties of the iron-carbohydrate complexes are, by themselves, practically noncarcingoenic (Haddow and Horning, 1960; Neukomm, 1969). It should be mentioned that Haddow and Horning (1960) succeeded in inducing sarcomas in rats when aluminum-dextran was substituted for iron-dextran. Roe and Haddow (1965) failed to induce sarcomas in rats after intramuscular administration of iron-sorbitol complex ("Jectofer") under identical conditions in which an iron-dextran complex ("Imferon") did produce local sarcomas. Likewise, Wrba and Mohr (1968) failed to induce any sarcomas in Sprague-Dawley rats after subcutaneous injection of iron-sorbitol complex. Neukomm (1969) observed a sarcoma at the injection site in one of 33 mice which were given massive intramuscular injections of iron-sorbitol. Roe and Carter (1967) observed that the incidence of tumors at locations distant from the injection site was slightly higher in rats which received injections of iron-dextran than in control rats, and that several of the distant neoplasms in

TABLE 5 Iron-Carbohydrate Carcinogenesis in Experimental Animals

Authors	Animals	Compounds and routes	Tumors
Richmond, 1959	Rats	Iron-dextran ("Imferon") (im)	Sarcomas
Haddow and Horning, 1960	Rats, hamsters, mice	Iron-dextran ("Imferon") (sc)	Sarcomas
Golberg et al., 1960; Baker et al., 1961; Muir and Golberg, 1961	Rats, mice	Iron-dextran ("Imferon") (sc)	Sarcomas
Lundin, 1961	Rats	Iron-dextran ("Imferon") and Iron-dextrin ("Ferrigen") (im)	Sarcomas
Fielding, 1962	Mice	Iron-dextran ("Imferon") and Iron-dextrin ("Astrafer") (sc)	Sarcomas
Zollinger, 1962	Rats	Iron-polyisomaltose ("Ferrum") (im)	Sarcomas
Kunz et al., 1963	Rats	Iron-dextran ("Ursoferran") (sc)	Sarcomas
Haddow et al., 1964	Rabbits	Iron-dextran ("Imferon") (im)	Sarcomas
Roe et al., 1964; Roe and Carter, 1967	Rats	Iron-dextran ("Imferon") (sc)	Sarcomas and distant varied neoplasms
Pai et al., 1967	Mice	Iron-dextran ("Imferon," "Muscularon") (im, sc)	Sarcomas
Braun and Kren, 1968; Kren et al., 1968, 1970	Rats	Iron-dextran ("Sopfa") (sc, ip)	Sarcomas
Carter et al., 1968	Rats, mice	Ferric sodium gluconate and iron-dextran glycerol glucoside (sc)	Sarcomas
Langvad, 1968	Rats, mice	Iron-dextran ("Imferon") and iron-carbohydrate ("Intrafer") (sc)	Sarcomas and distant varied neoplasms
Neukomm, 1969	Mice	Iron-sorbitol ("Jectofer") and iron-nitrilotriprionate (im)	Sarcomas

266

the iron-dextran treated rats were of unusual types. However, the risk of dying with a distant tumor was not significantly greater in the test group than in the control group, and they concluded that the relationship of distant tumors to treatment with iron-dextran remained uncertain. Langvad (1968) administered iron-dextran and iron-carbohydrate complexes to mice and observed increased incidences of distant nonlymphoreticular neoplasms. Langvad concluded that these iron-complexes induced distant tumors in mice at dosages below those accepted for human therapy.

Possible mechanisms of iron-dextran carcinogenesis. There is little experimental work which relates to the possible carcinogenic mechanisms of iron-dextran complexes.

LEAD

Epidemiological studies. There is no evidence that industrial lead poisoning is associated with increased incidence of cancers of any type (Dingwall-Fordyce and Lane, 1963; Hammond et al., 1972).

Experimental carcinogenesis. In 1953, Zollinger reported renal carcinomas in rats which had received injections of lead phosphate. This observation has been confirmed by numerous investigators (Table 6), who have observed renal carcinomas following oral or parenteral administration of lead phosphate, acetate, or basic lead acetate. Renal cancers have only been induced in rats and mice. Van Esch and Kroes (1969) did not succeed in inducing renal carcinomas in hamsters. Zawirska and Medras (1968) observed testicular Leydigiomas in rats after prolonged feeding with lead acetate, and they demonstrated that the Leydigiomas were readily transplantable and hormonally active (1972a). Kobayashi and Okamoto (1974) reported that lead oxide exerted a cocarcinogenic effect on the induction of lung tumors by benzo(a)pyrene, and they suggested that atmospheric lead may enhance the carcinogenicity of polycyclic aromatic hydrocarbons. Epstein and Mantel (1968) tested the carcinogenicity of tetraethyl lead following subcutaneous injection into neonatal Swiss mice. They obtained equivocal results, since enhanced incidence of lymphomas was noted after tetraethyl lead treatment in female mice but not in male mice.

Possible mechanisms of lead carcinogenesis. Several studies of lead induction of renal carcinomas have focused attention upon the possible pathogenic role of porphyrinuria (Boyland et al., 1962; Zawirska and Medras, 1972b; Coogan, 1973). Boyland et al. (1962) speculated that the porphyrins which pass through the kidney might be the intermediate carcinogen, rather than the administered lead. They tested this hypothesis by comparing the incidence of renal tumors in two groups of rats which were maintained on diets containing either 1% lead acetate or 0.5% sedormid. Both groups of rats developed pronounced coproporphyrinuria. Single or multiple kidney tumors developed in 15 of the 16 lead-treated rats which survived for 320 days. In

TABLE 6 Lead Carcinogenesis in Experimental Animals

Authors	Animals	Compounds and routes	Tumors
Zollinger, 1953	Rats	$Pb_3(PO_4)_2$ (sc)	Renal adenomas and carcinomas
Tönz, 1957	Rats	$Pb_3(PO_4)_2$, $Pb(C_2H_3O_2)_2$ (sc)	Renal adenomas and carcinomas
Boyland et al., 1962	Rats	$Pb(C_2H_3O_2)_2$ (po)	Renal adenomas and carcinomas
van Esch et al., 1962	Rats	$Pb(C_2H_3O_2)_2.2Pb(OH)_2$ (po)	Renal adenomas and carcinomas
Roe et al., 1965	Rats	$Pb(C_2H_3O_2)_2$ (sc, intraperitoneal)	Renal adenomas and carcinomas
Mao and Molnar, 1966	Rats	$Pb(C_2H_3O_2)_2.2Pb(OH)_2$ (po)	Renal adenomas
Zawirska and Medras, 1968, 1972a	Rats	$Pb(C_2H_3O_2)_2$ (po)	Renal and testicular carcinomas
van Esch and Kroes, 1969	Mice	$Pb(C_2H_3O_2)_2.2Pb(OH)_2$ (po)	Renal adenomas and carcinomas
Coogan, 1973	Rats	$Pb(C_2H_3O_2)_2$ (po)	Renal carcinomas
Stiller, 1973	Rats	$Pb(C_2H_3O_2)_2$ (po)	Renal adenomas and carcinomas
Kobayashi and Okamoto, 1974	Hamsters	PbO and benzo(*a*)pyrene (intratracheal)	Adenomas and adenocarcinoma

comparison, a kidney tumor developed in only one of the 17 sedormid-treated rats which survived for 320 days. Zawirska and Medras (1972b) observed no correlation between urinary porphyrin excretion and development of renal carcinomas in rats which were treated with lead acetate. Coogan (1973) administered lead acetate to rats alone or in combination with either phenylhydrazine or sedormid in order to aggravate coproporphyrinuria. Rather than increasing the incidence of renal cancer, both combined treatments resulted in substantially lower incidences of renal carcinomas than when inorganic lead was given alone. On the basis of all of these findings, it appears that the pathogenesis of lead-induced renal cancer is unrelated to the toxic effects of lead upon porphyrin metabolism.

ZINC

Epidemiological studies. There is no evidence that industrial zinc poisoning is associated with increased incidence of cancers of any type (Henkin et al., in press).

Experimental carcinogenesis. In 1926, Michalowsky reported that injections of zinc chloride into the testicles of roosters resulted in the development of testicular teratomas. This observation furnished the first experimental system for induction of tumors by administration of metallic compounds. As summarized in Table 7, Michalowsky's findings have been amply confirmed by other workers (Bagg, 1936; Falin, 1940, 1941; Carleton et al., 1953; Smith and Powell, 1957; Rivière et al., 1960; Bresler, 1964; Guthrie, 1964a, 1966, 1967, 1971; Guthrie and Guthrie, 1974). The greatest incidence of teratomas is achieved if the injection is made during the early spring, when there is active proliferation of testicular epithelium. Alternatively, testicular proliferation can be stimulated by administration of gonadotrophic hormone (Bagg, 1936) or by manipulation of artificial photoperiods (Guthrie, 1971). Rivière et al. (1960) and Bresler (1964) produced testicular tumors in rats by intratesticular injection of $ZnCl_2$. It may be mentioned that Bresler (1964) succeeded in inducing teratomas in mice by intratesticular injection of $CuCl_2$ with simultaneous hormonal stimulation by testosterone, but not in the absence of testosterone. Guthrie and Guthrie (1974) induced embryonal carcinomas in Syrian hamsters by intratesticular injection during the early spring, when spermatogonial division was active. There is no evidence that zinc compounds are carcinogenic after administration by any route except that of intratesticular injection. For example, Gunn et al. (1964) failed to observe any tumorigenicity of zinc acetate after subcutaneous injection, and Heath et al. (1962) failed to produce any tumors after intramuscular injection of zinc powder. Walters and Roe (1975) reported that long-term addition of $ZnSO_4$ to drinking water of mice did not lead to increased incidence of tumors at any site.

TABLE 7 Zinc Carcinogenesis in Experimental Animals

Authors	Animals	Compounds and routes	Tumors
Michalowsky, 1926, 1927, 1930	Chickens	$ZnCl_2$ (intratesticular)	Teratomas
Bagg, 1936	Chickens	$ZnCl_2$ (intratesticular)	Teratomas
Falin, 1940, 1941	Chickens	$ZnSO_4$, $ZnNo_3$ (intratesticular)	Teratomas
Carleton et al., 1953	Chickens	$ZnCl_2$, $ZnCl_2$ (intratesticular)	Teratomas
Smith and Powell, 1957	Chickens	$ZnCl_2$, $ZnSO_4$ (intratesticular)	Teratomas
Rivière et al., 1960	Rats	$ZnCl_2$ (intratesticular)	Teratomas
Bresler, 1964	Rats	$ZnCl_2$ (intratesticular)	Teratomas, adenocarcinoma, Leydigiomas, seminomas
Guthrie, 1964a, 1966, 1967	Chickens	$ZnCl_2$ (intratesticular)	Teratomas
Guthrie, 1971	Quail	$ZnCl_2$ (intratesticular)	Teratomas
Guthrie and Guthrie, 1974	Hamsters	$ZnCl_2$ (intratesticular)	Embryonal carcinomas

Possible mechanisms of zinc carcinogenesis. Falin and Anissimova (1940) and Guthrie (1967) attempted to induce testicular tumors in cockerels by intratesticular injections of other caustic substances, including hydrochloric acid and 10% formalin, but they only obtained necrosis. Apparently, a metallic ion such as Zn(II), Cu(II), and Cd(II) is necessary for the experimental induction of testicular tumors (Rivière et al., 1960; Guthrie, 1967; Haddow et al., 1964a; Falin and Anissimova, 1940). Intratesticular injection of the metal salts causes partial castration, which is believed to stimulate release of pituitary gonadotropins. Most workers believe that pituitary stimulation is necessary for the induction of the testicular teratomas, although the available evidence is entirely indirect.

NICKEL

Epidemiological studies. More than 386 cases of lung cancer and 126 cases of cancer of the nasal cavities have been recognized among workmen who were occupationally exposed to nickel compounds (Table 8) (Sunderman and Mastromatteo, 1975). The data in Table 9 are illustrative of the increased incidence of cancers of the respiratory tract among nickel refinery workers in Wales (Doll et al., 1970) and Norway (Pedersen et al., 1973). Similar increases in incidence of cancers of the lung and nasal cavities have also been observed among nickel workers in Canada (Mastromatteo, 1967; Virtue, 1972) and the Soviet Union (Saknyn and Shabynina, 1970, 1973). In addition to respiratory cancers, the incidence of laryngeal cancer is slightly increased in Norwegian nickel workers (Pedersen et al., 1973) and the incidences of gastric cancer and of diverse sarcomas are increased in Russian nickel workers (Saknyn and Shabynina, 1973). Several studies suggest that

TABLE 8   Cancers of Lung and Nasal Cavities
in Nickel Workers[a]

Industrial process	Country	Years	Respiratory cancers	
			Lung	Nasal cavities
Nickel refining	Wales	1921–1971	174	78
Nickel refining	Canada	1930–1968	92	24
Nickel refining	Norway	1950–1971	51	14
Nickel refining	Germany	1932–1953	45	–
Nickel refining	USSR	1955–1967	>3	>6
Unspecified	Japan	1957–1959	19	–
Nickel plating	France	1960	1	1
Nickel plating	U.S.	1972	1	–
Totals			>386	>123

[a]Sunderman and Mastromatteo, 1975.

TABLE 9   Respiratory Cancers Among Workers in Two Nickel Refineries

Location of refinery	Year of first employment	Respiratory cancers (Observed/Expected)	
		Lung	Nasal cavities
Clydach, Wales (Mond process)[a]	prior to 1910	9.5	308
	1910–1914	10.5	870
	1915–1919	5.7	400
	1920–1924	6.3	116
	1925–1944	1.3	–
Kristiansand, Norway (Electrolytic process)[b]	1910–1929	10.4	100
	1930–1940	4.5	64
	1945–1954	4.4	4.3
	1955–1960	2.5	–

[a]Doll et al., 1970.
[b]Pedersen et al., 1973.

furnace operations in nickel refineries are associated with the greatest hazard of respiratory cancers (Doll et al., 1970; Pedersen et al., 1973; Mastromatteo, 1967; Virtue, 1972; Saknyn and Shabynina, 1970, 1973; Doll, 1958; Morgan, 1958; Sunderman, 1973). It appears that fresh nickel dusts from sintering or roasting processes are especially carcinogenic (Sunderman and Mastromatteo, in press). Current speculation suggests that respirable particles of metallic nickel, nickel subsulfide (Ni_3S_2), and nickel oxide (NiO) may be the principal respiratory carcinogens in nickel refineries (Sunderman and Mastromatteo, 1975).

Experimental carcinogenesis. The numerous investigations which are cited in Table 10 indicate that parenteral administration of metallic nickel dust or pellets to mice, rats, guinea pigs, and rabbits results in induction of malignant sarcomas at the injection sites. Nickel subsulfide (Ni_3S_2), when injected intramuscularly into rats, is a very potent inducer of rhabdomyo-sarcomas. Graphs in Figure 1 illustrate the cumulative incidence of sarcomas at the site of injection and the cumulative mortality in groups of 15 Fischer rats which were given single intramuscular injections of several dosages of Ni_3S_2, based upon unpublished studies in the author's laboratory. In Table 11 are given data for the relative frequencies of sarcomas at the injection sites after similar intramuscular injections of several nickel compounds (as well as Fe), as observed at 88 weeks after injection, based upon a study in progress in the author's laboratory. Induction of sarcomas in Fischer rats by intramuscular injection of nickel subsulfide has proven to be an exceptionally simple and convenient system for experimental carcinogenesis, and it has been employed for studies of endocrine effects (Jasmin, 1963, 1965; Jasmin et al., 1963) and of cancer chemotherapy (Furst et al., 1972). Several cell lines derived from Ni_3S_2-induced sarcomas have been successfully propagated in tissue culture (Basrur and Gilman, 1963, 1967; Corbeil, 1967, 1969; Nath et al., 1971; Shvemberger et al., 1972; Sykes and Basrur, 1971). From a methodological viewpoint, Ni_3S_2 carcinogenesis is an especially attractive experimental model, in as much as the compound is inexpensively available in high purity and is readily labeled with ^{63}Ni, a beta-emitting radioisotope with a long half-life that is ideally suited for liquid scintillation spectrometry and autoradiography. Moreover, the carcinogenic potency of nickel subsulfide is greater than any other metallic compound which has been investigated. Induction of pulmonary carcinomas in guinea pigs has been reported after inhalation of nickel dust (Hueper, 1958a) and in rats after inhalation of nickel carbonyl (Sunderman et al., 1959; Sunderman and Donnelly, 1965; Sunderman, 1966). Lau et al. (1972) have reported the occurrence of carcinomas and sarcomas in diverse organs (including liver and kidney) of rats that received multiple intravenous injections of nickel carbonyl. Toda (1962), Maenza et al. (1971) and Kasprzak et al. (1973) have found carcinogenic synergism between some nickel compounds (NiO and Ni_3S_2) and polycyclic aromatic hydrocarbons (methylcholanthrene and benzo(*a*)pyrene). There is no experimental evidence that nickel compounds are carcinogenic in animals when administered by oral or cutaneous routes.

Possible mechanisms of nickel carcinogenesis. Elucidating the mechanisms whereby nickel enters the target cells is an important initial step in understanding the mechanisms of nickel carcinogenesis. Owing to its lipid solubility, nickel carbonyl is able to pass across cell membranes without metabolic alteration (Kasprzak and Sunderman, 1969; Sunderman et al., 1968a; Sunderman and Selin, 1968). The ability of nickel carbonyl to penetrate intracellularly is presumed to be responsible for its extreme toxicity. Nickel carbonyl decomposes within cells to liberate carbon monoxide and Ni°, which is oxidized to Ni(II) by intracellular oxidation sytems (Sunderman et al., 1968; Sunderman and Selin, 1968). On the basis of the studies of Buu-Hoï et al. (1970), it appears likely that nickelocene is also able to penetrate cellular membranes without decomposition and then exert its pharmacologic effects. A different mechanism may be postulated for the intracellular transport of insoluble inorganic carcinogens, such as nickel dust and nickel subsulfide. After intramuscular injection, these compounds are presumed to be deposited extracellularly and to dissolve slowly in the extracellular fluid and muscle autolysate. Singh and Gilman (1973) have studied the interaction between rat rhabdomyocytes and nickel subsulfide by use of double-diffusion chambers that were implanted intraperitoneally in adult rats. Explants of embryonic rat skeletal muscle were cultured in one compartment of the double-diffusion chamber, and nickel subsulfide was placed in the adjacent compartment, separated from the muscle cells by a 0.1 μm-pore membrane. Cytologic effects of nickel were detected throughout the period from 2–24 days. This study suggests that a diffusible soluble

TABLE 10 Nickel Carcinogenesis in Experimental Animals

Authors	Animals	Compounds and routes	Tumors
Campbell, 1943	Mice	Ni dust (inhalation)	Not specified
Hueper, 1952, 1955	Rats, rabbits	Ni dust (intrapleural)	Sarcomas
Hueper, 1958	Guinea pigs	Ni dust (inhalation)	Anaplastic and adenocarcinomas
Sunderman et al., 1959, 1965, 1966	Rats	Ni(CO)$_4$ (inhalation)	Squamous cell and adenocarcinomas
Mitchell et al., 1960	Rats	Ni pellets (sc)	Sarcomas
Hueper and Payne, 1962	Rats	Ni dust (intrapulmonary)	Sarcoma
Gilman, 1962	Mice, rats	Ni$_3$S$_2$, NiO (im)	Sarcomas
Toda, 1962	Rats	NiO and methylcholanthrene (intratracheal)	Squamous cell carcinomas
Noble and Capstick, 1963	Rats	Ni$_3$S$_2$ (im)	Sarcomas
Gilman and Basrur, 1963; Gilman and Herchen, 1963; Herchen and Gilman, 1964; Gilman et al., 1966; Hebert et al., 1970	Rats	Ni$_3$S$_2$ (im)	Sarcomas
Jasmin, 1963, 1965; Jasmin et al., 1963	Rats	Ni$_3$S$_2$ (im)	Sarcomas
Heath and Daniel, 1964; Heath and Webb, 1967; Daniel et al., 1967; Webb et al., 1972	Rats	Ni dust (im)	Sarcomas

274

Reference	Species	Compound (route)	Tumor type
Payne, 1964	Rats	Ni_3S_2, NiO, $NiCO_3$ (im)	Sarcomas
Daniel, 1967	Rats	Ni_3S_2 (im)	Sarcomas
Corbeil, 1967, 1968, 1969	Rats	Ni_3S_2 (im)	Sarcomas
Haro et al., 1968; Furst and Haro, 1970; Furst and Schlauder, 1971	Rats, hamsters	$Ni(C_5H_5)_2$ (im)	Sarcomas
Friedmann and Bird, 1969	Rats	Ni (im)	Sarcomas
Mason, 1970, 1972; Mason et al., 1971	Rats	Ni_3S_2 (im, sc)	Sarcomas
Sanina, 1971	Rats	$Ni(CO)_4$ (inhalation)	Diverse neoplasms
Kasprzak et al., 1971, 1972, 1974	Rats	Ni_3S_2 (im)	Sarcomas
Nath et al., 1971	Rats	Ni_3S_2 (im)	Sarcomas
Maenza et al., 1971	Rats	Ni_3S_2 and benzo(a)pyrene (im)	Sarcomas
Lau et al., 1972	Rats	$Ni(CO)_4$ (iv)	Carcinomas and sarcomas
Furst et al., 1972, 1973	Rats	Ni dust (ip, im, intrapleural)	Sarcomas, mesotheliomas
Shvemberger et al., 1973	Rats	Ni_3S_2 (im)	Sarcomas
Geissinger et al., 1973	Rats	Ni_3S_2 (im)	Sarcomas
Olinici et al., 1973	Mice	Ni dust (im)	Sarcomas
Jasmin, 1973	Rats	Ni_3S_2 (intrarenal)	Renal carcinomas
Kasprzak et al., 1973	Rats	Ni_3S_2 and benzo(a)pyrene (intratracheal)	Squamous cell carcinoma
Sunderman et al., 1974, 1975	Rats	Ni_3S_2 (im)	Sarcomas

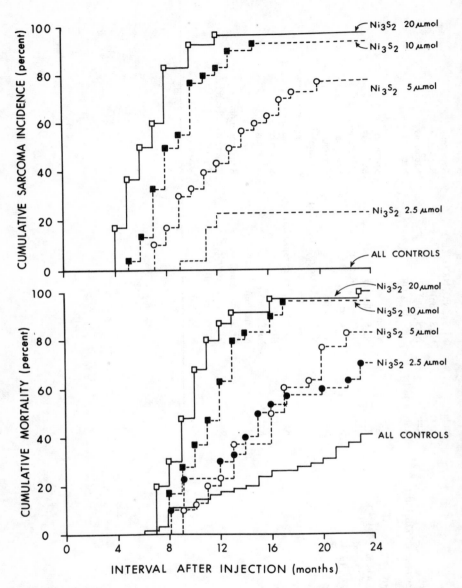

FIGURE 1 Cumulative incidence of sarcomas at the injection site and cumulative mortality in 4 groups of 30 ♂ Fischer rats which received a single im injection of Ni_3S_2 in dosages of 2.5, 5, 10, or 20 μmol, and in a control group of 180 ♂ Fischer rats. The materials and methods are the same as previously described (Sunderman et al., 1974, 1975).

TABLE 11 Comparisons of Carcinogenicities of Metallic Compounds Following Intramuscular Injection in Fischer Rats[a]

Group	Compound	Dosage (μmol)	Sarcoma incidence at injection site by 88 weeks
A	None	0	0/20
B	Ni_3S_2	20	8/10[b]
C	Ni_3S_2	80	9/10[b]
D	NiS	60	0/10
E	NiS	240	0/10
F	Ni^0	60	0/10
G	Ni^0	240	1/10
H	Fe^0	60	0/10
I	Fe^0	240	0/10
J	NiFeS	60	1/10
K	NiFeS	240	7/10[b]

[a]Single im injection in thigh musculature in 0.5 ml of penicillin suspension.
[b]$p < 0.0005$ versus Group A by χ^2 test.

intermediate complex is involved in the intracellular transport of nickel subsulfide. Heath et al. (1969), Webb et al. (1972), Webb and Weinzierl (1972), and Weinzierl and Webb (1972) have shown that nickel dust gradually dissolves when incubated aseptically with horse serum to form complexes with serum proteins and with ultrafiltrable molecules (primarily amino acids, such as histidine). Heath and associates have advanced two alternative hypotheses to account for the cellular penetration of nickel, cobalt, and other metallic carcinogens. In 1969, Heath et al. suggested that metal-serum protein complexes enter the cells by endocytosis and that later hydrolysis of the carrier proteins may lead to intracellular release of the electrophilic metal ion. In 1972, Webb et al. and Webb and Weinzierl suggested as an alternative hypothesis that complexes of nickel with small molecules may play key roles as intermediates in the intracellular transport of nickel. They found that nickel dust slowly dissolves when incubated with rat muscle homogenates and that the nickel becomes complexed almost entirely (90%) with ultrafiltrable molecules. Weinzierl and Webb (1972) showed that the ultrafiltrable nickel complexes obtained on dissolution of nickel dust in muscle homogenates *in vitro* were similar to those formed when nickel implants slowly dissolved in muscle *in vivo*. They speculated that myoblasts involved in the attempted repair of muscle injury may take up the diffusible nickel complexes and, under the influence of the intracellular nickel, may undergo neoplastic transformation. In support of this speculation, Webb and Weinzierl (1972) demonstrated the uptake of diffusible ^{63}Ni-complexes by mouse dermal fibroblasts in tissue culture. A second step in understanding the mechanisms of nickel carcinogenesis is elucidation of the intracellular biochemical and biologic effects of the Ni(II) ions. The biochemical alterations that develop in rats after administration of nickel carbonyl have been investigated by Beach and Sunderman (1970) and by Sunderman (1973) in an attempt to identify possible mechanisms of neoplastic transformation. Nickel carbonyl was found to have an inhibitory effect on the induction of several enzymes in lung and liver (Sunderman, 1967a, 1967b, 1968; Sunderman and Liebman, 1970). Nickel carbonyl did not affect substrate (tryptophan) induction of hepatic tryptophan pyrrolase, but did impair cortisone induction of tryptophan pyrrolase; this suggests that nickel carbonyl may produce a metabolic block at the level of messenger RNA (Sunderman, 1967b). Nickel carbonyl also inhibited phenothiazine induction of hepatic benzopyrene hydroxylase (Sunderman, 1967a) and phenobarbital induction of hepatic cytochrome P_{450} and aminopyrine demethylase (Sunderman and Liebman, 1970). These findings led to studies of the effects of nickel carbonyl on hepatic synthesis of RNA and proteins. At 1 day after injection of a dose of nickel carbonyl equivalent to LD_{50}, there was 60% inhibition of DNA-dependent RNA polymerase activity in hepatic nuclei (Sunderman and Esfahani, 1968) and 75% inhibition of RNA synthesis, as measured by incorporation of ^{14}C-orotic acid into RNA (Beach and Sunderman, 1969). Under identical experimental conditions, nickel carbonyl produced only 18% reduction of hepatic protein synthesis, as measured by incorporation of ^{14}C-leucine into microsomal proteins (Sunderman, 1970). Beach and Sunderman (1970) showed that exposure of rats to nickel carbonyl inhibited RNA synthesis *in vitro* by a chromatin-RNA polymerase complex that was prepared from hepatic nuclei. This study demonstrated that nickel carbonyl inhibition of RNA synthesis persists after disruption of the nuclei and thereby excluded inhibition, owing to impaired transport of RNA precursors across the nuclear membrane. Independent confirmation of the inhibitory effect of nickel carbonyl on hepatic RNA synthesis has been furnished by Witschi (1972). Beach (1969) has found that administration of nickel carbonyl did not significantly impair the template activity of isolated rat liver chromatin or DNA for transcription by RNA polymerase from *Micrococcus lysodeikticus*. The lack of an inhibitory effect of nickel carbonyl on the template activities of rat liver chromatin and DNA may possibly be ascribed to elution of nickel during isolation of the chromatin and DNA (Beach, 1969).

Webb et al. (1972) have studied the intracellular distribution of nickel in nickel-induced rhabdomyosarcomas and have found that a major portion (70–90%) of the nickel is within the nucleus. Furthermore, subfractionation indicated that an average of 53% (range: 41–63%) of nuclear nickel is present on the nucleolar fraction (Webb et al., 1972). The remainder of the nuclear nickel is distributed approximately equally between the nuclear sap and the chromatin fractions. Nucleolar localization of nuclear nickel has also been observed by Webb and Weinzierl (1972) in mouse dermal fibroblasts grown *in vitro* in the presence of [63]Ni-complexes. Intracellular [63]Ni in the fibroblasts was predominantly within the nuclei, and half the nuclear [63]Ni was associated with the nucleolar fraction (Webb and Weinzierl, 1972). Webb et al. (1972) emphasized the possible relations between their findings of nucleolar localization of nickel in rhabdomyoblasts and fibroblasts and the findings of Beach and Sunderman (1970) that nickel is bound to an RNA polymerase-chromatin complex isolated from hepatocyte nuclei of rats that were treated with nickel carbonyl. Buu-Hoï et al. (1970) have shown that administration of nickelocene in rats prolongs paralysis induced by zoxazolamine and potentiates the anticoagulant effects of tromexan. The mechanism of nickelocene inhibition of metabolism of zoxazolamine and tromexan has not been explained (Buu-Hoï et al., 1970), but it is presumed to resemble the inhibitory effects of nickel carbonyl on hepatic enzyme induction (Sunderman, 1967a, 1967b, 1968; Sunderman and Leibman, 1970). Treagan and Furst (1970) have shown that addition of nickel chloride to tissue cultures of mouse L-929 cells inhibits their capacity to synthesize interferon and antiviral protein in response to inoculation with Newcastle disease virus. From these observations, they have speculated that nickel might also inhibit the synthesis of repressors of tumor virus replication. Basrur and Gilman (1967) and Swierenga and Basrur (1968) have shown that addition of nickel subsulfide to cultured embryonic muscle cells inhibits mitotic activity and induces abnormal mitotic figures. Their findings suggest that nickel may interfere with gene replication and with the control of cell division. The studies of Swierenga and Basrur (1968) suggest that nickel subsulfide arrests mitoses in telophase and posttelophase, indicating that nickel may be interfering with the dissolution of the mitotic spindle. Olinici et al. (1973) reported the presence of marker chromosomes in two mouse cell lines derived from Ni-induced rhabdomyosarcomas. In both cell lines, chromosome aberrations persisted through many transplant generations.

Swierenga (1970) has found that, in tissue cultures of rat embryo muscle cells, Ni_3S_2 profoundly inhibits glyceraldehyde-3-phosphate dehydrogenase activity, whereas $NiCl_2$ and $NiSO_4$ cause only slight inhibition of this important glycolytic enzyme. Inhibition of glyceraldehyde-3-phosphate dehydrogenase results in a shift of energy metabolism to the alternative hexosemonophosphate pathway. Swierenga has proposed that this alteration of muscle glycolysis may be a factor in tumor induction.

A new avenue for investigations of the mechanisms of nickel carcinogenesis has been furnished by the recent observations of Sunderman et al. (1974) that manganese inhibits the carcinogenicity of nickel subsulfide. Fischer rats in five experimental groups were given a single im injection of penicillin suspension containing Ni_3S_2 dust, alone or in combination with equimolar amounts of aluminum, copper, chromium, or manganese dusts. Rats in five control groups were treated identically, except that the Ni_3S_2 dust was omitted. After 24 months, the incidence of sarcomas at the injection site was 63% in the group that received the combination of Ni_3S_2 and Mn^0, compared with incidences of 96–100% in the groups that received Ni_3S_2 alone or in

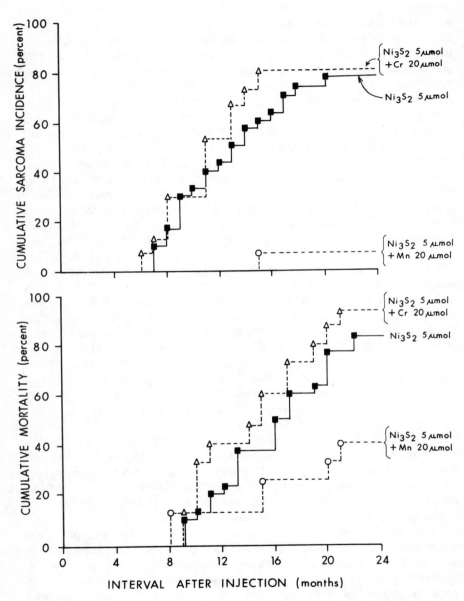

FIGURE 2 Cumulative incidence of sarcomas at the injection site and cumulative mortality in a group of 30 ♂ Fischer rats which received a single im injection of Ni_3S_2 (5 μmol), and in two groups of 15 ♂ Fischer rats which received Ni_3S_2 (5 μmol) in combination with 20 μmol of Cr^0 or Mn^0 dusts (Sunderman et al., 1975). The materials and methods are the same as previously described (Sunderman et al., 1974).

combination with Al^0, Cu^0, or Cr^0 dusts, $(p < 0.001)$. No sarcomas occurred at the injection site in control groups that did not receive Ni_3S_2. The inhibitory effect of Mn upon Ni_3S_2 tumorigenesis has been confirmed by another investigation in our laboratory (Sunderman et al., 1975). As illustrated in Figure 2, the incidence of sarcomas at 2 years after *im* administration of 5 μmol of Ni_3S_2 in Fischer rats was 23/30, compared with a sarcoma incidence of 0/180 in control rats. When 5 μmol of Ni_3S_2 was administered in combination with 20 μmol of Cr^0, there was no significant change in tumor incidence (12/15). In contrast, when 5 μmol of Ni_3S_2 was administered in combination with 20 μmol of Mn^0, the tumor incidence was profoundly reduced (1/15, $p < 0.001$). Mn^0 did not affect the mobilization or excretion of nickel nor alter the acute pathological reactions at the injection site. Sunderman et al. (1975) concluded that their findings are consistent with the hypothesis that Mn^0 may antagonize Ni-inhibition of RNA polymerase activity in rhabdomyocytes.

OTHER METALS

As summarized in Table 12, various other metallic compounds have been reported to produce tumors in experimental animals, including metallic mercury (Druckrey et al., 1957), colloidal silver (Schmähl and Steinhoff, 1960), aluminum-dextran (Haddow and Horning, 1960), copper sulfate (Bresler, 1964), sodium chlorostannate (Roe et al., 1965b), rare earth pellets (Ball et al., 1970), and titanocene (Furst and Haro, 1970; Furst and Schlauder, 1971). The equivocal increase in incidence of various tumors which was found by Roe et al. (1965b) after dietary administration of sodium chlorostannate to rats is probably of no practical significance.

SUMMARY

Epidemiological studies have demonstrated that certain occupational exposures of workmen to arsenic, chromium, and nickel compounds are associated with increased incidences of specific types of cancer.

Compounds of eight metals (beryllium, cadmium, chromium, cobalt, lead, nickel, zinc and iron-carbohydrate complexes) have been shown to induce cancers in experimental animals. The carcinogenic potency of nickel subsulfide, Ni_3S_2, is greater than any other metallic compound which has been investigated. A single im injection of 5 μmol of Ni_3S_2 in Fischer rats induces 77% incidence of malignant sarcomas within 20 months.

Following administration of carcinogenic nickel and beryllium compounds to rats, these metals become localized in cell nuclei, and they exert acute inhibitory effects upon mitosis and upon the expression of genetic information. These effects may be related to the unknown molecular mechanisms whereby the metallic compounds induce malignant transformation.

TABLE 12 Carcinogenesis by Various Metals in Experimental Animals

Authors	Animals	Compounds and routes	Tumors
Druckrey et al., 1957	Rats	Metallic mercury (ip)	Sarcomas
Schmähl and Steinhoff, 1960	Rats	Colloidal silver (iv, sc)	Sarcomas
Haddow and Horning, 1960	Rats	Aluminum-dextran (im)	Sarcomas
Bresler, 1964	Mice	$CuSO_4$ (intratesticular)	Teratomas, seminoma, chorioepithelioma
Roe et al., 1965	Rats	Sodium chlorostannate (po)	Varied tumors
Ball et al., 1970	Mice	Ytterbium, gadolinium (im)	Sarcomas
Furst and Haro, 1970; Furst and Schlauder, 1971	Rats	Titanocene (im)	Sarcomas

REFERENCES

Araki, M., Sanya, O. and Mikio, M. 1954. Experimental studies on beryllium-induced malignant tumors of rabbits. *Gann* 45:449–451.

Baetjer, A. M. 1950. Pulmonary carcinoma in chromate workers. I. Review of the literature and report of cases. *Arch. Industr. Hyg. Occup. Med.* 2:487–504.

Baetjer, A. M., Birmingham, D. J., Enterline, P. E., Mertz, W. and Pierce, J. V., II. 1974. *Chromium* Washington, D.C.: National Academy of Sciences.

Bagg, H. J. 1936. Experimental production of teratoma testis in the fowl. *Amer. J. Cancer* 26:69–84.

Baker, S. B. de C., Golberg, L., Martin, L. E. and Smith, J. P. 1961. Tissue changes following injection of iron-dextran complex. *J. Path. Bact.* 82:453–470.

Ball, R. A., Van Gelder, G., Green, J. W., Jr. and Reece, W. O. 1970. Neoplastic sequelae following subcutaneous implantation of mice with rare earth metals. *Proc. Soc. Exp. Biol. Med.* 135:426–430.

Barnes, J. M., Denz, F. A. and Sissons, H. A. 1950. Beryllium bone sarcomata in rabbits. *Brit. J. Cancer* 4:212–222.

Baroni, C., van Esch, G. J. and Saffiotti, U. 1963. Carcinogenesis tests of two inorganic arsenicals. *Arch. Environ. Health* 7:668–674.

Basrur, P. K. and Gilman, J. P. W. 1963. Behavior of two cell strains derived from rat rhabdomyosarcomas. *J. Nat. Cancer Inst.* 30:163–200.

Basrur, P. K. and Gilman, J. P. W. 1967. Morphologic and synthetic response of normal and tumor muscle cultures to nickel sulfide. *Cancer Res.* 27: 1168–1177.

Bayliss, D. 1972. Expected and observed deaths by selected causes occurring to beryllium workers. In National Insititue for Occupational Health and Safety. *Criteria for a recommended standard: Occupational exposure to beryllium*, pp. IV–22 to IV–23, Tables VII–X. Washington, D.C.: U.S. Dept. of Health, Education and Welfare.

Beach, D. J. 1969. Some aspects of the molecular biology of nickel carbonyl. Doctoral dissertation, University of Florida, Gainesville.

Beach, D. J. and Sunderman, F. W., Jr. 1969. Nickel carbonyl inhibition of ^{14}C-orotic acid incorporation into rat liver RNA. *Proc. Soc. Exp. Biol. Med.* 131:321–322.

Beach, D. J. and Sunderman, F. W., Jr. 1970. Nickel carbonyl inhibition of RNA synthesis by a chromatin-RNA polymerase complex from hepatic nuclei. *Cancer Res.* 30:48–50.

Bidstrup, P. L. 1951. Carcinoma of the lung in chromate workers. *Brit. J. Industr. Med.* 8:302–305.

Bidstrup, P. L. and Case, R. A. M. 1956. Carcinoma of the lung in workmen in the biochromates-producing industry in Great Britain. *Brit. J. Industr. Med.* 13:260–264.

Boyd, J. T., Doll, R., Faulds, J. S. and Leiper, J. 1970. Cancer of the lung in iron ore (haematite) miners. *Brit. J. Industr. Med.* 27:97–105.

Boyland, E., Dukes, C. E., Grover, P. L. and Mitchley, C. B. V. 1962. The induction of renal tumours by feeding lead acetate to rats. *Brit. J. Cancer* 16:283–288.

Braun, A. and Kren, V. 1968. Attempt to induce tumors by subcutaneous and intraperitoneal administration of ferridextran. *Neoplasma* 15:21.

Braun, P., Guillerm, B., Pierson, B. and Sadoul, P. 1960. A propos du cancer bronchique chez les mineurs de fer (Bronchial cancer among iron miners). *Rev. Med. (Nancy)* 85:702–708.

Bresler, W. M. 1964. On the dynamics of blastomogenesis in the testis. *Acta Unio Intern. Contra Cancrum* 20:1501–1503.

Bühler, R. H. O. and Kägi, J. H. R. 1974. Human hepatic metallothioneins. *FEBS Letters* 39:229–234.

Buu-Hoï, N. P., Hien, D-P. and Hieu, H-T. 1970. Effets des métallo-cènes sur la métabolisation des certains médicaments chez le rat (Effects of metallocenes on the metabolism of some medications in the rat). *C. R. Acad. Sci. (D)* 270:217–219.

Campbell, J. A. 1943. Lung tumours in mice and men. *Brit. Med. J.* 1:179–183.

Carleton, R. L., Freedman, N. B. and Bomze, E. J. 1953. Experimental teratomas of the testis. *Cancer* 6:464–473.

Carter, R. L., Mitchley, B. C. V. and Roe, F. J. C. 1968. Induction of tumours in mice and rats with ferric sodium gluconate and iron-dextran glycerol glycoside. *Brit. J. Cancer* 22:521–526.

Chen, R. W., Eakin, D. J. and Whanger, P. D. 1974. Biological function of metallothionein. II. Its role in zinc metabolism in the rat. *Nutr. Reports Intern.* 10:195–200.

Cloudman, A. M., Vining, D., Barkulis, S. and Nickson, J. J. 1949. Bone changes observed following intravenous injections of beryllium. *Amer. J. Path.* 25:810–811.

Coogan, P. S. 1973. Lead-induced renal carcinoma. *Proc. Inst. Med. Chicago*, 29:309.

Corbeil, L. B. 1967. Differentiation of rhabdomyosarcoma and neonatal muscle cells. *Cancer* 20:572–578.

Corbeil, L. B. 1968. Antigenicity of rhabdomyosarcomas induced by nickel sulfide, Ni_3S_2. *Cancer* 21:184–189.

Corbeil, L. B. 1969. Observations on multinucleation of rhabdomyosarcoma and neonatal muscle cells *in vitro. Life Sci.* 8:651–656.

Daniel, M. R. 1966. Strain differences in the response of rats to injection of nickel sulphide. *Brit. J. Cancer* 20:886–895.

Daniel, M. R., Heath, J. C. and Webb, M. 1967. Respiration of metal induced rhabdomyosarcomata. *Brit. J. Cancer* 21:780–786.

Dingwall-Fordyce, I. and Lane, R. E. 1963. A follow-up study of lead workers. *Brit. J. Industr. Med.* 20:313–315.

Dobson, R. L. and Pinto, J. S. 1966. Arsenical carcinogenesis. In *Advances in biology of skin*, ed. W. Montagna and R. L. Dobson, vol. 8, pp. 237–245. New York: Permagon Press.

Doll, R. 1958. Cancer of the lung and nose in nickel workers. *Brit. J. Industr. Med.* 15:217–223.

Doll, R., Morgan, L. G. and Speizer, F. E. 1970. Cancers of the lung and nasal sinuses in nickel workers. *Brit. J. Cancer* 24:623–632.

Druckrey, H., Hamperl, H. and Schmahl, D. 1957. Cancerogene Wirkung von metallischem Quecksilber nach intraperitonealer Gabe bei Ratten. *Ztschr. Krebsforsch.* 61:511–519.

Dutra, F. R. and Largent, E. J. 1950. Osteosarcoma induced by beryllium oxide. *Amer. J. Path.* 26:197–209.

Dutra, F. R., Largent, E. J. and Roth, J. 1951. Osteogenic sarcoma after inhalation of beryllium oxide. *Arch. Path.* 51:473–479.

Dvizhkov, P. P. and Fedorova, V. I. 1967. Cancerogenic properties of chromic oxide. *Vop. Onkol.* 13:57–62.

Ehlers, G. 1974. Klinische und histologische Untersuchungen zur Frage arzneimittelbedingter Arsen-tumoren. *Z. Haut-Geschlkr* 43:763–774.

Enterline, P. E. 1974. Respiratory cancer among chromate workers. *J. Occup. Med.* 16:523–526.

Epstein, S. S. and Mantel, N. 1968. Carcinogenicity of tetraethyl lead. *Experientia* 24:580–581.

Falin, L. I. 1940. Experimental teratoma testis in the fowl. *Amer. J. Cancer* 38:199–211.

Falin, L. I. 1941. Morphologie und Differenczcerung der Nervenelement im der experimentellen Teratoiden. *Z. Mikroscop Anat. Forsch.* 49:193–224.

Falin, L. I. and Anissimova, W. W. 1940. Zur Pathogenese der experimentellen teratoiden Geschwülste der Geschlechtsdrüsen. Teratoide Hodengeschwulst beim Hahn, erzeugt durch Einführung von $CuSO_4$-Lösung. *Z. Krebsforsch.* 50:339–351.

Faulds, J. S. and Stewart, M. J. 1956. Carcinoma of the lung in hematite miners. *J. Path. Bact.* 72:353–366.

Favino, A., Cavalleri, A., Nazari, G. and Tilli, M. 1968. Testosterone excretion in cadmium chloride induced testicular tumours in rats. *Med. Lavoro* 59:36–40.

Fielding, J. 1962. Sarcoma induction by iron-carbohydrate complexes. *Brit. Med. J.* 1:1800–1803.

Flick, D. F., Kraybill, H. F. and Dimitroff, J. M. 1971. Toxic effects of cadmium: A review. *Environ. Res.* 4:71–85.

Friedmann, I. and Bird, E. S. 1969. Electron microscope investigation of experimental rhabdomyosarcoma. *J. Path.* 97:375–382.

Friedrich, E. G., Jr. 1972. Vulvar carcinoma in situ in identical twins—an occupational hazard. *Obstet. Gynec.* 39:837–841.

Frost, D. V. 1967. Arsenicals in biology—retrospect and prospect. *Fed. Proc.* 26:194–208.

Furst, A. 1971. Trace elements related to specific chronic diseases: Cancer. In *Environmental geochemistry*, ed. H. L. Cannon and H. C. Hopps, pp. 109–129. Boulder, Colo.: Geological Society of America.

Furst, A. and Cassetta, D. M. 1972. Failure of zinc to negate cadmium carcinogenesis. *Proc. Amer. Ass. Cancer Res.* 13:62.

Furst, A. and Cassetta, D. M. 1973. Carcinogenicity of nickel by different routes. *Proc. Amer. Ass. Cancer Res.* 14:31.

Furst, A. and Haro, R. T. 1969a. Possible mechanisms of metal ion carcinogenesis. In *Jerusalem symposium on quantum chemistry and biochemistry*, ed. E. D. Bergmann and B. Pullman, pp. 310–320. Jerusalem: Israel Academy of Sciences and Humanities.

Furst, A. and Haro, R. T. 1969b. A survey of metal carcinogenesis. *Prog. Exp. Tumor Res.* 12:102–133.

Furst, A. and Haro, R. T. 1970. Carcinogenicity of metal pi-complex compounds: Metallocenes. In *Abstracts of the tenth international cancer congress, Houston*, p. 28. Austin: University of Texas Press.

Furst, A. and Schlauder, M. C. 1971. The hamster as a model for metal carcinogenesis. *Proc. West. Pharmacol. Soc.* 14:68–71.

Furst, A., Haro, R. T. and Schlauder, M. C. 1972. Experimental chemotherapy of nickel-induced fibrosarcomas. *Oncology* 26:422–426.

Furst, A., Cassetta, D. M. and Sasmore, D. P. 1973. Rapid induction of pleural mesotheliomas in the rat. *Proc. West Pharmacol. Soc.* 16:150–153.

Gardner, L. U. and Heslington, H. F. 1946. Osteosarcoma from intravenous beryllium compounds in rabbits. *Fed. Proc.* 5:221.

Geissinger, H. D., Basrur, P. K. and Yamishiro, S. 1973. Fast scanning electron microscopic and light microscopic correlation of paraffin sections and chromosome spreads of nickel-induced tumor. *Trans. Amer. Microscop. Soc.* 92:209–217.

Gilman, J. P. W. 1962. Metal carcinogenesis. II. A study on the carcinogenic activity of cobalt, copper, iron and nickel compounds. *Cancer Res.* 22:158–162.

Gilman, J. P. W. 1965. Muscle tumorigenesis. In *Proceedings of the sixth Canadian cancer research conference, Honey Harbor, Ontario*, pp. 209–223. Elmsford, N.Y.: Pergamon Press.

Gilman, J. P. W. and Basrur, P. K. 1963. Precancerous changes in muscle cells exposed to nickel sulphide. *Proc. Amer. Ass. Cancer Res.* 4:23.

Gilman, J. P. W. and Herchen, H. 1963. The effect of physical form of implant on nickel sulphide tumorigenesis in the rat. *Acta Unio Intern. Contra Cancrum* 19:615–619.

Gilman, J. P. W. and Ruckerbauer, G. 1962. Metal carcinogenesis. I. Observations on the carcinogenicity of a refinery dust, cobalt oxide, and colloidal thorium dioxide. *Cancer Res.* 22:152–157.

Gilman, J. P. W., Daniel, M. R. and Basrur, P. K. 1966. Observations of tissue selectivity in nickel tumorigenesis. *Proc. Amer. Ass. Cancer Res.* 7:24.

Golberg, L., Martin, L. E. and Smith, J. P. 1960. Iron overloading phenomena in animals. *Toxic Appl. Pharmacol.* 2:683–707.

Goldman, A. L. 1973. Lung cancer in Bowen's disease. *Amer. Rev. Resp. Dis.* 108:1205–1207.

Gunn, S. A., Gould, T. C. and Anderson, W. A. D. 1963. Cadmium-induced interstitial cell tumors in rats and mice and their prevention by zinc. *J. Nat. Cancer Inst.* 31:745–759.

Gunn, S. A., Gould, T. C. and Anderson, W. A. D. 1964. Effect of zinc on cancerogenesis by cadmium. *Proc. Soc. Exp. Biol. Med.* 115:653–657.

Gunn, S. A., Gould, T. C. and Anderson, W. A. D. 1965. Comparative study of interstitial cell tumors of rat testis induced by cadmium injection and vascular ligation. *J. Nat. Cancer Inst.* 35:329–337.

Gunn, S. A., Gould, T. C. and Anderson, W. A. D. 1967. Specific response of mesenchymal tissue to cancerogenesis by cadmium. *Arch. Path.* 83: 493–499.

Guthrie, J. 1964a. Observations on the zinc induced testicular teratomas of fowl. *Brit. J. Cancer* 18:130–142.

Guthrie, J. 1964b. Histological effects of intra-testicular injections of cadmium chloride in domestic fowl. *Brit. J. Cancer* 18:255–260.

Guthrie, J. 1966. Effects of a synthetic antigonadotrophin (2-amino,5-nitrothiazole) on growth of experimental testicular teratomas. *Brit. J. Cancer* 20:582–587.

Guthrie, J. 1967. Specificity of the metallic ion in the experimental induction of teratomas in the fowl. *Brit. J. Cancer* 21:619–622.

Guthrie, J. 1971. Zinc induction of testicular teratomas in Japanese quail (Coturnix japonica) after photoperiodic stimulation of testes. *Brit. J. Cancer* 25:311–314.

Guthrie, J. and Guthrie, O. A. 1974. Embryonal carcinomas in Syrian hamsters after intratesticular inoculation of zinc chloride during seasonal testicular growth. *Cancer Res.* 34:2612–2614.

Haddow, A. and Horning, E. S. 1960. On the carcinogenicity of an iron-dextran complex. *J. Nat. Cancer Inst.* 24:109–147.

Haddow, A., Roe, F. J. C., Dukes, C. E. and Mitchley, B. C. V. 1964a.

Cadmium neoplasia: Sarcomata at the site of injection of cadmium sulphate in rats and mice. *Brit. J. Cancer* 18:667–673.

Haddow, A., Roe, F. J. C. and Mitchley, B. C. V. 1964b. Induction of sarcomata in rabbits by intramuscular injection of iron-dextran ("Imferon"). *Brit. Med. J.* 1:1593–1594.

Hammond, P. B., Aronson, A. L., Chisolm, J. J., Jr., Falk, J. L., Keenan, R. G. and Sandstead, H. H. 1972. *Airborne lead in perspective.* Washington, D.C.: National Academy of Sciences.

Haro, R. T., Furst, A., Payne, W. W. and Falk, H. 1968. A new nickel carcinogen. *Proc. Amer. Ass. Cancer Res.* 9:28.

Hasan, F. M. and Kazemi, H. 1974. Chronic beryllium disease: A continuing epidemiologic hazard. *Chest* 65:289–293.

Heath, J. C. 1956. The production of malignant tumours by cobalt in the rat. *Brit. J. Cancer* 10:668–673.

Heath, J. C. 1960. The histogenesis of malignant tumours induced by cobalt in the rat. *Brit. J. Cancer* 14:478–482.

Heath, J. C. and Daniel, M. R. 1964a. The production of malignant tumours by cadmium in the rat. *Brit. J. Cancer* 18:124–129.

Heath, J. C. and Daniel, M. R. 1964b. The production of malignant tumours by nickel in the rat. *Brit. J. Cancer* 18:261–264.

Heath, J. C., Daniel, M. R., Dingle, J. T. and Webb, M. 1962. Cadmium as a carcinogen. *Nature* 193:592–593.

Heath, J. C., Freeman, M. A. R. and Swanson, S. A. V. 1971. Carcinogenic properties of wear particles from prostheses made in cobalt-chromium alloy. *Lancet* 1:564–566.

Heath, J. C. and Webb, M. 1967. Content and intracellular distribution of the inducing metal in the primary rhabdomyosarcomata induced in the rat by cobalt, nickel and cadmium. *Brit. J. Cancer* 21:768–779.

Heath, J. C., Webb, M. and Caffrey, M. 1969. The interaction of carcinogenic metals with tissues and body fluids. Cobalt and horse serum. *Brit. J. Cancer* 23:153–166.

Hebert, G. J., Basrur, P. K. and Gilman, J. P. W. 1970. Arginase activity in nickel sulfide-induced rat tumors. *Cancer* 25:1134–1141.

Henkin, R., Apgar, J., Cole, J., Cotterill, C., Goyer, R. and Knezek, B. In press. *Zinc.* Report of the Committee on Medical and Biologic Effects of Environmental Pollutants. Washington, D.C.: National Academy of Sciences.

Herchen, H. and Gilman, J. P. W. 1964. Effect of duration of exposure on nickel sulphide tumorigenesis. *Nature* 202:306–307.

Hoagland, M. B., Grier, R. S. and Hood, M. B. 1950. Beryllium and growth. I. Beryllium-induced osteogenic sarcomata. *Cancer Res.* 10:629–635.

Holden, H. 1969. Cadmium toxicology. *Lancet* 2:57.

Hueper, W. C. 1952. Experimental studies in metal carcinogenesis. I. Nickel cancer in rats. *Texas Rep. Biol. Med.* 10:167–186.

Hueper, W. C. 1955. Experimental studies in metal carcinogenesis. IV. Cancer produced by parenterally introduced metallic nickel. *J. Nat. Cancer Inst.* 16:55–67.

Hueper, W. C. 1958a. Experimental studies in metal carcinogenesis. IX. Pulmonary lesions in guinea pigs and rats exposed to prolonged inhalation of powdered metallic nickel. *Arch. Path.* 65:600–607.

Hueper, W. C. 1958b. Experimental studies in metal carcinogenesis. X. Cancerigenic effects of chromite ore deposited in muscle tissue and pleural cavity of rats. *Arch. Industr. Health* 18:284–291.

Hueper, W. C. 1966. *Occupational and environmental cancers of the respiratory system.* New York: Springer-Verlag.

Hueper, W. C. and Payne, W. W. 1959. Experimental cancers in rats produced by chromium compounds and their significance to industry and public health. *Amer. Industr. Hyg. Ass. J.* 20:274–280.

Hueper, W. C. and Payne, W. W. 1962. Experimental studies in metal carcinogenesis. Chromium, nickel, iron, arsenic. *Arch. Environ. Health* 5:445–462.

International Agency for Research on Cancer 1973. *Evaluation of carcinogenic risk of chemicals to man.* Vol. 2: *Some Inorganic and Organo-metallic Compounds.* Geneva: World Health Organization.

Janes, J. M., Higgins, G. M. and Herrick, J. F. 1954. Beryllium-induced osteogenic sarcoma in rabbits. *J. Bone Joint Surg.* 36B:543–552.

Janes, J. M., Higgins, G. M. and Herrick, J. F. 1956. The influence of splenectomy on the induction of osteogenic sarcoma in rabbits. *J. Bone Joint Surg.* 38A:809–816.

Jasmin, G. 1963. Effects of methandrostenolone on muscle carcinogenesis induced in rats by nickel sulphide. *Brit. J. Cancer* 17:681–686.

Jasmin, G. 1965. Influence of age, sex and glandular extirpation on muscle carcinogenesis in rats. *Experientia* 21:149–150.

Jasmin, G. 1973. Alterations cellulaires rénales induites par le sulfure de nickel chez le rat. *J. Microscop.* 17:68a–69a.

Jasmin, G., Bajusz, E. and Mongeau, A. 1963. Influence du sexe et de la castration sur la production de tumeurs musculaires chez le rat par le sulfure de nickel. *Revue Can. Biol.* 22:113–114.

Jung, E. G. and Trachsel, B. 1970. Molekularbiologische Untersuchungen zur Arsencarcinogenese. *Arch. klin. exp. Derm.* 237:819–826.

Kägi, J. H. R., Himmelhoch, S. R., Whanger, P. D., Bethune, J. L. and Vallee, B. L. 1974. Equine hepatic and renal metallothioneins. *J. Biol. Chem.* 249:3537–3542.

Kägi, J. H. R., Whanger, P. D. and Bethune, J. L. 1973. Purification and characterization of metallothionein. *Fed. Proc.* 32:942 ABS.

Kanisawa, M. 1971. Aspects of metal carcinogenesis. *Chiba Daigaku Kogakubu Kenkyn Hokuku* 24:1–35.

Kasprzak, K. S. 1974. An autoradiographic study of nickel carcinogenesis in rats following injection of $^{63}Ni_3S_2$ and $Ni_3{}^{35}S_2$. *Res. Commun. Chem. Path. Pharmacol.* 8:141–150.

Kasprzak, K. S. and Marchow, L. 1972. Nickel subsulfide in muscle carcinogenesis. *Patol. Polska* 23:135–142.

Kasprzak, K. S., Marchow, L. and Breborowicz, J. 1971. Parasites and carcinogenesis. *Lancet* 2:106–107.

Kasprzak, K. S., Marchow, L. and Breborowicz, J. 1973. Pathological reactions in rat lungs following intratrachial injection of nickel subsulfide and 3,4-benzpyrene. *Res. Commun. Chem. Path. Pharmacol.* 6:237–245.

Kasprzak, K. S. and Sunderman, F. W., Jr. 1969. The metabolism of nickel carbonyl-^{14}C. *Toxicol. Appl. Pharmacol.* 15:295–303.

Kazantzis, G. 1963. Induction of sarcoma in the rat by cadmium sulphide pigment. *Nature* 198:1213–1214.

Kazantzis, G. and Hanbury, W. J. 1966. The induction of sarcoma in the rat by cadmium sulphide and by cadmium oxide. *Brit. J. Cancer* 20:190–199.

Kelly, P. J., Janes, J. M. and Peterson, L. F. A. 1961. The effect of beryllium on bone. *J. Bone Joint Surg.* 43A:829–844.

Kipling, M. D. and Waterhouse, J. A. H. 1967. Cadmium and prostatic cancer. *Lancet* 1:730–731.

Kjeldsberg, C. R. and Ward, H. P. 1972. Leukemia in arsenic poisoning. *Ann. Intern. Med.* 77:935–937.

Knorre, D. 1970. Ortliche Hautschädigungen an der Albinoratte in der Latenzperiode der Sarkomentwicklung nach Kadmiumchloridinjection. *Zbl. Allg. Path.* 113:192–197.

Knorre, D. 1971. Zur Induktion von Hodenzwischenzelltumoren an der Albinoratte durche Kadmiumchlorid. *Arch. Geschwulstforsch.* 38:257–263.

Kobayashi, N. and Okamoto, T. 1974. Effects of lead oxide on the induction of lung tumors in Syrian hamsters. *J. Nat. Cancer Inst.* 52:1605–1610.

Komitowski, D. 1968. Experimental beryllium-induced bone tumors as a model of osteogenic sarcoma. *Chirugia. Narz. Ruchu Ortop. Pol.* 33:237.

Kren, V., Braun, A. and Krenova, D. 1968. The transplantability of the tumor induced in rats by ferridextran. *Neoplasma* 15:29.

Kren, V., Krenova, D. and Stark, O. 1970. Properties of sarcomas induced by ferridextran Sopfa in the inbred rat strain AVN-1. *Neoplasma* 17:329–337.

Kunz, J., Shabad, L., Henze, K. and David, H. 1963. The carcinogenic effect of iron-dextran (Ursoferran). *Acta Biol. Med. Germ.* 10:602–614.

Kuschner, M. and Laskin, S. 1971. Experimental models in environmental carcinogenesis. *Amer. J. Path.* 64:183–191.

Lacassagne, A. 1971. Revue critique des tumeurs expérimentales des cellules de Leydig, plus particulièrement chez le rat. *Bull. Cancer* 58:235–276.

Langvad, E. 1968. Iron-dextran induction of distant tumours in mice. *Int. J. Cancer* 3:415–423.

Laskin, S., Kuschner, M. and Drew, R. T. 1970. Studies in pulmonary carcinogenesis. In *Inhalation carcinogenesis*, ed. M. G. Hanna, Jr., P. Nettesheim and J. R. Gilbert, pp. 321–351. Washington, D.C.: U.S. Atomic Energy Commission.

Lau, T. J., Hackett, R. L. and Sunderman, F. W., Jr. 1972. The carcinogenicity of intravenous nickel carbonyl in rats. *Cancer Res.* 32:2253–2258.

Lee, A. M. and Fraumeni, J. F., Jr. 1969. Arsenic and respiratory cancer in man: An occupational study. *J. Nat. Cancer Inst.* 42:1045–1052.

Lucis, O. J., Lucis, R. and Aterman, K. 1972. Tumorigenesis by cadmium. *Oncology* 26:53–67.

Lundin, P. M. 1961. The carcinogenic action of complex iron preparations. *Brit. J. Cancer* 15:838–847.

Machle, W. and Gregorius, F. 1948. Cancer of the respiratory system in the United States chromate-producing industry. *Publ. Health Reports* 63:1114–1127.

MacKinnon, A. E. and Bancewicz, J. 1973. Sarcoma after injection of intramuscular iron. *Brit. Med. J.* 2:277–279.

Maenza, R. M., Pradhan, A. M. and Sunderman, F. W., Jr. 1971. Rapid induction of sarcomas in rats by a combination of nickel sulfide and 3,4-benzpyrene. *Cancer Res.* 31:2067–2071.

Malcolm, D. 1972. Potential carcinogenic effect of cadmium in animals and man. *Ann. Occup. Hyg.* 15:33–36.

Mancuso, T. F. 1970. Relation of duration of employment and prior respiratory illness to respiratory cancer among beryllium workers. *Environ. Res.* 3:251–275.

Mancuso, T. F. and Hueper, W. C. 1951. Occupational cancer and other health hazards in a chromate plant: A medical appraisal. I. Lung cancers in chromate workers. *Industr. Med. Surg.* 20:358–363.

Mao, P. and Molnar, J. J. 1966. Fine structure of lead induced renal tumors. *Amer. J. Path.* 48:9a.

Marcotte, J. and Witschi, H. P. 1972. Synthesis of RNA and nuclear proteins in early regenerating rat livers exposed to beryllium. *Res. Commun. Chem. Path. Pharmacol.* 3:100–104.

Mason, M. M. 1970. Nickel sulfide, a model carcinogen. In *Abstracts of the tenth international cancer congress, Houston*, p. 27. Austin: University of Texas Press.

Mason, M. M. 1972. Nickel sulfide carcinogenesis. *Environ. Physiol. Biochem.* 2:137–141.

Mason, M. M., Cate, C. C. and Baker, J. 1971. Toxicology and carcinogenesis of various chemicals used in the preparation of vaccines. *Clin. Toxicol.* 4:185–204.

Mastromatteo, E. 1967. Nickel: A review of its occupational health aspects. *J. Occup. Med.* 9:127–136.

McLaughlin, A. I. G. and Harding, H. E. 1956. Pneumoconiosis and other causes of death of iron and steel foundry workers. *Arch. Industr. Health* 14:350–378.

Michalowsky, I. 1926. Die experimentelle Erzeugung einer teratoiden Neubildung der Hoden beim Hahn. Vorlaufige Mitteilung. *Zentbl. Allg. Path. Anat.* 38:585–587.

Michalowsky, I. 1928. Eine experimentelle Erzeugung teratoider Geschwülste der Hoden beim Hahn. *Virchows Arch. Path. Anat.* 267:27–62.

Michalowsky, I. 1930. Das 10 experimentelle Zink-Teratom. II. Mitteilung. *Virchows Arch. Path. Anat.* 274:319–325.

Milner, J. E. 1969. The effect of ingested arsenic on methylcholanthrene-induced skin tumors in mice. *Arch. Environ. Health* 18:7–11.

Minkowitz, S. 1964. Multiple carcinomata following ingestion of medicinal arsenic. *Ann. Intern. Med.* 61:296–299.

Mitchell, D. F., Shankwalker, G. B. and Shazer, S. 1960. Determining the tumorigenicity of dental materials. *J. Dent. Res.* 39:1023–1028.

Monlibert, L. and Roubille, R. H. 1960. A propos du cancer bronchique chez les mineurs de fer. *J. Franc. Med. Chir. Thor.* 14:435–439.

Morgan, J. G. 1958. Some observations on the incidence of respiratory cancer in nickel workers. *Brit. J. Industr. Med.* 15:224–234.

Muir, A. R. and Golberg, L. 1961. The tissue response to iron-dextran: An electron-microscope study. *J. Path. Bact.* 82:471–482.

Myhre, E. 1973. Sarkom etter intramuskulaer injeksjon av jern. *Tidsskr. Norsk. Laegef.* 93:2500–2501.

Nash, P. 1950. Experimental production of malignant tumors by beryllium. *Lancet* 1:519.

Nath, N., Basrur, P. K. and Limebeer, R. 1971. A new cell line derived from nickel-sulfide-induced rat rhabdomyosarcoma. *In Vitro* 7:158–160.

National Institute for Occupational Safety and Health 1973. *Criteria for a recommended standard: Occupational exposure to inorganic arsenic.* Washington, D.C.: U.S. Dept. of Health, Education and Welfare.

Nazari, G., Favino, A. and Pozzi, U. 1967. Effetti di un'unica iniezione sottocutanea di cadmio cloruro nel ratto maschio. *Riv. Anat. Pathol. Oncol.* 31:251–270.

Nettesheim, P., Hanna, M. G., Jr., Dohertz, D. G., Newell, R. F. and Hellman, A. 1971. Effect of calcium chromate dust, influenza virus and 100 R whole-body x-radiation on lung tumor incidence in mice. *J. Nat. Cancer Inst.* 47:1129–1144.

Neukomm, S. 1969. Recherches sur le pouvoir cancérigène et co-cancérigène de diverses préparations à base de fer. *Proc. Europ. Soc. Drug Toxicity* 2:174–177.

Noble, R. L. and Capstick, V. 1963. Rhabdomyosarcomas induced by nickel sulphide in the rat. *Proc. Amer. Ass. Cancer Res.* 4:48.

Nordberg, G. F. 1971. Effects of acute and chronic cadmium exposure on the testicles of mice with special reference to protective effects of metallothionein. *Exper. Physiol.* 1:171–187.

Nordberg, G. F., Nordberg, M., Piscator, M. and Vesterberg, O. 1972. Separation of two forms of rabbit metallothionein by isoelectric focusing. *Biochem. J.* 126:491–498.

Olinici, C. D., Risca, R. and Todorutiu, C. 1973. Cytogenic observations of nickel-induced tumors in mice. *Oncologia Radiologia* 12:41–46.

Osswald, H. and Goerttler, K. 1971. Arsenic-induced leukoses in mice after diaplacental and postnatal application. *Verb. Dtsch. Ges. Path.* 55:289–293.

Ott, M., Holder, B. B. and Gordon, H. L. 1974. Respiratory cancer and occupational exposure to arsenicals. *Arch. Environ. Health* 29:250–255.

Pai, S. R., Gothoskar, S. V., and Ranadive, K. J. 1967. Testing of iron complexes. *Brit. J. Cancer* 21:448–451.

Paton, G. R. and Allison, A. C. 1972. Chromosome damage in human cell cultures induced by metal salts. *Mutation Res.* 16:332–336.

Payne, W. W. 1960. Production of cancers in mice and rats by chromium compounds. *Arch. Industr. Health* 21:530–535.

Payne, W. W. 1964. Carcinogenicity of nickel compounds in experimental animals. *Proc. Amer. Ass. Cancer Res.* 5:50.

Pedersen, E., Hogetveit, A. C. and Canderson, A. 1973. Cancer of respiratory organs among workers at a nickel refinery in Norway. *Int. J. Cancer* 12:32–41.

Petres, J., Schmid-Ullrich, K. and Wolf, U. 1970. Chromosomenaberrationen an menschlichen Lymphozyten bei chronischen Arsenschaden. *Deut. Med. Wschr.* 95:79–82.

Piotrowski, J. K., Trojanowska, B., Wisniewska-Knype, J. M. and Bolanorska, W. 1974. Mercury binding in the kidney and liver of rats repeatedly exposed to mercuric chloride: Induction of metallothionein by mercury and cadmium. *Toxicol. Appl. Pharmacol.* 27:11–19.

Potts, C. L. 1965. Cadmium proteinuria. The health of battery workers exposed to cadmium oxide dust. *Ann. Occup. Hyg.* 8:55–61.

Pulido, P., Kägi, J. H. R. and Vallee, B. L. 1966. Isolation and some properties of human metallothionein. *Biochemistry* 5:1768–1777.

Reddy, J., Svoboda, D., Azarnoff, D. and Dawas, R. 1973. Cadmium-induced Leydig cell tumors of rat testis: Morphologic and cytochemical study. *J. Nat. Cancer Inst.* 51:891–903.

Reeves, A. L. 1967. Isoenzymes of lactate dehydrogenase during beryllium carcinogenesis in the rat. *Cancer Res.* 27:1895–1899.

Reeves, A. L., Deitch, D. and Vorwald, A. J. 1967. Beryllium carcinogenesis. I. Inhalation exposure of rats to beryllium sulfate aerosol. *Cancer Res.* 27:439–445.

Reeves, A. L. and Vorwald, A. J. 1967. Beryllium carcinogenesis. II. Pulmonary deposition and clearance of inhaled beryllium sulfate in the rat. *Cancer Res.* 27:446–451.

Richmond, H. G. 1959. Induction of sarcoma in the rat by iron-dextran complex. *Brit. Med. J.* 1:947–949.

Rivière, M. R., Chouroulinkov, I. and Guérin, M. 1960. The production of tumours by means of intratesticular injections of zinc chloride in the rat. *Bull. Ass. Franc. Étude Cancer* 47:55–87.

Robinson, C. E. G., Bell, D. N. and Sturdy, J. H. 1960. Possible association of malignant neoplasm with iron-dextran: A case report. *Brit. Med. J.* 2:648–650.

Roe, F. J. C. 1967. On the potential carcinogenicity of the iron macromolecular complexes. In *Potential carcinogenic hazards from drugs*, ed. R. Truhaut, pp. 105–118. Berlin: Springer-Verlag.

Roe, F. J. C., Boyland, E., Dukes, C. E. and Mitchley, B. C. V. 1965a. Failure of testosterone or xanthopterin to influence the induction of renal neoplasms by lead in rats. *Brit. J. Cancer* 19:860–866.

Roe, F. J. C., Boyland, E., and Millican, K. 1965b. Effect of oral administration of two tin compounds to rats over prolonged periods. *Food Cosmet. Toxicol.* 3:277–280.

Roe, F. J. C. and Carter, R. L. 1967. Iron-dextran carcinogenesis in rats: Influence of dose on the number and types of neoplasm induced. *Int. J. Cancer* 2:370–380.

Roe, F. J. C. and Carter, R. L. 1969. Chromium carcinogenesis: Calcium chromate as a potent carcingoen for the subcutaneous tissues of the rat. *Brit. J. Cancer* 23:172–176.

Roe, F. J. C., Dukes, C. E., Cameron, K. M., Pugh, R. C. B. and Mitchley, B. C. V. 1964a. Cadmium neoplasia: Testicular atrophy and Leydig cell hyperplasia and neoplasia in rats and mice following the subcutaneous injection of cadmium salts. *Brit. J. Cancer* 18:674–681.

Roe, F. J. C. and Haddow, A. 1965. Test of an iron-sorbitol-citric acid complex ("Jectofer") for carcinogenicity in rats. *Brit. J. Cancer* 19:855–859.

Roe, F. J. C., Haddow, A., Dukes, C. E. and Mitchley, B. C. V. 1964b. Iron-dextran carcinogenesis in rats: Effect of distributing injected material between one, two, four, or six sites. *Brit. J. Cancer* 18:801–808.

Roe, F. J. C. and Lancaster, M. C. 1963. Natural, metallic and other substances as carcinogens. *Brit. Med. Bull.* 19:127–133.

Rosen, P. 1971. Theoretical significance of arsenic as a carcinogen. *J. Theor. Biol.* 32:425–426.

Roussel, J., Pernot, C., Schoumacher, P., Pernot, M. and Kessler, Y. 1964. Considerations statistiques sur le cancer bronchique du mineur de fer au bassin de Lorraine. *J. Radiol. Electrol.* 45:541–546.

Saknyn, A. V. and Shabynina, N. K. 1970. Some statistical data on carcinogenous hazards for workers engaged in the production of nickel from oxidized ores. *Gig. Trud. Prof. Zabol.* 14:10–13.

Saknyn, A. V. and Shabynina, N. K. 1973. Epidemiology of malignant neoplasms in nickel plants. *Gig. Trud. Prof. Zabol.* 17:25–28.

Sanina, Y. P. 1968. Toxicology of nickel carbonyl. *Toksikol Nov. Prom. Khim. Veshchestv.* 10:144–149.

Schepers, G. W. H. 1961. Neoplasia experimentally induced by beryllium compounds. *Prog. Exp. Tumor Res.* 2:203–244.

Schepers, G. W. H. 1964. Biological action of beryllium. Reaction of the monkey to inhaled aerosols. *Industr. Med. Surg.* 33:1–16.

Schinz, H. R. and Uehlinger, E. 1942. Der Mettallkrebs; ein neues Prinzip der Krebserzeugung. *Z. Krebsforsch.* 52:425–437.

Schmähl, D. and Steinhoff, D. 1960. Versuche zur Krebserzeugung mit kolloidolen Silber-und Goldlösungen an Ratten. *Z. Krebsforsch.* 63:586–591.

Shvemberger, I. N., Ivashkevitch, L. G. and Yu, B. V. 1972. Variability of cell elements in myogenic and connective tissue rat tumors. *Tsitologiia* 14:1405–1413.

Singh, A. and Gilman, J. P. W. 1973. Use of the double diffusion chamber for an analysis of muscle-nickel sulfide interaction. *Indian J. Med. Res.* 61:704–707.

Sissons, H. A. 1950. Bone sarcomas produced experimentally in the rabbit, using compounds of beryllium. *Acta Unio Intern. Contra Cancrum* 7:171.

Smith, A. G. and Powell, L. 1957. Genesis of teratomas of the testis. A study of normal and zinc-injected testes of roosters. *Amer. J. Path.* 33:653–669.

Stiller, D. 1973. Topochemie tubulärer Enzymanuster während der experimentellen Cancerisierung. Untersuchungen zur Enzymhistochemie der chronischen Bleinephropathie der Ratte. *Exp. Path.* 8:137–153.

Sunderman, F. W. 1966. Metastasizing pulmonary tumors in rats induced by the inhalation of nickel carbonyl. In *Lung tumors in animals,* ed. L. Severi, pp. 551–564. Proceedings of the Third Quadrennial Conference on Cander, Perugia, Italy. Perugia: Division of Cancer Research.

Sunderman, F. W. and Donnelly, A. J. 1965. Studies of nickel carcinogenesis. Metastasizing pulmonary tumors in rats induced by the inhalation of nickel carbonyl. *Amer. J. Path.* 46:1027–1041.

Sunderman, F. W., Donnelly, A. J., West, B. and Kincaid, J. F. 1959. Nickel poisoning. IX. Carcinogenesis in rats exposed to nickel carbonyl. *Arch. Indust. Health* 20:36–41.

Sunderman, F. W., Jr. 1967a. Inhibition of induction of benzopyrene hydroxylase by nickel carbonyl. *Cancer Res.* 27:950–955.

Sunderman, F. W., Jr. 1967b. Nickel carbonyl inhibition of cortisone induction of hepatic tryptophan pyrrolase. *Cancer Res.* 27:1595–1599.

Sunderman, F. W., Jr. 1968. Nickel carbonyl inhibition of phenobarbital induction of hepatic cytochrome P-450. *Cancer Res.* 28:465–470.

Sunderman, F. W., Jr. 1970. Effect of nickel carbonyl upon incorporation of ^{14}C-leucine into hepatic microsomal proteins. *Res. Commun. Chem. Path. Pharmacol.* 1:161–168.

Sunderman, F. W., Jr. 1971. Metal carcinogenesis in experimental animals. *Food Cosmet. Toxicol.* 9:105–120.

Sunderman, F. W., Jr. 1973. The current status of nickel carcinogenesis. *Ann. Clin. Lab. Sci.* 3:156–180.

Sunderman, F. W., Jr. and Esfahani, M. 1968. Nickel carbonyl inhibition of RNA polymerase activity in hepatic nuclei. *Cancer Res.* 28:2565–2567.

Sunderman, F. W., Jr., Lau, T. J. and Cralley, L. J. 1974. Inhibitory effect of manganese upon muscle tumorigenesis by nickel subsulfide. *Cancer Res.* 34:92–95.

Sunderman, F. W., Jr., Lau, T. J., Minghetti, P. P., Maenza, R., Becker, N., Onkelinx, C. and Goldblatt, P. 1975. Effects of manganese on tumorigenesis and metabolism of nickel subsulfide. *Proc. Amer. Ass. Cancer Res.* 16:554.

Sunderman, F. W., Jr. and Leibman, K. C. 1970. Nickel carbonyl inhibition of induction of aminopyrine demethylase activity in liver and lung. *Cancer Res.* 30:1645–1650.

Sunderman, F. W., Jr. and Mastromatteo, E. 1975. Nickel carcinogenesis. In *Nickel*, ed. F. W. Sunderman, Jr., F. Coulston, G. L. Eichorn, J. A. Fellows, E. Mastromatteo, H. T. Reno, and M. H. Samitz. Washington, D.C.: National Academy of Sciences.

Sunderman, F. W., Jr., Roszel, N. O. and Clark, R. J. 1968. Gas chromatography of nickel carbonyl in blood and breath. *Arch. Environ. Health* 16:836–843.

Sunderman, F. W., Jr. and Selin, C. E. 1968. The metabolism of nickel-63 carbonyl. *Toxicol. Appl. Pharmacol.* 12:207–218.

Swanson, S. A. V., Freeman, M. A. R. and Heath, J. C. 1973. Laboratory tests on total joint replacement protheses. *J. Bone Joint Surg.* 55B:759–773.

Swierenga, S. H. H. 1970. The role of nickel in the induction of muscle tumors. Doctoral dissertation, University of Guelph, Ontario.

Swierenga, S. H. H. and Basrur, P. K. 1968. Effect of nickel on cultured rat embryo muscle cells. *Lab. Invest.* 19:663–674.

Sykes, A. K. and Basrur, P. K. 1971. A melinex coverslip method for ultrastructural studies of monolayer cultures. *In Vitro* 7:68–73.

Taylor, F. H. 1966. The relationship of mortality and duration of employment, as reflected by a cohort of chromate workers. *Amer. J. Pub. Health* 56:218–229.

Thomas, J. A. and Thiery, J. P. 1953. Production elective de liposarcoma chez les lapins par les oligoelements zinc et cobalt. *Compt. Rendu Hebd. Séanc. Acad. Sci. (Paris)* 236:1387.

Toda, M. 1962. Experimental studies of occupational lung cancer. *Bull. Tokyo Med. Dent. Univ.* 9:440–441.

Tönz, O. 1957. Nierenveränderungen bei experimenteller chronischer Bleivergiftung (Ratten). *Z. Ges. Exp. Méd.* 128:361–377.

Treagan, L. and Furst, A. 1970. Inhibition of interferon synthesis in mammalian cell cultures after nickel treatment. *Res. Commun. Chem. Path. Pharmacol.* 1:395–402.

Turner, H. M. and Grace, H. G. 1938. An investigation into cancer mortality among males in certain Sheffield trades. *J. Hyg.* 38:90–103.

Van Esch, G. J. and Kroes, R. 1969. The induction of renal tumours by feeding basic lead acetate to mice and hamsters. *Brit. J. Cancer* 23:765–770.

van Esch, G. J., van Genderen, H. and Vink, H. H. 1962. The induction of renal tumours by feeding basic lead acetate to rats. *Brit. J. Cancer* 16:289–297.

Venitt, S. and Levy, L. S. 1974. Mutagenicity of chromates in bacteria and its relevance to chromate carcinogenesis. *Nature* 250:493–495.

Virtue, J. A. 1972. The relationship between the refining of nickel and cancer of the nasal cavity. *Canad. J. Otolaryngol.* 1:37–42.

Vollmann, J. 1940. Tierexperimente mit intraossärem Arsen-, Chrom- und Kobaltdepot. *Schweiz. A. Allg. Path. Bakt.* 2:440–443.

Vorwald, A. J. and Reeves, A. L. 1959. Pathologic changes induced by beryllium compounds. *Arch. Industr. Health* 19:190–199.

Vorwald, A. J., Reeves, A. L. and Urban, E. C. J. 1966. Experimental beryllium toxicology. In *Beryllium—Its industrial hygiene aspects*, ed. H. E. Stokinger, pp. 201–234. New York: Academic Press.

Walters, M. and Roe, F. J. C. 1975. A study of the effects of zinc and tin administered orally to mice over a prolonged period. *Food Cosmet. Toxicol.* 3:271–276.

Webb, M. 1972. Binding of cadmium ions by rat liver and kidney. *Biochem. Pharmacol.* 21:2751–2765.

Webb, M., Heath, J. C. and Hopkins, T. 1972. Intranuclear distribution of the inducing metal in primary rhabdomyosarcomata induced in the rat by nickel, cobalt and cadmium. *Brit. J. Cancer* 26:274–278.

Webb, M. and Weinzierl, S. M. 1972. Uptake of ^{63}Ni^{2+} from its complexes with proteins and other ligands by mouse dermal fibroblasts *in vitro. Brit. J. Cancer* 26:292–298.

Weinzierl, S. M. and Webb, M. 1972. Interaction of carcingoenic metals with tissue and body fluids. *Brit. J. Cancer* 26:279–291.

Williams, D. R. 1972. Metals, ligands and cancer. *Chem. Rev.* 72:203–213.

Winge, D. R. and Rajagopolan, K. V. 1972. Purification and some properties of Cd-binding protein from rat liver. *Arch. Biochem. Biophys.* 153:755–762.

Winkelstein, W. and Kantor, S. 1969. Prostatic cancer: Relationship to suspended particulate air pollution. *Amer. J. Pub. Health* 59:1134–1138.

Witschi, H. P. 1968. Inhibition of DNA synthesis in regenerating rat liver by beryllium. *Lab. Invest.* 19:67–70.

Witschi, H. P. 1970. Effects of beryllium on deoxyribonucleic acid-synthesizing enzymes in regenerating rat liver. *Biochem. J.* 120:623–634.

Witschi, H. P. 1972. A comparative study of *in vivo* RNA and protein synthesis in rat liver and lung. *Cancer Res.* 32:1686–1694.

Witschi, H. P. and Aldridge, W. N. 1968. Uptake, distribution and binding of beryllium to organelles of the rat liver cell. *Biochem. J.* 106:811–820.

Witschi, H. P. and Marchand, P. 1971. Interference of beryllium with enzyme induction in rat liver. *Toxicol. Appl. Pharmacol.* 20:565–572.

Wrba, H. and Mohr, U. 1968. Krebs durch Injektionen von Eisenkomplexen. Langzert-Tierversuche zur Frage der kanzerogenen Wirkung von Eisen-Sorbitol-Zitronensäure. *Münch. Med. Wschr.* 110:139–140.

Yamaguchi, S. 1963. Study of beryllium-induced osteogenic sarcoma. *Nagasaki Iggaki Zasshi* 38:127.

Yeh, S. 1973. Skin cancer in chronic arsenicalism. *Human Path.* 4:469–485.

Zachariae, H. 1972. Arsenik og cancerrisko. *Ugeskr. Laeg.* 134:2720–2721.

Zawirska, B. and Medras, K. 1968. Tumors and disorders of porphyrin metabolism in rats with chronic experimental lead poisoning. *Zentbl. Allg. Path.* 111:1–12.

Zawirska, B. and Medras, K. 1972a. Morphology and biological activity of transplantable interstitioma testis induced with lead acetate. *Arch. Immunol. Therap. Exp.* 20:243–256.

Zawirska, B. and Medras, K. 1972b. Role of the kidneys in disorders of porphyrin metabolism during carcinogenesis induced with lead acetate. *Arch. Immunol. Therap. Exp.* 20:257–272.

Zollinger, H. U. 1953. Durch chronische Bleivergiftung erzeugte Nierenadenome und -carcinome bei Ratten und ihre Beziehungen zu den entsprechenden Neubildungen des Menschen. *Virchows Arch. Path. Anat.* 323:694–710.

Zollinger, H. U. 1962. Weichteiltumoren bei Ratten nach sehr massiven Eiseninjektionen. *Schweiz Med. Wschr.* 92:130–134.

INDEX

Absorption (*see* Gastrointestinal absorption)
Air:
    arsenic in, 85
    vanadium in, 154–155
δ-ALA (*see* δ-Aminolevulinic acid)
ALA-D (*see* δ-Aminolevulinic acid
        dehydratase)
Albumin, 123
Alcohol, in nickel toxicity, 141
Alkylmercury:
    other alkylmercury compounds, 9
    (*See also* Methylmercury, Ethylmercury)
Allergic response, due to nickel, 134–135
δ-Aminolevulinic acid (δ-ALA):
    dehydratase, 50
    lead metabolism, 50
    lead poisoning, 56–57
    method of measurement in urine, 70–71
    relationship to blood lead, 56
δ-Aminolevulinic acid dehydratase
        (ALA-D), methods of measurement,
        65–66
Anemia:
    copper, 125
    lead, 49–50
Anticarcinogenic effects, of selenium, 211
Aorta, 248
Apoferritin, effect of lead on, 50
Arsanilic acid, used to promote growth, 85
Arsenic:
    air content, 85
    biliary excretion, 87
    blood cells, 87
    blood levels, 85
    cardiac effects, 92–93
    coal, 81
    epidemiological studies, 257
    fingernails, 90
    goiter, 88
    mechanisms of toxicity, 258
    pentavalent, 87
    promote growth, 85
    protein binding, 88

selenium interrelationships, 209
    trivalent, 86
    urine content, 85
    (*See also* specific topic)
Arsenic acid, 85
Arsenic exposure, 79–86
    human, 85–86
    poultry and meat products, 85
Arsenic poisoning(s):
    accidental deaths, 89
    hemolysis, 90, 93
    hepatotoxicity, 92
    Manchester, England, 89
    Morinaga milk incident, 89
    nervous system effects, 92
    Niigata Prefecture of Japan, 90
    renal failure, 90
Arylmercury (*see* Phenylmercury)
Ascorbate, 178
Ascorbic acid, 179, 185
Atherosclerosis, 247
Atomic absorption, for measurement of
        lead, 42, 60–61

BAL (British antilewisite):
    and biliary excretion of mercury, 14
    for treatment of arsine poisoning, 91–92
Basophilic stippling, 49
Berylliosis, 258
Beryllium, 170, 258
    carcinogenicity of, 258
    and osteogenic sarcomas, 258
    and pulmonary carcinomas, 258
Beta cells, of pancreas, 249
Biliary excretion:
    of arsenic, 87
    of mercury, 12–14
        BAL, 13
        cysteine, 12
        penicillamine, 14
        Sephadex, 12